Developments in Soil Science 5A

SOIL CHEMISTRY
A. Basic Elements

Further Titles in this Series

1. I. VALETON
BAUXITES

2. IAHR
FUNDAMENTALS OF TRANSPORT PHENOMENA
IN POROUS MEDIA

3. F.E. ALLISON
SOIL ORGANIC MATTER AND ITS ROLE IN
CROP PRODUCTION

4. R.W. SIMONSON (Editor)
NON-AGRICULTURAL APPLICATIONS OF SOIL SURVEYS

Developments in Soil Science 5A

SOIL CHEMISTRY
A. Basic Elements

EDITED BY

G.H. BOLT and M.G.M. BRUGGENWERT

CONTRIBUTING AUTHORS:

J. BEEK
G.H. BOLT
M.G.M. BRUGGENWERT
F.A.M. DE HAAN
A. KAMPHORST
I. NOVOZAMSKY

Department of Soil Science and Plant Nutrition,
Agricultural University of Wageningen, The Netherlands

N. VAN BREEMEN   R. BRINKMAN

Department of Soil Science and Geology,
Agricultural University of Wageningen, The Netherlands

P.J. ZWERMAN

Department of Agronomy, Cornell University, Ithaca, N.Y.,
U.S.A.

SECOND REVISED EDITION

ELSEVIER SCIENTIFIC PUBLISHING COMPANY
Amsterdam — Oxford — New York   1978

ELSEVIER SCIENCE PUBLISHERS B.V.
Sara Burgerhartstraat 25
P.O. Box 211, 1000 AE Amsterdam, The Netherlands

Distributors for the United States and Canada:

ELSEVIER SCIENCE PUBLISHING COMPANY INC.
52, Vanderbilt Avenue
New York, NY 10017, U.S.A.

First edition 1976
Second revised edition 1978
Second impression 1981
Third impression 1986
Fourth impression 1988

ISBN 0-444-41435-5 (Vol. 5A)
ISBN 0-444-40882-7 (Series)

© Elsevier Science Publishers B.V., 1976

All rights reserved. No part of this publication may be reproduced, stored in a retrieval system or transmitted in any form or by any means, electronic, mechanical, photocopying, recording or otherwise, without the prior written permission of the publisher, Elsevier Science Publishers B.V./ Physical Sciences & Engineering Division, P.O. Box 330, 1000 AH Amsterdam, The Netherlands.

Special regulations for readers in the USA – This publication has been registered with the Copyright Clearance Center Inc. (CCC), Salem, Massachusetts. Information can be obtained from the CCC about conditions under which photocopies of parts of this publication may be made in the USA. All other copyright questions, including photocopying outside of the USA, should be referred to the publisher.

No responsibility is assumed by the Publisher for any injury and/or damage to persons or property as a matter of products liability, negligence or otherwise, or from any use or operation of any methods, products, instructions or ideas contained in the material herein.

Printed in The Netherlands

PREFACE

The present text is primarily meant for student instruction at advanced undergraduate level. It grew during a number of years from a set of lecture syllabi covering courses in soil chemistry given to students of the State Agricultural University at Wageningen, majoring in the fields of Soils, Drainage and Irrigation Engineering and Environmental Sciences. Because of such multipurpose usage care has been taken to make certain chapters sufficiently self-supporting in order to allow skipping others. In this manner the chapter sequences 1-3-4-5-(7)-9, 1-2-6-8 and 1-4-5-7-10 could be used more or less independently of the other chapters.

In selecting a method of approach to the subject material, it was decided to use the basic sciences as point of departure. This is obviously in contrast to the historic development of Soil Science, in which practical experience came first. Application of knowledge derived from the basic sciences occurred gradually, usually as a result of difficulties encountered when it was attempted to interpret and generalize practical experience.

For a basic text in Soil Chemistry it appears that following the historic development is rather inefficient, as it implies a 'repeated, incidental digging back' into the fundamental background of the phenomena observed. In such a process only parts of this background are encountered, often in a rather disconnected manner. In selecting the reverse approach it is felt that there is a better chance to give to the student a coherent insight in the fundamental aspects of soil science. Such insight should then serve as a skeleton to which later more factual knowledge may be attached; it should, because of its comparatively high generalization value, also serve as a dependable guideline when attempting to interpret practical observations.

While stressing these fundamental aspects of Soil Chemistry, the factual information of this text has been kept brief in order to stay within a reasonable length for teaching purposes. Admitting that a fair background in basic chemistry has been preassumed, particular care was taken to add small print sections meant to bring back to memory - or render plausible - those aspects of basic chemistry needed at that point. These small print sections should thus be viewed as an additional - sometimes cautioning - note, which could be skipped by those with a good background in chemistry. A few sample problems have been included to show the type of estimates one should be able to make in practice.

The approach chosen explains why in this text (with the exception of chapter 10) only very seldom direct reference is made to the scientists that were involved in the development of the particular subject: the 'pioneers' in Soil Science were seldom discoverers of new laws of science, but rather the ones that recognized the significance of existing knowledge in the basic sciences for solving the problems in soil science. In some instances a list of standard works used extensively in preparing the relevant text has been added. In other instances certain selected sections of other texts have been recommended for reading, in part to supplement the factual information but also to confront the student with the viewpoint of other authors given in a text of comparable length.

Finally it should be pointed out that the 'application' chapters 9 and 10 have by necessity a slightly different character from the foregoing ones, being somewhat more detailed. Particularly with respect to chapter 10 it was envisaged that with the comparatively recent awakening of interest in environmental sciences many outsiders from the field of Soil Science have come into contact with soil as a biotope which could become spoiled. For those a summary of factual information may be the prime interest, while the time needed to go through a regular study of Soil Science as it developed in the last 50 years is lacking. Accordingly this last chapter has been supplied with an extensive list of references supporting the information presented. It is hoped nevertheless, that some of those who start reading this text with the last chapter will become enticed to leaf through the earlier chapters to see a bit more of the pattern of thinking in Soil Chemistry.

Returning to the first sentence, the scope of the present text precluded the inclusion of advanced model theories that have been used to describe the transport and accumulation phenomena occurring in soil. As has been indicated locally in this volume 5A of the present series of texts, these will be covered in a separate volume 5B.

It is a pleasure to acknowledge all those who contributed to this book: Dr. A.R.P. Janse and Dr. F.F.R. Koenigs, who were involved in the first issue of the lecture syllabus mentioned; Dr. W.L. Lindsay of Colorado State University, who gave a course on solubility equilibria in soils while spending his sabbatical leave at Wageningen in 1972, and whose ideas on the presentation of this subject material are clearly reflected in chapter 6; Mr. W.H. van Riemsdijk, who read and commented on the contents of chapters 2 and 6; Dr. J. van Schuylenborgh of the University of Amsterdam, whose course on the chemical aspects of soil formation given at the time at Wageningen University set the stage for chapter 8; Mr. B.W. Matser, who made the majority of the drawings of the figures; Mr. T. Klaassen and co-workers of the Institute for Land and Water Management Research, who prepared the figures of chapter 10; Miss G.G. Gerding and Miss D.J. Hoftijser, who did the (repeated) draft typing and particularly Miss A.H. Kap, who prepared the camera-ready copy as printed.

Wageningen, 1975   G.H.B. and M.G.M.B.

CONTENTS

PREFACE V

LIST OF SYMBOLS XI

PREFACE TO THE SECOND EDITION X

CHAPTER 1. COMPOSITION OF THE SOIL 1
  1.1. Solid phase components 2
      1.1.1. Inorganic components 2
      1.1.2. The organic components. 8
  1.2. The liquid phase 10
  1.3. The gas phase 11

CHAPTER 2. CHEMICAL EQUILIBRIA 13
  2.1. The condition for equilibrium 13
  2.2. Standard states and activities. 14
  2.3. Activity coefficients of ions in aqueous solutions. 16
      2.3.1. Activity coefficients in mixed aqueous solutions at high ionic strength. 20
  2.4. Calculation of equilibrium constants from thermodynamic data 21
  2.5. Some thermodynamic considerations 22
  2.6. Illustrative calculations 23
      2.6.1. Calculation of the thermodynamic equilibrium constant 23
      2.6.2. Calculation of the equilibrium solution composition at low electrolyte level 26
      2.6.3. Calculation of the equilibrium solution composition at 'high' electrolyte level 27
  2.7. Reactions involving the transfer of protons and/or electrons 30
      2.7.1. Acid - base equilibria 30
      2.7.2. Oxidation - reduction equilibria 32
      2.7.3. The electrometric determination of pH and pe 35
  2.8. Graphical presentation of solubility equilibria. 36
  2.9. Surface structure and solubility 40
      Literature consulted. 41

CHAPTER 3. SURFACE INTERACTION BETWEEN THE SOIL SOLID PHASE AND THE SOIL SOLUTION. 43
  3.1. The surface charge of the solid phase. 43
  3.2. Properties of the liquid layer adjacent to the solid phase 45
      3.2.1. The extent of the diffuse double layer at high water content 47
      3.2.2. The diffuse double layer at low liquid content of the system 50
  3.3. The influence of the interaction between solid and liquid phase on soil properties 52
      Recommended literature. 53

CHAPTER 4. ADSORPTION OF CATIONS BY SOIL. 54
  4.1. Qualitative description of the exchange reaction 54
  4.2. Experimental approach. 56
      4.2.1. Interpretation of the analysis-data 56
      4.2.2. Some experimental data 62
  4.3. Model considerations 63
  4.4. The exchange equilibrium 65
      4.4.1. Exchange equations 66

		4.4.2. Application of the exchange equations in estimating changes in composition of solution and complex 69
4.5.	Highly selective adsorption of cations by soil 72	
	4.5.1. Fixation of cations in clay lattices 73	
	4.5.2. Complex formation of cations by organic matter ligands. 75	
4.6.	The adsorption of H- and Al-ions by soil constituents. 76	
	4.6.1. Analysis of the different types of adsorption mechanisms 76	
	4.6.2. The titration curve of soil constituents 82	
	4.6.3. Correction of the soil pH 86	
	4.6.4. Measurement of pH in soil; the suspension effect. 87	
	Illustrative problems 89	
	Recommended literature. 90	

CHAPTER 5.   ADSORPTION OF ANIONS BY SOIL 91
 5.1. Anion exclusion at negatively charged surfaces 91
 5.2. The positive adsorption of anions. 92
 5.3. Phosphate 'fixation'. 94
      Illustrative problems 95
      Recommended literature. 95

CHAPTER 6.   COMMON SOLUBILITY EQUILIBRIA IN SOILS. 96
 6.1. Carbonate equilibria. 96
      6.1.1. The CO_2-H_2O system. 96
      6.1.2. Systems containing $CaCO_{3(s)}$ 100
 6.2. Iron oxides and hydroxides. 105
      6.2.1. Ferrous compounds. 105
      6.2.2. Redox reactions involving iron compounds. 106
      6.2.3. pe-pH diagrams for the system hematite-magnetite-siderite-H_2O . . . 111
 6.3. Aluminum 113
      6.3.1. Al_2O_3-H_2O system. 114
      6.3.2. Al_2O_3-SiO_2-H_2O system 116
 6.4. Phosphorus 118
      6.4.1. Solubility of phosphates in soils 119
      6.4.2. Phosphate solubility diagram in the system Al_2O_3-Fe_2O_3-CaO-P_2O_5-H_2O 120
 6.5. Relevant thermodynamic data of the systems discussed 121
      Illustrative problems 124
      Recommended and Consulted literature. 125

CHAPTER 7.   TRANSPORT AND ACCUMULATION OF SOLUBLE SOIL
COMPONENTS. 126
 7.1. Transport with and in the liquid phase. 126
 7.2. Solute displacement in soil 127
      7.2.1. Displacement in case of complete exchange 128
      7.2.2. Displacement in case of incomplete exchange 130
      7.2.3. The influence of the exchange isotherm on solute displacement . . . 131
 7.3. The penetrating solute front 132
      7.3.1. Influence of the exchange isotherm. 132
      7.3.2. Influence of diffusion and dispersion. 134
      7.3.3. Order of magnitude of the front spreading effects 135
 7.4. Some practical examples. 135
      7.4.1. Reclamation of Na-soils 135
      7.4.2. The sodication process 137
      7.4.3. The penetration of trace components into soil 137

7.5.	Some cautioning remarks		139
	Illustrative problems		139

CHAPTER 8.   CHEMICAL EQUILIBRIA AND SOIL FORMATION 141
8.1. Introduction... 141
    8.1.1. Soil formation and soil forming factors 141
    8.1.2. The use of water analyses in the study of soil formation 142
    8.1.3. A landscape model..................................... 143
8.2. Weathering of soil minerals 145
    8.2.1. Congruent and incongruent dissolution 145
    8.2.2. Solubility and stability relationships 146
    8.2.3. The concept of partial equilibrium 153
    8.2.4. Weathering products 154
    8.2.5. Decay of organic matter, humification and chelation 155
    8.2.6. Composition of the soil solution....................... 156
8.3. Soil reduction and oxidation 158
    8.3.1. Environmental requirements for soil reduction 158
    8.3.2. The sequential appearance of reduction products upon flooding... 160
    8.3.3. Soil reaction and production of alkalinity during reduction 162
    8.3.4. Water regimes in hydromorphic soils.................... 163
    8.3.5. Weathering under seasonally reduced conditions 164
8.4. Reverse weathering 166
    8.4.1. Vertisols, calcium carbonate, salinity and high pH 166
    8.4.2. Absolute accumulation of iron oxide..................... 169
    Illustrative problems 170
    Recommended literature...................................... 170

CHAPTER 9.   SALINE AND SODIC SOILS............................. 171
9.1. Chemical characterization of saline and alkali soils................ 173
9.2. Salinization of soils upon irrigation............................. 175
9.3. Sodication of soils upon irrigation 179
9.4. Alkalinization under irrigation................................. 183
9.5. Chemical aspects of the reclamation of saline and sodic soils 186
    Illustrative problems 189
    Consulted literature.. 190
    Recommended literature 191

CHAPTER 10.   POLLUTION OF SOIL................................. 192
10.1. Soil as an environmental component........................... 193
10.2. Recognition and prediction of soil pollution..................... 194
10.3. Nitrogen and phosphorus in soil............................... 197
    10.3.1. Pollution effects involving nitrogen..................... 198
    10.3.2. Sources of (excess) nitrogen in soil..................... 199
    10.3.3. Forms of organic nitrogen in soil 201
    10.3.4. Forms of inorganic nitrogen in soil 203
    10.3.5. The pathway of nitrogen through soil 204
    10.3.6. Pollution effects involving phosphates.................. 210
    10.3.7. Sources of phosphates in soil.......................... 211
    10.3.8. The interaction between phosphates and soil 213
    10.3.9. A characteristic phosphate distribution profile as found on a
               sewage farm... 216
10.4. Heavy metals and trace elements 218
    10.4.1. As, arsenic... 221
    10.4.2. Cd, cadmium 222

	10.4.3.	Co, cobalt	223
	10.4.4.	Cr, chromium	224
	10.4.5.	Cu, copper	225
	10.4.6.	Hg, mercury	227
	10.4.7.	Mo, molybdenum	230
	10.4.8.	Ni, nickel	231
	10.4.9.	Pb, lead	232
	10.4.10.	Se, selenium	234
	10.4.11.	V, vanadium	235
	10.4.12.	Zn, zinc	236
	10.4.13.	Chelation and metal mobility	238
10.5.	Organic pesticides in soil	239
	10.5.1.	Bonding by soil constituents	242
	10.5.2.	Decomposition of pesticides in soil	251
10.6.	Miscellaneous soil pollution sources	255
	10.6.1.	Oil spills and oil sludge disposal	255
	10.6.2.	Gas leakages	256
	10.6.3.	Sanitary landfills	259
10.7.	Positioning the present treatise with respect to adjacent areas of interest	262
	Literature	264

SUBJECT INDEX 272

PREFACE TO THE SECOND EDITION

A number of irritating misprints, "dutchisms" and a few errors were removed: thanks are due to readers who pointed out some of these to us. After ample consideration the term Free Enthalpy was retained — in favor of the more ambiguous term Free Energy as still often used in Anglo-Saxon literature — when referring to the thermodynamic state function named Gibbs Energy or Gibbs Function in the I.U.P.A.C. recommendations.

Wageningen, 1978   G.H.B. and M.G.M.B.

LIST OF SYMBOLS

As to the units used, SI has been the guideline, with some notable exceptions in order to maintain the necessary links with practical usage. Thus the kcal/mole has been used because available tables of thermodynamic data are largely expressed in this unit. After some hesitance the exchange capacity and related quantities were still expressed in the practical unit of meq per 100 g. Combined with the moisture content in ml or cm^3 per 100 g of dry soil this allows one to express the concentration in $keq/m^3 \equiv meq/cm^3$. In a few instances other units have been used very locally but then these units are mentioned at the spot. This is particularly the case in chapter 9 which was meant to reflect existing 'field' practice. In those instances were certain system parameters were used solely in a formal context, without specifying numerical values, units are not specified in the present list.

		unit
A, B	: chemical species	
(A)	: activity of species A	
[A]	: concentration of species A	
$\tilde{a}$	: activity in solution	mol/l
c	: concentration in solution	defined locally
C	: total electrolyte concentration	$keq/m^3 = eq/l$
C	: electrical capacity of the double layer	
$d_{ex,an}$	: effective (equivalent) distance of exclusion of anions from a double layer	$Å = 10^{-10}$ m
d_l	: (mean) thickness of the liquid layer on a solid surface	m or Å
E	: electrode potential	volt
e	: charge of a monovalent electron	Coulomb
e^-	: electron	
pe	: minus logarithm of the relative electron activity, cf p 33	---
F	: Faraday constant	Coulomb/eq
f	: activity coefficient in solution	---
f	: fractional concentration (c/C)	---
G	: free enthalpy	kcal
$\bar{G}$	: partial molar free enthalpy	kcal/mole
$\bar{G}_f^o$	: free enthalpy of formation per mole	kcal/mole
ΔG_r^o	: standard free enthalpy of a reaction	kcal/mole
I	: ionic strength	
J^v	: volume flux of the soil solution per unit cross-section of a soil column	m/sec
j	: (total) solute flux per unit cross-section of a soil column	keq/m^2 sec
j^D	: autonomous flux of a solute with respect to the carrier solution (due to diffusion or dispersion cf. chapter 7)	keq/m^2 sec

K	: equilibrium 'constant' of a reaction	
K°	: thermodynamic equilibrium constant of a reaction	
K_A°	: thermodynamic acidity constant	
K_D°	: thermodynamic dissociation constant	
K_K	: Kerr exchange constant'	---
K_G	: Gapon exchange 'constant'	$(mol/l)^{-\frac{1}{2}}$
K_{SO}°	: solubility product	
N	: equivalent fraction adsorbed (of an ion, relative to the CEC)	
N	: slope of the normalized adsorption isotherm (dN/df)	
pH_o	: pH-value at zero point of charge (of a solid surface)	
Q	: cation exchange capacity per unit bulk volume of soil (cf. chapter 7)	keq/m^3
q	: amount adsorbed of an ion per unit bulk volume of soil, $q \equiv N \times Q$	keq/m^3
R_D	: distribution ratio, amount adsorbed relative to the amount in the solution phase	
$\bar{R}_D$	: overall distribution ratio $(Q/\theta C_o \equiv \gamma/WC_o)$	
S	: entropy	
S	: specific surface area	$m^2/kg, m^2/g$
T_{cat}, T_{an}	: total amount of cations and anions, respectively, present per unit weight of soil	$meq/100\ g$
V	: volume of solution fed into a unit cross-section of soil column feed volume (cf. chapter 7)	m
W	: moisture content	$cm^3/100\ g$
x	. distance coordinate	
x_p	: mean depth of penetration	m
z	: ionic valence	

GREEK LETTERS

γ : adsorption capacity of the soil solid phase (constituents), particularly for exchangeable cations and then referred to as Cation Exchange Capacity, CEC $\qquad meq/100\ g$

$\overset{+}{\gamma}_k$	: amount of ion species k present in excess of the product of moisture content, W, and its equilibrium concentration, $c_{o,k}$	meq/100 g
$\bar{\gamma}_l$	: deficit of ion species l, defined in analogy with the above	meq/100 g
Γ	: surface density of charge of the soil solid phase (constituents) as derived from $\gamma_{cat}/10^5 S$	keq/m²
F_k	: excess of ion species k per unit surface ($\overset{+}{\gamma}_k/10^5 S$)	keq/m²
F_l	: deficit of ion species l per unit surface ($\bar{\gamma}_l/10^5 S$)	keq/m²
ψ	: electrostatic potential, i.c. in the DDL	Volt
ρ_b	: bulk density of the soil	kg/m³
θ	. volume fraction of bulk soil occupied by the soil solution	---
μ_k	: chemical potential of species k	kcal/mole
ν	: stoichiometric equivalence number	---

LETTER COMBINATIONS

AEC	: Anion Exchange Capacity	meq/100 g
Alk	: Alkalinity (concentration)	meq/m³ or eq/m³
CEC	: Cation Exchange Capacity	meq/100 g
DL	: Double layer	
DDL	. Diffuse Double Layer	
EC	. Electrical Conductivity	mmho/cm
EC_e	: Electrical Conductivity of the saturation extract	mmho/cm
EC_{FC}	: Electrical Conductivity of the soil solution at Field Capacity	mmho/cm
EC_{iw}, EC_{dw}	: Electrical Conductivity of the irrigation and drainage water, resp.	mmho/cm
EMF	: Electromotive force	Volt
ESP	: Exchangeable Sodium Percentage	---
FC	. moisture content at Field Capacity	cm³/100 g
LR	: Leaching Requirement	---
RSC	: Residual Sodium Concentration	eq/m³ = meq/l
SAR	: Sodium Adsorption Ratio	(mmol/l)^½
SP	: moisture content at saturation	cm³/100 g

SUBSCRIPTS

+	: monovalent cation species, positive charge
2+	: divalent cation species

−	:	monovalent anion species, negative charge
cat	:	cations
an	:	anions
a,b,k,l, 1,2	:	chemical species a, b, k, l, 1, 2
dw,iw,cw, rw	:	refers to drainage, irrigation, consumptive and rain water, respectively (cf. chapter 9)
i	:	initial
f	:	final
ox	:	oxidized state
red	:	reduced state
(aq)	:	solvated species in aqueous solution
(g)	:	gas phase
(s)	:	solid phase
s	:	refers to the solid surface
o	:	refers to the bulk equilibrium solutions
r	:	refers to a reaction

CHAPTER 1

COMPOSITION OF THE SOIL

G.H. Bolt and M.G.M. Bruggenwert

A discussion of the chemical behavior of soil, as is the purpose of this text, is logically preceded by a description of the chemical composition of the constituents of the soil. The chemical behavior is then defined as the totality of (physico-) chemical reactions occurring between these constituents (and possibly other materials added to the soil in situ). The difficulty with this reasoning is to allocate a suitable starting point. Strictly speaking one must view the soil as a reaction intermediate between some parent material, often consisting of certain rock formations, and a 'dead' end of the weathering processes (acting on these parent materials), comprising some extremely resistant components like e.g. quartz and some iron and aluminum oxides, plus the wash-out present in the oceans as solutes. This encompassing approach to soil chemistry offers little perspective, as the rate factors of all chemical reactions in soil vary from a half-time of minutes for e.g. certain adsorption reactions to one of centuries for others. If furthermore the chemical reactions occurring in soil are viewed as induced by the action of certain 'environmental' factors, it is clear that for the very slow reactions mentioned above the external action may vary much faster than the rate at which the reaction in the soil is proceeding. It thus seems reasonable to ignore at this stage the reactions of the very slow type comprising the processes that are usually indicated with the name soil formation, and treat these in a separate chapter (cf. chapter 8).

Narrowing down the definition of the chemical behavior of soil to the relatively fast reactions (e.g. with a half time of less than a growing season) one may try to enumerate the composition of the soil in situ. In view of the above, however, it is understandable that 'the' composition of the soil will comprise an enormous list of different components. While referring to existing texts for such listings, here it will only be attempted to present an overall picture based primarily on the reactivity on short term basis, as particularly governed by the solubility and adsorption behavior of the solid phase components. Recognizing that the composition of the mobile phases of the soil, i.e. liquid and gas phase, is determined largely by the interaction of the solid phase with the inflowing mobile phases, these phases will be treated thereafter.

1.1. SOLID PHASE COMPONENTS

1.1.1. Inorganic components

A broad listing of the types of inorganic components found in soil is given in table 1.1., grouped according to the anionic constituents. These will be discussed according to their significance with respect to adsorption behavior and solubility. In this respect one must distinguish between materials with a high specific surface area, S, and the coarse materials with a very low value of S, which are of no importance with regard to adsorption reactions. For practical purposes it is convenient to use a value of S equal to 10^3 m^2/kg (= 1 m^2/g) as a rough division line between coarse materials and fine materials accounting for the surface phenomena treated in chapter 3.

> As is usually treated in Soil Physics texts, the value of S depends on the particle size of the solid phase grains. Using a limit of 2 μm 'equivalent' diameter of the grains, as determined with 'mechanical analysis' employing sedimentation in water, one finds that the value of S corresponding to this diameter is about 1 m^2/g.

As to solubility it is pointed out that materials with a high value of S almost invariably have a very low solubility, since the comparatively rapid translocation of the liquid phase in soil would normally lead to a fast disappearance of very small particles of a readily soluble salt. The solubility of certain salts, or better the solubility of a solid phase salt in soil, depends on the composition of the liquid phase, e.g. notably the pH-value of the latter, and hence division lines remain somewhat arbitrary. From the viewpoint of soil as a biotope, however, it is practical to make a division into three classes, i.c. 'high' solubility covering those salts that in a saturated solution will inhibit plant growth because of the high osmotic pressure (e.g. in excess of 5-10 bars), 'intermediate' solubility, i.e. the saturated solutions not exhibiting osmotic inhibition of plant growth but the solubility being high enough to play a significant role in determining the ionic composition of the soil solution (around 0.01 molar), and the remaining class of 'low' solubility.

Looking at table 1.1 with the above classification in mind one finds as coarse materials with a high solubility the nitrates, most halides and some sulfates. These solid phase components are only present in soil in sizable amounts under exceptional conditions, i.e. in so-called saline soils (at low moisture contents). As large areas on earth are (adversely) affected by the presence of such salts, a separate treatment of these soils (saline- and sodic--soils) is given in chapter 9.

Intermediate solubility could be assigned to a range of salts, the most common ones being gypsum ($CaSO_4 \cdot 2H_2O$) and several carbonates (e.g. calcite which dissolves as bicarbonate at favorable conditions with respect to pH and CO_2-pressure and ferrous sulfate under reduced conditions, cf. chapter 6).

TABLE 1.1.

Some common minerals

Oxides/Hydroxides	
Si-oxides	: quartz, tridymite
Fe-oxides/hydroxides	: goethite, hematite, limonite
Al-oxides/hydroxides	: gibbsite, boehmite, diaspore
Silicates	
Nesosilicates	: olivine (Mg)*, garnet (Ca, Mg, Mn^{2+}, Ti, Cr), tourmaline (Na, Ca, Li, Mg, BO_3), Zircon (Zr)
Inosilicates	: augite (Ca, Mg), hornblende (Na, Ca, Mg, Ti)
Phyllosilicates	: talc (Mg), biotite (K, Mg, F), muscovite (K, F), clay minerals : illite (K), kaolinite, montmorillonite, vermiculite (Mg)
Tectosilicates	: albite (Na), anorthite (Ca), orthoclase (K), zeolites (Ca, Na, K, Ba)
Carbonates	: calcite ($CaCO_3$), dolomite ($MgCa(CO_3)_2$)
Sulfates	: gypsum ($CaSO_4.2H_2O$)
Halides	: halite (NaCl), sylvine (KCl), carnallite ($KMgCl_3.6H_2O$), ($CaCl_2 nH_2O$)
Sulphides	: pyrite (FeS_2)
Phosphates	: apatite ($Ca_5(F,Cl,OH)(PO_4)_3$), vivianite ($Fe_3(PO_4)_2.8H_2O$)
Nitrates	: soda-nitre ($NaNO_3$), nitre (KNO_3)

The remainder of the components of table 1.1 belong to the group with low solubility. Particularly the many different types of silicates and some phosphates should be considered as the (metastable) left-overs from parent materials, i.e. minerals derived from rock formations. While often being characteristic of this parent material, these residual minerals usually play only a very minor role in the incidental chemical behavior of the soil, at least as far as they combine very low solubility with a low specific surface area. Seen with some longer range perspective they are of importance in slowly delivering to the solution phase certain elements necessary for plant nutrition, usually at a very low to extremely low concentration level (notably some minor elements). In table 1.1 some characteristic cations, occurring in

---

*As an indication of the chemical composition of the silicates the cations except Si, Al and Fe, are mentioned.

silicates, (aside from Al and/or Fe which are abundantly present) are indicated. The oxides listed, i.e. those of Si, Al and Fe, may be considered as the stable endpoint of the weathering processes. In chapter 8 the conditions favoring the accumulation of either one will be discussed. For the present purpose it suffices to state that if coarse grained they constitute an inert skeleton of the soil which is hardly interesting from a soil chemical standpoint.

Switching to the inorganic components with a moderately high to high specific surface area, one finds here again the oxides and hydroxides of Fe and Al, and particularly the so-called clay minerals (cf. phyllosilicates). As the mineral structure of the oxides is often fairly simple, reference is made to mineralogy texts for details. It suffices here to mention that they consist 'superficially seen' of a fairly dense packing of O-ions, held together in specific coordination by the metal cations. This internal coordination varies from perfectly regular to rather irregular (specifically if impurities are present), leading to a distinction between crystalline and amorphous oxides and hydroxides. In the hydroxides of Fe and Al, OH-ions take the place of part or all the O-ions present in the oxides, giving rise to cleavage planes in the crystalline forms (cf. also below). Depending on the conditions prevailing during their formation, the specific surface area of these oxides and hydroxides may range from several tens of square meters per gram to the very low values of macroscopic crystals and concretions. In the latter case, given the low solubility, their role is mainly that of contributing to the soil skeleton (at least for the very common Fe- and Al-oxides).

The clay minerals deserve separate treatment as they play a dominant role in many soils. This is mainly due to their (often) very large surface area, which is connected in turn with their lattice structure. These silicate minerals belong to the phyllosilicates which have a layered structure. Two main types may be distinguished:

a. The 2:1 type consisting of two layers of SiO_4-tetrahedrons, all tetrahedrons sharing corners with each other and with an octahedral layer of e.g. $AlO_4(OH)_2$, situated in between. The thickness of this three-layered unit, to be referred to as platelet, is about 10 Å.

b. The 1:1 type, in which one tetrahedral SiO_4-layer shares corners with an octahedral layer of e.g. $AlO_2(OH)_4$. The thickness of this two-layered unit is about 7 Å.

In both types three Mg-ions may take the place of two Al-ions in the octahedral layer, leading to the distinction tri-octahedral and di-octahedral clays. This layered structure explains that the clay minerals occur in plate-shaped crystals. Inasfar as these plates may be extremely thin, the specific surface area may amount to hundreds of square meters per gram.

In discussing further the lattice structure of clay minerals (which turns out to be of decisive importance with regard to the adsorption behavior of these materials) it is easier to consider first the composing units, i.e. the Si- and Al-layers. The Si-layer consists of SiO_4-tetrahedrons arranged in a similar manner as in tridymite, a mineral form of SiO_2. As is shown in figure 1.1 the SiO_4-tetrahedrons share three out of four corners with each other, the fourth one sticking out towards one side. It should be noted that the upper layer of O-ions is densely packed, but with hexagonal 'holes' i.e. one out of four O-ions is missing. The Al-layer consists of a layer of $Al(OH)_6$ octahedrons as in gibbsite, sharing all corners with each other. Here the OH-ions are not quite densely packed, and as is shown in figure 1.1, two thirds of the available positions in the octahedrons are filled with Al-atoms. A similar arrangement with all available positions filled with Mg-ions is the structure of the mineral brucite. Writing out the composition of these constituting layers of the clay minerals as $Si_2O_3(OH)_2$ (using a proton to balance the charge at the fourth corner of the SiO_4-tetrahedron) and as $Al_2(OH)_6$ respectively, one may now visualize clay minerals as condensates of the above layers, according to figure 1.2.

Remembering that the Si- and Al-ions are situated in interstitial holes between the large O-ions, the individual platelet of the 1:1 type consists, superficially seen, of three layers of fairly densely packed O-ions (partly as OH), while the 2:1 type consists of four layers of O-ions (partly as OH).

The above description, although correct in principle, is still incomplete. Clay minerals have another important characteristic, viz. part of the Si and/or Al (or Mg)-atoms have been replaced by cations of lower valence. This isomorphic substitution has taken place during the formation of the clay minerals because the Si- and Al-atoms were not present in exactly the correct ratio. Replacements were then accepted, provided the replacing ion was roughly of the correct size, e.g. Al for Si and Mg for Al. A direct consequence of this substitution is a deficit of positive charge of the clay lattice. As the solution or molten mixture, from which the clay minerals were formed, was electrically neutral, the deficit of positive charge of the clay crystal was compensated for by the adsorption of an equivalent amount of other cations (e.g. Ca, Mg, K, Na etc.) on the exterior surface.

Whereas in the above the lattice structure of the individual platelets of the clay minerals was discussed, it is particularly the combination of these platelets into a larger unit which is decisive with respect to the physico--chemical and physical behavior of the clay. In this respect a main distinction between the 1:1 and 2:1 clays is the 'polar' structure of the 1:1 type (upper side differs from lower side of the platelets). Platelets of this type are bonded together rather tightly via H-bonds between the octahedral OH--groups of one platelet and the O-ions of the tetrahedral layer of the next one. Thus, at least for the dioctahedral forms, very large crystals (up to truly macroscopic dimensions) are common. The most abundant clay mineral of this type is kaolinite, with a specific surface area ranging from tens of square meters to negligible values, depending on the degree to which the platelets are bonded together as multilayer plates. Adding to this the low degree of substitution (if present at all), this type of clay is relatively 'inert' from a physico-chemical point of view.

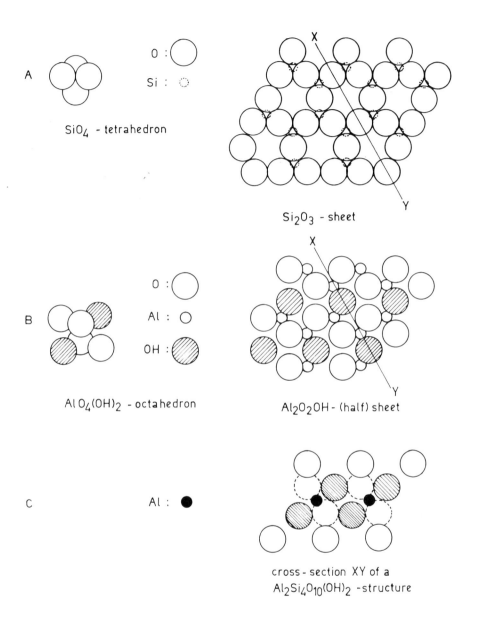

Fig. 1.1. Scale drawing of the structure of a 2 : 1 clay mineral and its composing elements.

A: SiO_4-tetrahedrons may form Si_2O_5-sheets, analogous to the Si_2O_3-structure shown, but containing O atoms underneath every Si-atom.

B: $Al(OH)_6$-octahedrons may form $Al_2(OH)_6$-sheets (gibbsite), analogous to the $Al_2O_2(OH)$-structure shown, with the same amount of O and OH underneath.

C: A cross-section of the condensate of the above sheets in a 2 : 1 arrangement.

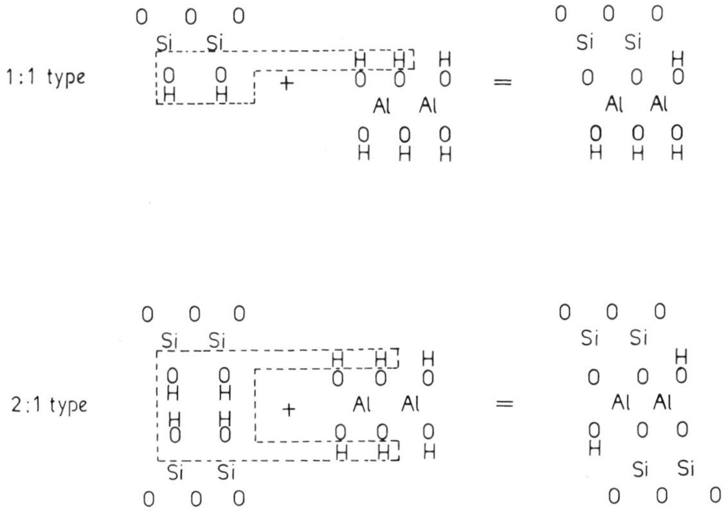

Fig. 1.2. Clay minerals as condensates of $Si_2O_3(OH)_2$ and $Al_2(OH)_6$

Much in contrast the 2:1 type platelets are mirror symmetric and bonding between them depends on specific conditions. The muscovites (mica), with a high degree of substitution, show a fairly perfect coupling mechanism between platelets via (dehydrated) K-ions partly sunken into the hexagonal holes of the tetrahedral layers. This again leads to a low value of the specific surface area. The illites have a similar arrangement, but due to partial replacement of the K-ions by H_3O-ions the binding is less strong and the number of individual platelets combined into one unit is only e.g. 5-10. Accordingly S is large, e.g. 80 m²/g for a unit of 10 platelets with total thickness of about 100 Å. These clays, also named 'hydrous mica' thus show the typical behavior associated with large S and a considerable substitution charge (cf. section 4.1). The tri-octahedral chlorites form larger units again, consisting of platelets bonded together mainly with Mg-ions.

Finally the montmorillonites (and to a lesser extent the vermiculites) are the extreme examples with respect to the value of S. Probably due to the deeper seating of the substitution charges, bonding between montmorillonite platelets is 'incidental and variable'. This means that one may easily obtain a dispersion of individual platelets of 10 Å thickness (Na-montmorillonite, cf. chapter 4), but at low moisture content and with di- or trivalent cations larger units ('polyplates') are formed with a variable spacing between the individual platelets. Because of this these clays are often referred to as 'expanding lattice' clays. The highly dispersed Na-montmorillonite system mentioned above has a specific surface area of 800 m²/g, and, having again a considerable amount of substitution, this clay mineral shows all properties characteristic for clays (adsorption, swelling, etc., cf. following chapters) to an extreme degree. This explains the widespread usage of this material in many studies of clay behavior. The vermiculites exhibit bonding between the platelets via Mg-ions, but these ions may be displaced without too much difficulty.

A summarizing list of the mentioned types of clay minerals is presented in table 1.2, the underlined ones being the most widespread. In this respect illite is typical for many river deposits all over the world but specifically in the temperate regions. Montmorillonites are of local importance, mainly in the tropics and subtropics (black cotton soils or Vertisols) and are known for their extreme and often adverse behavior (cf. chapter 9). Kaolinites are dominant in many tropical areas with lateritic weathering.

TABLE 1.2.

Some clay minerals

1:1 type :	probably low degree of substitution
	    a. di-octahedral: <u>kaolinite</u>
	                      halloysite
	    b. tri-octahedral: antigorite (tubular)

2:1 type :	mainly substitution in tetrahedral layer
	    a. di-octahedral: muscovite, <u>illite</u> (decreasing interlayer K)
	    b. tri-octahedral: chlorite (interlayer Mg)
	mainly substitution in octahedral layer
	    a. di-octahedral: <u>montmorillonite</u>
	    b. tri-octahedral: <u>vermiculite</u> (interlayer Mg)

1.1.2. The organic components

While the inorganic components may be traced back to parent rocks - - though often via translocation (usually by water) - the organic components all stem from the biomass which is characteristic for a 'living' soil. The paramount importance of the biosphere in the weathering processes acting on the inorganic parent material mentioned above is digressed upon in chapter 8. While leaving a discussion of the nature of biochemical processes taking place in soil to other texts in the fields of Soil Microbiology and Soil Fertility, the outcome of the metabolic activities of soil organisms will in this text be acknowledged summarily as resulting in the production and/or consumption of CO_2, O_2 and sometimes H_2S, plus the left-over in the form of (metastable) residues (cf. see also section 8.2.5).

Though strictly speaking both living organisms and the organic components formed by decay of the latter could be considered as organic components of the soil, only the non-living components will be considered here. These are organic compounds formed by chemical and biological decay of mainly plant materials.

One should distinguish between materials in which the anatomy of the parent substance may still be recognized and materials which have decayed completely. The first group is usually of minor concern from a soil chemical point of view, because the intact structure has a relatively small specific surface area and thus is rather inactive as adsorbens. From the viewpoint of soil physics it may be of greater significance (e.g. protection of the soil surface by a leaf-mulch), whereas for organic soils (peat soils) the incompletely decayed organic material often forms the bulk of the soil skeleton.

The 'end' product of the decay of plant material in soil is sometimes referred to as humus. It has been found that often the humus content of the soil stabilizes at a fairly definite amount, depending on climate and cultural practices. The ratio between carbon and nitrogen (C/N-ratio) usually narrows down in a humification process from a value in excess of 20 for 'fresh' material to a value of 8-15 for the rather stable endproduct. The chemical composition and structure of humus cannot be specified precisely (chemical instability of the material), so a generalized chemical formula does not exist. Considering the diversity of the parent material, the superficial uniformity of many types of humus is actually remarkable. This may be caused by the fact that only few components of the parent material survive the decay process (possibly lignin compounds). It is also possible that humification is in fact an 'ad random' resynthesis of organic compounds based on fairly elementary break-down products of the parent material (e.g. polysaccharides, amino acids, phenols and lignin fragments). Schematizing the situation it could perhaps be stated that different types of humus vary mainly in the arrangement and frequency of characteristic groups, but not so much in overall structure. According to this scheme humus could be seen as a branched, coiled polymer with certain functional groups. It should be noted here that the external shape of these polymers is flexible, e.g. the coils may become stretched upon adsorption on clay minerals. Even if coiled up the internal space remains accessible, so the specific surface area (internal and external) is large.

A number of the characteristic groups of the polymer chains contains hydroxyls, many of which may dissociate depending on the pH of the system. Inasfar as one could study the dissociation equilibrium for the different types of OH-groups without interference from neighboring OH-groups (e.g. at high electrolyte level of the system, cf. chapter 3) one should expect probably a fairly continuous range of pK-values. Even in simple organic acids the dissociation constant varies greatly depending on the chemical composition of adjacent groups (cf. acetic acid with a pK-value of about 5 and trichloroacetic acid with a pK-value of about 1). Considering a full range of carboxyl groups present in many different configurations along the polymer chain, as well as phenolic OH-groups, it is clear that particular pK-values of humic materials can not be given. Another important feature of the chemical structure of the hetero-polymer chains is the occurrence of particular combinations of different groups (involving the OH-groups discussed) which are able to form complexes with certain metal cations, notably the transition metals (cf. section 4.5.2). These complexing sites are usually referred to as ligands. Experimental evidence indicates that these ligands consist mainly, if not entirely, of particular structural combinations of the mentioned acidic OH-groups.

Summarizing the situation from a physico-chemical point of view, humus may be considered as an amorphous compound with varying external dimensions (coiling and stretching of the polymer chains), a large internal surface area (in the coil), a variable charge and a strong tendency to form complexes with certain cations. It should also be noted that in soil the organic matter fraction will contain humus compounds of varying molecular weight and that polymerization (or condensation) and depolymerization reactions may occur continually, i.e. the frequency distribution of molecular sizes may change with conditions.

In view of these considerations it appears that a description of the composition of soil organic matter is more or less limited to an inventory of the types of groups that are likely to be present, and perhaps a fractionation on the basis of solubility in different solvents.

In this context the names Fulvic, Hymatomelanic and Humic acids have been used in this order for the fractions with decreasing solubility and, presumably, increasing molecular weight.

1.2. THE LIQUID PHASE

In general terms the liquid phase of the soil (often referred to as the soil solution) is an usually dilute, aqueous solution of common salts from the ions Na, K, Mg, Ca, Cl, NO_3, SO_4, HCO_3 etc. In addition it may contain small amounts of many other ions, depending on the presence in the solid phase of compounds with low solubility. Also organic compounds will be present, usually stemming from breakdown (or decay) of the soil organic matter. As some of these organic compounds form complexes with e.g. heavy metal ions, the organic constituents of the soil solution may contribute significantly to the mobility of such ions in soil (cf. below). Finally all types of organic and inorganic 'pollutants' may be present in the soil solution as a result of human activities in, on or near the soil.

Any attempt to specify the composition of 'the' soil solution is somewhat futile, the soil water being highly mobile as part of the hydrological cycle. The latter involves additions of water at the soil surface (rain, irrigation water, waste water) and disappearance from the soil via plants and the vapor phase, and as drainage water joining the groundwater or surface water.

Leaving the discussion of soil water movement to Soil Physics it is pointed out here that the soil water is *the* carrier for transport in the soil system. It provides a connecting pathway for diffusion of solutes towards, or away from, adsorbing surfaces and plants, while also dragging along the solutes in its flow (cf. chapter 7). Admitting that nothing is static as far as the soil solution is concerned it is nevertheless of interest to study the solubility and adsorption equilibria involving the soil solution, as these equilibria constitute conditions which are always being approached and often reached for limited time periods. Special attention is given to these aspects in chapter 2 and 3.

While foregoing any attempt to specify the composition of the soil solution with any precision it may still be stated that in 'normal' soils in the temperate region the soil solution at field capacity is around 0.01 normal in total electrolyte, with roughly equal amounts of mono- and divalent cations. Around the wilting point the concentration may increase to about 0.1 normal. In contrast, so-called saline soils (cf. chapter 9) will contain a soil solution with at least about 0.1 normal salt at field capacity.

1.3. THE GAS PHASE

The volume occupied by the gas phase in soil is complementary to the volumetric moisture content, the sum of both phases constituting the pore volume of the soil. Admitting beforehand that a discussion of the significance of the volume ratio of liquid and gas phase belongs to Soil Physics, it is mentioned here that probably around a gas-filled pore volume of about one tenth of the total soil volume, the connections between gas-filled pores become severely hampered. As furthermore diffusion in the gas phase is relatively rapid it may be said that unless a particular volume of soil is separated from the atmosphere by a layer having less than 10 % gas-filled pores, the composition of the soil gas phase strives continually towards that of the earth's atmosphere, i.e. 78 % N_2, 21 % O_2 and 1 % rare gases, via diffusion processes.

The second process governing the composition of the soil gas phase is the consumption of O_2 and production of CO_2 by living organisms. The two processes together lead towards a roughly stationary state, corresponding to a gradual increase of the CO_2-concentration with depth from about 0.03 % at the soil surface to e.g. 1-5 % underneath the plant rooting zone. The increase of CO_2 is then accompanied by a decrease of O_2 with depth. Notable exceptions to this pattern arise under a dense surface crust or other dense layers. Once the diffusion towards the atmosphere is seriously impeded the O_2-concentration may drop to values close to zero. Both the O_2- and the

CO_2-concentration influence chemical equilibria in the soil as will be discussed in chapters 2 and 6.

Finally it is mentioned that several organic pesticides exhibit fairly high vapor pressures. Thus occasionally noticeable amounts of these compounds appear in the soil gas phase (fumigation procedures !). For completeness the presence of water vapor at about 1 % of the total gas volume is noted.

CHAPTER 2

CHEMICAL EQUILIBRIA

I. Novozamsky, J. Beek, G.H. Bolt.

In chapter 1 the nature and composition of the different components of the soil were discussed. Naturally conditions may arise under which certain components will react with each other to form other chemical components. Trivial examples of this situation are e.g. the formation of solid salts in a soil solution upon disappearance of water, the precipitation of certain hydroxides upon a rise of pH and the escape of CO_2 from $CaCO_3$ present in soil upon a decrease of pH.

2.1. THE CONDITION FOR EQUILIBRIUM

Limiting the discussion here to reactions involving crystalline solids, ions in solution and gases in the soil atmosphere, some general rules governing chemical equilibria between these mentioned components will be given. Of prime importance here is the reversibility of the reactions studied. If this condition is fulfilled there will always exist an equilibrium state corresponding to a particular composition of the system with regard to reactants and reaction products. At this equilibrium state the reaction rates in forward and backward direction just compensate each other and the composition remains constant.

Thus for a reaction of the type:

$$aA + bB = cC + dD \tag{2.1}$$

equilibrium is achieved when the forward reaction rate, s_1, given by:

$$s_1 = k_{r1} [A]^a [B]^b \tag{2.2}$$

equals the backward reaction rate, s_2, according to:

$$s_2 = k_{r2} [C]^c [D]^d \tag{2.3}$$

where brackets denote the concentration of the reacting species and k_{r1} and k_{r2} are rate coefficients. So at equilibrium the following relationship can be derived:

$$\frac{k_{r1}}{k_{r2}} = \frac{[C]^c [D]^d}{[A]^a [B]^b} = K \tag{2.4}$$

where K is the equilibrium constant of the reaction, which is still a function of temperature and pressure. This so-called mass-action principle is only approximate at best.

Application of thermodynamics to chemical reactions yields exact information with regard to conditions necessary for equilibrium. This condition is then formulated in terms of the chemical activities of the reactants and products rather than concentrations. The condition for thermodynamic equilibrium applied to (2.1) gives:

$$\frac{(C)^c (D)^d}{(A)^a (B)^b} = K^\circ \tag{2.5}$$

where the parentheses refer to chemical activities of the species and K° represents the thermodynamic equilibrium constant. This thermodynamic equilibrium constant may be expressed in terms of the standard free enthalpy of the reaction, ΔG_r° specified in units of energy per mole (e.g. kcal/mole), according to:

$$K^\circ = \exp(-\Delta G_r^\circ / RT) \tag{2.6}$$

in which R and T are the gas constant and the absolute temperature, respectively. As will be elaborated on in section 2.4, the value of ΔG_r° may be derived in principle from tabulated values of the standard free enthalpy of the reactants and products of the reaction studied. Furthermore the activities of reactants and products in the reaction mixture are related to their concentrations (cf. following section). Accordingly equation (2.5) together with (2.6) prescribes a relationship between the concentrations at the reaction equilibrium. Some examples of such relationships are given in section 2.6.

2.2. STANDARD STATES AND ACTIVITIES

To each chemical species in a reaction mixture a certain amount of ('free') energy can be ascribed. This amount of energy, expressed per unit amount of species, is called chemical potential (potential ≡ energy per unit amount of matter). This entity which is indicated with the symbol μ_k depends on the pressure, P, and temperature, T, of the system, on the chemical nature of the species and eventually on its mixing ratio with other species in the mixture. As the latter aspect is of prime concern for calculating mixture compositions, one may formally separate out this aspect by writing:

$$\mu_k \equiv \mu_k^o + \Delta\mu_k^c, \tag{2.7}$$

in which the last term then covers the concentration-dependent part of μ_k. The first term of the RHS of equation (2.7) is then referred to as the chemical potential of the species in its standard state. Often one uses the pure compound for this state, under standard pressure and temperature. For ideally behaving mixed systems (e.g. certain gases and solid solutions) it is now found that the concentration-dependent term is related to the mole-fraction of the species k in the mixture, M_k, according to:

$$\mu_k = \mu_k^o + RT \ln M_k \tag{2.8}$$

Accordingly one finds that for gas-mixtures at one bar total pressure (which behave practically ideally) $\Delta\mu_k^c$ equals $RT \ln P_k$, as the partial pressure, P_k, in this case equals the mole fraction.

Allowing for non-ideal behavior in mixtures equation (2.8) is formally replaced by:

$$\mu_k \equiv \mu_k^o + RT \ln a_k \tag{2.9}$$

in which a_k is named the activity of species k. Using again the pure compound as standard state one thus finds that the activity of a pure compound must be unity (as is its mole fraction). Where small amounts of impurities in a given compound usually follow ideal mixing behavior one may safely take the activity of the major compound equal to its mole fraction in such a case. Thus considering a solid solution of calcite-siderite with the composition $(Ca_{0.94} Fe_{0.06})CO_3$ the activity of calcite may be taken at 0.94 without introducing much error. In the same manner one may estimate the activity of water in a 3 molar solution at about $56/59 \approx 0.95$.

For dissolved substances it is rather impractical to use the pure compound as the standard state. Thus dissolution itself implies an important change in the properties of a compound, viz. the bonds between the molecules in the solid crystal are broken, and the individual molecules (or sometimes ions) swarm out in the solvent. Obviously the chemical potential of the crystalline compound is then a rather far-fetched standard of comparison for the value of the chemical potential of the dissolved species. As furthermore all calculations needed involve only differences in chemical potentials (cf. section 2.4) one has conveniently chosen a separate standard state for dissolved substances viz. the state corresponding to a (hypothetical) solution exhibiting ideal mixing behavior and with a concentration, c, of 1 mole of solute per liter. For such a hypothetical solution equation (2.8) would thus read:

$$\mu_k = \mu_k^\circ + RT \ln c_k \tag{2.10}$$

which implies that for ideal behavior in solution the chosen standard state makes a_k equal to the molar concentration. As all solutions approach ideality towards infinite dilution, equation (2.10) is indeed valid at very low concentrations. For higher concentrations one now introduces an activity coefficient f_k, according to $a_k \equiv f_k \cdot c_k$, such that

$$\mu_k = \mu_k^\circ + RT \ln f_k \cdot c_k \tag{2.11}$$

As will be shown in the next section the value of f_k may be estimated from certain equations, or from tables based on experimental observations, once the composition of the mixture is known. Obiously $f_k \to 1$ for $c_k \to 0$. The standard state for dissolved substances is thus defined as a solution for which $f_k \cdot c_k = 1$.

Because the value of f_k depends on the nature of the solute and the total solution composition, it is evident that the standard state of a specific solute in a given solvent corresponds to a particular concentration. For instance the standard states of the solutions of NaCl and KCl, respectively, in H_2O correspond to concentrations of 1.50 normal NaCl and 1.74 normal KCl, respectively, the activity coefficients being $f_{NaCl} = 0.666$ and $f_{KCl} = 0.575$ at these concentrations.

2.3. ACTIVITY COEFFICIENTS OF IONS IN AQUEOUS SOLUTIONS

The existence of long range electrostatic interactions between ions is the main reason for non-ideal behavior of these ions in solutions. Debye and Hückel developed a model for the estimation of the activity coefficients of ions in solution based on the theory of the electrostatic field.

The ions become distributed in the solution in a semi-ordered manner, the immediate environment of a particular ion always containing an excess of ions of opposite sign. The free enthalpy change associated with this arrangement may then be calculated, which forms the basis for the computation of the activity coefficient. This may be demonstrated with equations (2.10) and (2.11), where the difference between μ_k (from (2.11) and μ_k (ideal) from (2.10) represents $\Delta\mu_k$ (interaction) $\equiv RT \ln f_k$. In all equations μ_k corresponds to the free enthalpy of the system per mole of solute (actually the partial molar free enthalpy).

The outcome of the above theory is the well-known (extended) D(ebye)--H(ückel) equation:

$$-\log f_k = \frac{A z_k^2 \sqrt{I}}{1+å_k B \sqrt{I}} \qquad (2.12)$$

in which I ($=\frac{1}{2}\Sigma_k c_k z_k^2$) indicates the ionic strength of the solution, i.e. a weighted mean molarity; $å_k$ is an ionic size parameter, that is the diameter of the solvated ion in Å units and A and B are constants for a given system (actually proportional to $(\epsilon T)^{-1.5}$ and $(\epsilon T)^{-0.5}$, respectively, with ϵ = dielectric constant, T = temperature). For aqueous solutions at 25°C the constant A equals 0.51 (mol/l)$^{-0.5}$ and B is 0.33 Å^{-1} mole$^{-0.5}$ l$^{0.5}$. As is shown in table (2.1), $å_k$ varies between about 3 and 11 Å. As for many common ions $å_k$ is roughly 3 Å, equation (2.12) is often simplified to:

$$-\log f_k = A z_k^2 \sqrt{I} / (1 + \sqrt{I}) \qquad (2.12a)$$

Table (2.2) lists the outcome of equation (2.12) for different ionic sizes up to I = 0.1 molar.

As experiments involving chemical reactions always imply the transfer of ions in electrically neutral combinations, the relevant equations describing such reactions contain only those combinations of single ionic activities that pertain to the (weighted geometric) mean activity of neutral electrolytes. This mean activity of an electrolyte containing per molecule ν_+ ions of the valence z_+ and ν_- ions of the valence z_- (such that $\nu_+ z_+ = -\nu_- z_-$) is then defined as:

$$a_\pm = (a_+^{\nu_+} \cdot a_-^{\nu_-})^{1/\nu}$$

with $\nu = \nu_+ + \nu_-$. In the same way one introduces the mean activity coefficient according to:

$$f_\pm = (f_+^{\nu_+} \cdot f_-^{\nu_-})^{1/\nu} \qquad (2.13)$$

Introduced into equation (2.12a), one thus finds:

$$-\log f_\pm = \frac{|z_+ \cdot z_-| A \sqrt{I}}{1 + \sqrt{I}} \qquad (2.12b)$$

Equation (2.12) changes accordingly, with the understanding that å should now be taken as the mean diameter of the ions involved, i.e. $å_\pm = (å_+ + å_-)/2$. The equations (2.12) are reasonably reliable for values of I up to 0.1 molar. In systems that contain particularly the common (i.e. small) ions one often uses the 'Davies' extension of the D-H equation according to:

$$-\log f_\pm = 0.5 |z_+ \cdot z_-| \left(\frac{\sqrt{I}}{1+\sqrt{I}} - 0.3\, I \right) \qquad (2.12c)$$

TABLE 2.1.

Values of the parameter å used in the extended Debye-Hückel equation (2.12) for 130 selected ions (from Kielland, 1937).

Charge 1

9	H^+
8	$(C_6H_5)_2CHCOO^-$, $(C_3H_7)_4N^+$
7	$OC_6H_2(NO_3)_3^-$, $(C_3H_7)_3NH^+$, $CH_3OC_6H_4COO^-$
6	Li^+, $C_6H_5COO^-$, $C_6H_4OHCOO^-$, $C_6H_4ClCOO^-$, $C_6H_5CH_2COO^-$, $CH_2CHCH_2COO^-$, $(CH_3)_2CCHCOO^-$, $(C_2H_5)_4N^+$, $(C_3H_7)_2NH_2^+$
5	$CHCl_2COO^-$, CCl_3COO^-, $(C_2H_5)_3NH^+$, $(C_3H_7)NH_3^+$
4	Na^+, $CdCl^+$, ClO_2^-, IO_3^-, HCO_3^-, $H_2PO_4^-$, HSO_3^-, $H_2AsO_4^-$, $Co(NH_3)_4(NO_2)_2^+$, CH_3COO^-, CH_2ClCOO^-, $(CH_3)_4N^+$, $(C_2H_5)_2NH_2^+$, $NH_2CH_2COO^-$, $^+NH_3CH_2COOH$, $(CH_3)_3NH^+$, $C_2H_5NH_3^+$
3	OH^-, F^-, CNS^-, CNO^-, HS^-, ClO_3^-, ClO_4^-, BrO_3^-, IO_4^-, MnO_4^-, K^+, Cl^-, Br^-, I^-, CN^-, NO_2^-, NO_3^-, Rb^+, Cs^+, NH_4^+, Tl^+, Ag^+, $HCOO^-$, $H_2(citrate)^-$, $CH_3NH_3^+$, $(CH_3)_2NH_2^+$

Charge 2

8	Mg^{2+}, Be^{2+}
7	$(CH_2)_5(COO)_2^{2-}$, $(CH_2)_6(COO)_2^{2-}$, $(congo\ red)^{2-}$
6	Ca^{2+}, Cu^{2+}, Zn^{2+}, Sn^{2+}, Mn^{2+}, Fe^{2+}, Ni^{2+}, Co^{2+}, $C_6H_4(COO)_2^{2-}$, $H_2C(CH_2COO)_2^{2-}$, $(CH_2CH_2COO)_2^{2-}$
5	Sr^{2+}, Ba^{2+}, Ra^{2+}, Cd^{2+}, Hg^{2+}, S^{2-}, $S_2O_4^{2-}$, WO_4^{2-}, Pb^{2+}, CO_3^{2-}, SO_3^{2-}, MoO_4^{2-}, $Co(NH_3)_5Cl^{2+}$, $Fe(CN)_5NO^{2-}$, $H_2C(COO)_2^{2-}$, $(CH_2COO)_2^{2-}$, $(CHOHCOO)_2^{2-}$, $(COO)_2^{2-}$, $H(citrate)^{2-}$
4	Hg_2^{2+}, SO_4^{2-}, $S_2O_3^{2-}$, $S_2O_8^{2-}$, SeO_4^{2-}, CrO_4^{2-}, HPO_4^{2-}, $S_2O_6^{2-}$

Charge 3

9	Al^{3+}, Fe^{3+}, Cr^{3+}, Se^{3+}, Y^{3+}, La^{3+}, In^{3+}, Ce^{3+}, Pr^{3+}, Nd^{3+}, Sm^{3+}
6	$Co(ethylenediamine)_3^{3+}$
5	$Citrate^{3-}$
4	PO_4^{3-}, $Fe(CN)_6^{3-}$, $Cr(NH_3)_6^{3+}$, $Co(NH_3)_6^{3+}$, $Co(NH_3)_5H_2O^{3+}$

Charge 4

11	Th^{4+}, Zr^{4+}, Ce^{4+}, Sn^{4+}
6	$Co(S_2O_3)(CN)_5^{4-}$

TABLE 2.2.

Single-ion activity coefficients calculated from the extended Debye-Hückel equation (2.12) at 25°C.

				Ionic strength			
a	0.001	0.0025	0.005	0.01	0.025	0.05	0.1
				Charge 1			
9	0.967	0.950	0.933	0.914	0.88	0.86	0.83
8	0.966	0.949	0.931	0.912	0.88	0.85	0.82
7	0.965	0.948	0.930	0.909	0.875	0.845	0.81
6	0.965	0.948	0.929	0.907	0.87	0.835	0.80
5	0.964	0.947	0.928	0.904	0.865	0.83	0.79
4	0.964	0.947	0.927	0.901	0.855	0.815	0.77
3	0.964	0.945	0.925	0.899	0.85	0.805	0.755
				Charge 2			
8	0.872	0.813	0.755	0.69	0.595	0.52	0.45
7	0.872	0.812	0.753	0.685	0.58	0.50	0.425
6	0.870	0.809	0.749	0.675	0.57	0.485	0.405
5	0.868	0.805	0.744	0.67	0.555	0.465	0.38
4	0.867	0.803	0.740	0.660	0.545	0.445	0.355
				Charge 3			
9	0.738	0.632	0.54	0.445	0.325	0.245	0.18
6	0.731	0.620	0.52	0.415	0.28	0.195	0.13
5	0.728	0.616	0.51	0.405	0.27	0.18	0.115
4	0.725	0.612	0.505	0.395	0.25	0.16	0.095
				Charge 4			
11	0.588	0.455	0.35	0.255	0.155	0.10	0.065
6	0.575	0.43	0.315	0.21	0.105	0.055	0.027
5	0.57	0.425	0.31	0.20	0.10	0.048	0.021

which relation remains valid to about I = 0.5 molar. The Davies equation is one of the many semi-empirical equations used for the calculation of the activity coefficients at 'medium' ionic strength. The coefficient in front of the last term is obtained by curve-matching on experimental data.

Equation (2.12c) becomes an inferior approximation if ions of extreme sizes are involved. In that case an extension of equation (2.12) similar to the one used to transform equation (2.12b) into (2.12c) can be considered. Values of the parameter å according to Kielland (1937) are listed in table 2.1.

2.3.1. Activity coefficients in mixed aqueous solutions at high ionic strength

If the ionic strength exceeds 0.5 molar, the activity coefficients of the ions involved should be derived from experimental data (e.g. from measured osmotic pressure or with the help of electrodes reversible to the species involved). For solutions of single salts such experimental data have been tabulated, obviously in the form of $f_\pm$, as single ionic activities cannot be determined experimentally. In order to estimate activity coefficients in mixed systems with the help of the mentioned tables, one usually assumes that the ratio of the activity coefficients of two salts with a common anion (or cation, respectively), as measured in the single salt solutions at the same ionic strength, I, may be equated to the ratio of the cationic (or anionic, respectively) activity coefficients at that value of I. Thus if the activity coefficients of the mono-monovalent salts KCl, MCl and KA at ionic strength I are given, this assumption implies that also in the mixture:

$$\frac{f_{+,M}}{f_{+,K}} = \frac{f^2_{\pm,MCl}}{f^2_{\pm,KCl}} \quad \text{and} \quad \frac{f_{-,A}}{f_{-,Cl}} = \frac{f^2_{\pm,KA}}{f^2_{\pm,KCl}} \tag{2.14}$$

In this manner the activity coefficients of all cations and anions present in the system at ionic strength I may be expressed in terms of $f_{+,K}$ and $f_{-,Cl}$ at this value of I, or in any other pair of reference ions, even if the latter are not present in the system.

Any equilibrium condition pertaining to one or more of the ions present in the system may then be written out in terms of concentrations multiplied with the activity coefficients calculated according to (2.14), leaving a factor $f^n_{\pm,KCl}$ (where n depends on the reaction involved). As follows from the above there is no need to attempt to introduce the absolute value of the single ionic activity coefficients $f_{+,M}$ and $f_{-,Cl}$. Nevertheless it is sometimes useful to do so, especially if one uses computer programs to calculate equilibrium concentrations in complicated mixtures. As these calculations generally involve simultaneously the use of equilibrium conditions in terms of activities and balance equations in terms of concentrations, the latter are conveniently expressed in terms of single ionic activities according to $c_k = a_k/f_k$. To this purpose one then introduces an (arbitrary) convention with respect to the single ionic activity coefficients of the pair of reference cations. One of these conventions is to use KCl as the reference salt and putting $f_{+,K} = f_{-,Cl} = f_{\pm,KCl}$. The calculation of single ionic activity coefficients with the help of this convention and the assumed relations (2.14) has been referred to as the 'mean salt method' (Garrels and Christ, 1965). Obviously the single ion activities found with this method have no significance beyond a 'conventional' one. On the other hand, the ionic concentrations calculated with this convention are correct within the range of validity of equation (2.14).

2.4. CALCULATION OF EQUILIBRIUM CONSTANTS FROM THERMODYNAMIC DATA

The standard free enthalpy of a chemical reaction, ΔG_r°, is defined as the change in free enthalpy if the minimum integral number of moles of reactants at their standard state are combined to deliver the ensuing reaction products again at their standard state. Thus for a reaction of the type:

$aA + bB \rightleftharpoons cC + dD$,

one finds because of the conservation of free enthalpy:

$$\Delta G_r^\circ = (c\bar{G}_C^\circ + d\bar{G}_D^\circ) - (a\bar{G}_A^\circ + b\bar{G}_B^\circ) \tag{2.15}$$

The free enthalpy (per mole) of a pure substance involved, at their standard state, $\bar{G}^\circ$, is often referred to as the free enthalpy of formation of the compound, $\bar{G}_f^\circ$, and then specified relative to the free enthalpy of the composing elements in their pure form (at the same standard values of P and T). It is also indicated with the symbol μ° (cf. section 2.2.) and then named the standard chemical potential. The value of $\bar{G}_f^\circ$ of many chemical species at 298 °K and 1 atm. has been listed in tables, thus allowing the calculation of ΔG_r° with the help of equation (2.15) and the corresponding value of $\log K^\circ = 0.43 \ln K^\circ$ with equation (2.6). At 25 °C, specifying $\bar{G}^\circ$ in kcal/mole, one thus finds:

$$\log K^\circ = -\Delta G_r^\circ/1.364 \tag{2.16}$$

In those cases where only the standard enthalpy of formation, $\bar{H}_f^\circ$ and the standard entropy, S°, are tabulated one must first calculate ΔH_r° and ΔS_r° with equations similar to (2.15), whereafter ΔG_r° is found from:

$$\Delta G_r^\circ = \Delta H_r^\circ - T \Delta S_r^\circ \tag{2.17}$$

The effect of a temperature change on K° may be estimated by means of the van 't Hoff equation:

$$\frac{d\ln K}{dT} = \frac{\Delta H_r^\circ}{RT^2} \tag{2.18}$$

Although in principle ΔH_r° is itself a function of the temperature, as the specific heat of reactants and products are not the same, this effect is often negligible (at least if ΔH_r° exceeds 10 kcal/mole). In that case (2.18) may be integrated to yield:

$$\log K_{T_2}/K_{T_1} = \frac{0.43 \Delta H_r^\circ}{R}\left(\frac{1}{T_1} - \frac{1}{T_2}\right) \tag{2.19}$$

2.5. SOME THERMODYNAMIC CONSIDERATIONS

In the preceding sections use was made of a number of expressions, which follow from thermodynamics. While referring the reader to standard texts on this subject, a limited effort will be made here to render part of the argumentation a bit more plausible. Thus the reasoning leading to equations (2.5) and (2.6) may be summarized roughly as follows.

a. Each (chemical) system possesses a certain amount of free enthalpy, G, which is related to the internal energy, U (consisting of the chemical energy stored in the components present), the pressure P, the volume of the system V, the temperature T and the entropy of the system S (in turn related to the probability of arrangement of all the atoms present in the system), according to:

$$G = U + PV - TS \qquad (2.20)$$

b. It follows from the second law of thermodynamics that for a system at equilibrium, at a given P and T, the value of G must be at a minimum. Thus any change in the system, e.g. a chemical reaction leading to a change in composition of the system, is accompanied by an increase of G_{PT}. The above may be summarized by the statement that at equilibrium $dG_{TP} = 0$.

c. The free enthalpy of the system, G, may be apportioned over the components present according to:

$$G = \Sigma_k n_k \bar{G}_k \qquad (2.21)$$

in which:

$$\bar{G}_k = \left(\frac{\partial G}{\partial n_k}\right)_{P,T,n_l}$$

is termed the partial molar free enthalpy of component k in the system, i.e. the increase of G if a unit amount of component k is added to the system, while keeping P, T and the number of moles of all other components, n_l, constant. As furthermore G is fully determined by the composition of the system (specified as n_k) once the pressure and temperature are fixed, one finds:

$$dG_{PT} = \frac{\partial G}{\partial n_1} dn_1 + \frac{\partial G}{\partial n_2} dn_2 \ldots \text{etc.} \equiv \Sigma_k \bar{G}_k dn_k \qquad (2.22)$$

d. Applying now the condition stipulated under b one finds that at equilibrium (at constant P, T):

$$\Sigma_k \bar{G}_k dn_k = 0 \qquad (2.23)$$

e. If a reaction is studied of the type as specified in equation (2.1), any shift from its equilibrium state implies that the changes dn_k are proportional to the reaction coefficients a, b, c, d, with the understanding that if one proceeds an infinitesimal amount from left to right the coefficients c and d obtain positieve signs (they refer to the products), while a and b are negative (the reactants are disappearing). Thus the equilibrium condition may be written as:

$$\Sigma'_k r_k \bar{G}_k = 0 \qquad (2.24)$$

where the reaction coefficients r_k are taken negative for the reactants and positive for the products.

f. The partial molar free enthalpy of a component $\bar{G}_k$ is often referred to shortly as the chemical potential of the component and then indicated with the symbol μ_k. This chemical potential of component k depends now on P and T (of the system), the chemical nature of the component *and* on its mixing ratio with other components in the system (cf. equation 2.7). Separating off this last aspect (as it refers to the composition of the system, which is of special interest if one studies equilibrium systems) one then introduces equation (2.9):

$$\mu_k \equiv \mu_k^o + RT \ln a_k \qquad (2.25)$$

The chemical potential of the component in its chosen standard state, μ_k^o, is then equal to the standard partial molar free enthalpy, $\bar{G}_f^o$ (introduced in section 2.4) if the system is studied at standard values of P and T.

g. Combining now (2.24) and (2.25) one finds at standard values of P and T:

$$\Sigma_k' r_k \bar{G}_k^o + RT \ln \Pi' a_k^{r_k} = 0 \qquad (2.26)$$

Making use of equation (2.15) the first term is identified as ΔG_r^o. The second term involves the 'reduced activity product' of all components, the coefficients r_k having positive signs for products and negative for reactants. This then delivers the equation:

$$\ln K^o \equiv \ln \Pi' a_k^{r_k} = -\Delta G_r^o / RT \qquad (2.27)$$

which is the same as (2.6).

2.6. ILLUSTRATIVE CALCULATIONS

The actual application of the theoretical considerations presented in the previous sections to the prediction of the composition of systems at equilibrium may be elucidated with some rather simple examples. The relevant thermodynamic data are listed in table 2.3. Pointing out that such data are derived from experiments, it is to be understood that (usually slight) differences are found depending on the source of the data. For the present purpose of demonstration of a procedure the precise value is rather irrelevant, however. Accordingly the data have been rounded off and the source is omitted.

2.6.1. Calculation of the thermodynamic equilibrium constant

As an example a system is considered containing both bayerite and gibbsite, crystalline solid phases of hydrated aluminum oxide. Upon addition of (pure) water dissolution reactions are initiated and as a result Al^{3+} and OH^--ions are dissolved as is illustrated by equations (2.28) and (2.29).

$$Al(OH)_{3,bayerite} \rightleftharpoons Al^{3+}_{(aq)} + 3OH^-_{(aq)} \qquad (2.28)$$

$$Al(OH)_{3,gibbsite} \rightleftharpoons Al^{3+}_{(aq)} + 3OH^-_{(aq)} \qquad (2.29)$$

The equilibrium constants, denoted by $K^\circ_{(2.28)}$ and $K^\circ_{(2.29)}$, describing the solution composition in equilibrium with each of the solid phases are derived with the help of equation (2.15). For bayerite this gives:

$$\Delta G^\circ_r = \bar{G}^\circ_{f,Al^{3+}} + 3\bar{G}^\circ_{f,OH^-} - \bar{G}^\circ_{f,bayerite}$$

Introducing the numerical values listed in table 2.3 gives:

$$\Delta G^\circ_r = 48.4 \text{ kcal/mole}$$

According to equation (2.11) one thus finds:

$$\log K^\circ_{(2.28)} = -\Delta G^\circ_r / 1.364$$

which states that at equilibrium

$$\frac{(Al^{3+})(OH^-)^3}{(bayerite)} = 10^{-35.5} \qquad (2.30)$$

According to the definition of the standard state of pure solid compounds the activity of the crystalline bayerite in equation (2.30) equals unity. In a comparable manner the equilibrium constant, $K^\circ_{(2.29)}$ for gibbsite is derived

$$K^\circ_{(2.29)} = (Al^{3+})(OH^-)^3 = 10^{-36.3} \qquad (2.31)$$

As in this case both constants represent the solubility products of bayerite and gibbsite, respectively, it follows from the derived values that bayerite is more soluble than gibbsite. Accordingly bayerite will dissolve with the simultaneous precipitation of gibbsite. Thus in this system eventually only solid gibbsite will be present in equilibrium with a solution composition as described by equation (2.31). Although this conclusion is correct for this system one should realize that thermodynamics does not allow any prediction concerning the rate of transformation of the unstable phase into the stable one. In the present case the rate of transformation is determined by the rate of dissolution of the bayerite.

The conclusion that bayerite is the unstable phase in this system follows directly if one compares the corresponding values of the free enthalpies of formation of the solid phases involved. Considering the reaction

$$Al(OH)_{3,bayerite} \rightarrow Al(OH)_{3,gibbsite}$$

it is found that transformation of bayerite into gibbsite is accompanied by a decrease in free enthalpy equal to:

$$G^°_{f,\text{gibbsite}} - G^°_{f,\text{bayerite}} = -1.1 \text{ kcal/mole}$$

As the free enthalpy of a system at equilibrium is at a minimum (cf. equation 2.5), this equilibrium is reached at the moment that all the bayerite has disappeared. This conclusion remains true regardless of the presence of the water, presupposed above. Thus at the values of P and T considered, also 'dry' solid bayerite must in principle recrystallize in the form of gibbsite. Again, however, nothing may be concluded about the rate of this process, which in the present case is practically nil.

TABLE 2.3.

Values of standard free enthalpies of formation, $\bar{G}^°_f$, in kilocalories per mole, for a number of substances at 25°C and 1 bar total pressure.

Species		State [*1]	$\bar{G}^°_f$
$Al(OH)_3$	(bayerite)	s	−276.2
$Al(OH)_3$	(gibbsite)	s	−277.3
$CaSO_4 \cdot 2H_2O$	(gypsum)	s	−429.2
H_2O		l	−56.7
$CaSO_4^°$	(uncharged ion pair in aqueous solution)	aq	−312.7
Al^{3+}		aq	−115.0
Ca^{2+}		aq	−132.2
SO_4^{2-}		aq	−177.3
H^+		aq	0 [*2]
OH^-		aq	−37.6

[*1] s = solid phase. l = liquid phase. aq = in aqueous solution at unit activity (cf. section 2.2).

[*2] by convention $\bar{G}^°_f$ is put at zero for the H^+-ion at unit activity in water, which convention is then used as a reference for the calculation of $\bar{G}^°_f$ for all other ions.

In an aqueous solution also the dissociation reaction of water should be considered, which may be written as:

$$H_2O_{(l)} \rightleftharpoons H^+_{(aq)} + OH^-_{(aq)} \tag{2.32}$$

The free enthalpy of this reaction amounts to 19.1 kcal/mole yielding the well-known value of the dissociation constant for water of 10^{-14} at 25°C and 1 bar pressure. Thus:

$$\frac{(H^+)(OH^-)}{(H_2O)} = (H^+)(OH^-) = 10^{-14} \tag{2.33}$$

where the activity of $H_2O_{(l)}$ is assumed to be unity, which is justifiable for dilute solutions.

Returning to the gibbsite system above, one may combine reaction (2.29) and (2.32), according to:

$$Al(OH)_{3, \text{gibbsite}} \rightleftharpoons Al^{3+}_{(aq)} + 3OH^-_{(aq)}$$

$$\underline{3OH^-_{(aq)} + 3H^+_{(aq)} \rightleftharpoons 3H_2O_{(l)}} \qquad +$$

$$Al(OH)_{3, \text{gibbsite}} + 3H^+_{(aq)} \rightleftharpoons Al^{3+}_{(aq)} + 3H_2O_{(l)} \qquad (2.34)$$

Besides the usual manner of deriving the standard free enthalpy for the combined reaction (according to equation 2.15) plain algebra shows that also (number in parentheses refers to equation !):

$$\Delta G^\circ_{r\,(2.34)} = \Delta G^\circ_{r\,(2.29)} + 3(-\Delta G^\circ_{r\,(2.32)})$$

Accordingly one finds that $\Delta G^\circ_{r\,(2.34)}$ equals -7.8 kcal/mole. It follows furthermore that:

$$\log K^\circ_{(2.34)} = \log K^\circ_{(2.29)} - 3\log K^\circ_{(2.32)} = 5.7$$

The 'combined' equation (2.34), with its constant $K^\circ_{(2.34)}$, is a type of relationship frequently encountered in chapter 6. It expresses the activity of a particular ion (i.e. Al^{3+}) in terms of the pH, a parameter that can be determined experimentally in a relatively easy manner. The particular relation between $-\log(Al^{3+})$ and $-\log(H^+)$ in the presence of solid $Al(OH)_3$ is presented graphically in fig. (6.9).

It is mentioned here that in analogy with the convention to indicate $-\log(H^+)$ as pH, one uses the notation pK° and e.g. pAl^{3+} for $-\log K^\circ$ and $-\log(Al^{3+})$, respectively.

Accordingly the above calculations for the gibbsite system may be summarized as:

$$pK^\circ_{(2.34)} = -5.7 = (pAl^{3+} - 3pH) \qquad (2.35)$$

2.6.2. Calculation of the equilibrium solution composition at low electrolyte level.

Because of the very low solubility of gibbsite in neutral or slightly acidic solutions the above example lends itself very well to the calculation of the equilibrium concentrations of all constituents if (an excess of) chemically pure gibbsite is shaken with, say, 1 l. of water containing a (small) amount of HCl, e.g. 10^{-4} mole. The ionic species present in the system are then Al^{3+}, H^+, Cl^-, OH^-. Electroneutrality now requires that:

$$3[Al^{3+}] + [H^+] = [Cl^-] + [OH^-], \qquad (2.36)$$

where $[Cl^-] = 10^{-4}$, as it remains in solution. If one puts $[H^+]$ equal to $10^{-4}.x$ (where x must be slightly smaller than unity, as the small amount of gibbsite that dissolves will produce OH^--ions that will combine in part with the H^+-ions present due to the HCl), one may use equations (2.33) and (2.35) to express the other terms of (2.36) in x (replacing activities by concentrations because of the low value of the ionic strength). This gives:

$$3.10^{-12}.x^3.10^{5.7} + 10^{-4}.x = 10^{-4} + 10^{-10}/x.$$

As x is close to unity, the last term is obviously negligible, thus one finds, after multiplying with 10^4:

$$0.015\, x^3 + x = 1, \text{ or } x = 0.986 \quad \text{and}$$

$$[Al^{3+}] = 10^{+5.7} \times 10^{-12} \times 0.96 = 0.48 \times 10^{-6}$$

Thus about one half micromole of gibbsite has dissolved. As a check it may be verified that the amount of OH^--ions produced in this manner, i.e. 1.4 mmole has reduced the H^+-ion concentration from the original 10^{-4} molar to 0.986×10^{-4}.

Referring to chapter 6 it is pointed out that in reality the above calculation does not entirely apply to this system, as Al^{3+}-ions will form an association complex with OH^--ions. At the present pH of about 4 the concentration of these $Al(OH)^{2+}$-ions is about 10% of that of the Al^{3+}-ions, leading to a slightly higher pH value.

2.6.3. Calculation of the equilibrium solution composition of a moderately soluble salt.

The use of activity coefficients becomes necessary at higher electrolyte levels. This will be demonstrated by calculating the equilibrium concentrations of solid gypsum in water. The composition of a solution saturated with respect to gypsum is defined by equation (2.37):

$$CaSO_4.2H_2O_{gypsum} \rightleftharpoons Ca^{2+} + SO_4^{2-} + 2H_2O \qquad (2.37)$$

Using the data listed in table 2.3 the solubility product of gypsum equals $10^{-4.6}$ or 2.5×10^{-5}. Hence at equilibrium:

$$(Ca^{2+})(SO_4^{2-}) = 2.5 \times 10^{-5} \text{ moles}^2/l^2 \qquad (2.38)$$

In the present system only Ca- and SO_4-ions are introduced into the solution upon dissolution of solid gypsum. Neglecting the H- and OH-ions (present at a level around 10^{-7}), the electroneutrality condition of the solution yields:

$$2[Ca^{2+}] = 2[SO_4^{2-}] \qquad (2.39)$$

In order to combine equation (2.39), expressed in concentrations, with equation (2.38) in terms of activities, the activity coefficients must be calculated. The latter are, however, determined by the ionic strength, which depends in turn on the concentration. Thus in principle an iteration method could be used, e.g. first assuming $f_\pm = 1$, giving a first estimate of the concentrations and the ionic strength. The ensueing value of $f_\pm$ is then used to recalculate a second estimate, and so on. Thus the first estimate (assuming $f_+ = f_- = 1$) yields:

$[Ca] = [SO_4] = (2.5 \times 10^{-5})^{0.5} = 5 \times 10^{-3}$ mole/l

The corresponding value of I is then four times the above molar concentration (i.e. $z_+ = -z_- = 2$). The parameter å for the ions involved is 6 Å for the Ca-ions and 4 Å for the SO_4-ions (cf. table 2.1). Taking the numerical values of A and B at 0.51 and 0.33, respectively, equation (2.12) gives $f_+ = 0.60$ and $f_- = 0.57$ at $I = 2.10^{-2}$ mole/l.

Application of equation (2.38) now gives:

$[Ca] = [SO_4] = [2.5 \times 10^{-5}/(0.60 \times 0.57)]^{0.5} = 8.5 \times 10^{-3}$ mole/l

Repeating the above procedure gives the following list of consecutive estimates:

Approximation	assumed value		$c_{Ca} = c_{SO_4}$	I	Calculated value	
	f_+	f_-			f_+	f_-
1	1	1	0.0050	0.020	0.60	0.57
2	0.60	0.57	0.0085	0.034	0.53	0.50
3	0.53	0.50	0.0097	0.039	0.51	0.48
4	0.51	0.48	0.0101	0.040	0.51	0.48

The fourth approximation is satisfactory for the chosen accuracy of calculation, yielding:

$[Ca] = [SO_4] = 1.01 \times 10^{-2}$ mole/l.

As in the present case the electroneutrality condition is extremely simple, one may also combine equations (2.38) and (2.39) directly, putting:

$[Ca] = [SO_4] = x$ mole/l

Then:

$f_\pm^2 x^2 = 10^{-4.6}$, or $2 \log x + 4.6 = -2 \log f_\pm$

with:

$\log f_\pm = -4A\sqrt{I}/(1+å_\pm B\sqrt{I})$

Taking $A = 0.51$, $f_\pm$ as $\frac{1}{2}(f_{Ca} + f_{SO_4}) = 5\text{Å}$ (cf. table 2.1), and $B = 0.33$, while $I = \frac{1}{2}(4x + 4x) = 4x$, this gives:

$$2.30 + \log x = (4.08\sqrt{x})/(1 + 3.3\sqrt{x})$$

This equation is easily solved by plotting left hand side and right hand side on semi log paper or by trial and error. The result is again $x = 1.01 \times 10^{-2}$.

Calculation of the solubility would give: $1.01 \times 10^{-2} \times 172$ (molecular weight of gypsum) = 1.74 g/l soluble gypsum.

Experimental results, however, indicate that the amount of soluble gypsum is higher (about 2.64 g/l). This is so because also a soluble complex (or ion-pair) is involved. The formation reaction of the uncharged ion-pair $CaSO_4^\circ$ reads:

$$Ca^{2+} + SO_4^{2-} \rightleftharpoons CaSO_4^\circ \tag{2.40}$$

The formation constant is defined as $(CaSO_4^\circ)/[(Ca^{2+})(SO_4^{2-})]$ and equals $10^{2.31}$. In this system where also solid gypsum is present the product $(Ca^{2+})(SO_4^{2-})$ has a constant value, i.e. determined by the solubility product of gypsum (cf. equation 2.38). Therefore the activity of the ion-pair is fixed. This is also illustrated by combining equations (2.37) and (2.40) according to:

$$\begin{aligned}
CaSO_4 \cdot 2H_2O_{gypsum} &\rightleftharpoons Ca^{2+} + SO_4^{2-} + 2H_2O \\
Ca^{2+} + SO_4^{2-} &\rightleftharpoons CaSO_4^\circ \\
\hline
CaSO_4 \cdot 2H_2O_{gypsum} &\rightleftharpoons CaSO_4^\circ + 2H_2O
\end{aligned} \tag{2.41}$$

The numerical value of the equilibrium constant of equation (2.41) follows from the values of the constants of the composing reactions (cf. equation 2.34) and equals $10^{-2.29}$ mole/l. As for uncharged species the activity coefficient in aqueous solution is usually taken as unity (provided the total concentration is not excessive), the concentration of $CaSO_4^\circ$ amounts to $10^{-2.29}$ mole/l. Expressing this in terms of soluble gypsum yields 0.88 g/l. Thus the total concentration equals $1.74 + 0.88 = 2.62$ g/l gypsum which agrees with the experimentally determined value, considering the chosen accuracy of calculation.

2.7. REACTIONS INVOLVING THE TRANSFER OF PROTONS AND/OR ELECTRONS

As was discussed in the previous sections the application of thermodynamics to reaction equilibria leads to the definition of the thermodynamic equilibrium constant K°, which then prescribes certain relations between the activities of the reactants and reaction products if the system is at equilibrium. Knowledge of such a K°-value then allows at least the prediction of the activity of any one component, once the activities of the other species involved are known.

2.7.1. Acid — base equilibria

An important group of reactions involve the transfer of protons from a so-called proton-donor to a proton-acceptor. The proton donor is often referred to as an 'acid', the acceptor as a 'base'. Thus for each acid there exists a (conjugated) base, the two being connected by means of the general relationship:

$$\text{acid} \rightleftharpoons \text{base} + \text{proton} \tag{2.42}$$

Particular examples of this relationship are e.g.

$$HCl \rightleftharpoons Cl^- + H^+ \tag{2.42a}$$

$$H_2PO_4^- \rightleftharpoons HPO_4^{2-} + H^+ \tag{2.42b}$$

$$Al(H_2O)_6^{3+} \rightleftharpoons Al(H_2O)_5(OH)^{2+} + H^+ \tag{2.42c}$$

$$H_3O^+ \rightleftharpoons H_2O + H^+ \tag{2.42d}$$

$$H_2O \rightleftharpoons OH^- + H^+ \tag{2.42e}$$

Considering the above examples as proton-dissociation 'reactions' one may introduce corresponding equilibrium constants of the form:

$$K_A^\circ = \frac{(H^+)(\text{base})}{(\text{acid})} \tag{2.43}$$

where K_A° is then called the acidity constant. Recognizing that the activity of unsolvated protons (in aqueous sytems) cannot be determined as such (presumably unsolvated protons exist only at immeasurably low concentrations), the equation (2.43) is of little help in a quantitative sense; K_A° is accordingly experimentally inaccessible. It does stipulate, however, that for a given ratio of the activities of a particular acid and its conjugated base, the activity of protons has a fixed value.

If, however, a system is considered containing two acid-base couples, following (2.42) according to:

$acid_1 \rightleftharpoons H^+ + base_1$, with $K^\circ_{A,1}$

$acid_2 \rightleftharpoons H^+ + base_2$, with $K^\circ_{A,2}$

one may follow the proton exchange between e.g. $acid_1$ and $base_2$. Adding the above reaction equations (writing the second one in backward direction) thus gives:

$$acid_1 + base_2 \rightleftharpoons base_1 + acid_2, \qquad (2.44)$$

signifying the exchange of a proton from $acid_1$ (yielding $base_1$) to $base_2$ (yielding $acid_2$). The corresponding equilibrium constant then reads:

$$K^\circ_{1-2} = K^\circ_{A,1}/K^\circ_{A,2} = \frac{(base_1)(acid_2)}{(acid_1)(base_2)}$$

where now presumably the activities of the chemical species involved are experimentally accessible.

In aqueous systems it is convenient to use the solvent H_2O as the ever present proton acceptor (according to (2.42d), written in backward direction). Combining (2.42d) with (2.42a) then gives:

$$HCl + H_2O \rightleftharpoons H_3O^+ + Cl^- \qquad (2.45)$$

with the corresponding 'dissociation' constant of HCl in H_2O:

$$K^\circ_{D,HCl} = \frac{K^\circ_{A,HCl}}{K^\circ_{A,H_3O^+}} = \frac{(H_3O^+)(Cl^-)}{(HCl)(H_2O)} \qquad (2.46)$$

Defined in this manner, the dissociation constant of HCl in H_2O thus characterizes the 'acid strength' of HCl relative to that of H_3O^+, in terms of a measurable parameter.

If now, by convention, the experimentally inaccessible value of K°_{A,H_3O^+} – i.e. the acid strength of the reference acid in aqueous systems (2.42d) is set at unity, equation (2.46) yields:

$$pK^\circ_D = pK^\circ_A \qquad (2.47)$$

Moreover, the above convention implies that in dilute aqueous systems the activity of the unsolvated protons equals that of H_3O^+.

In actuality the symbol H_3O^+ comprises the totality of hydrated protons, $H(H_2O)_n^+$. The total titratable concentration of this species, $\Sigma_n[H(H_2O)_n^+]$, then approaches the activity of H_3O^+ in infinitely dilute aqueous systems, via the D.H. limiting law.

Where $-\log(H_3O^+)$ is indicated as pH, a corollary of the mentioned convention is to state that the pH-value in dilute aqueous solution characterizes the hydrogen ion activity, irrespective of the state of solvation of these ions.

Substitution of pH into (2.43) gives:

$$pH = pK_A^\circ + \log\frac{(base)}{(acid)} = pK_D + \log\frac{(base)}{(acid)} \qquad (2.48)$$

Referring to section 2.4, pK_A° may also be expressed as $\Delta G_r^\circ/2.3RT$, where ΔG_r° is the standard free enthalpy of the particular acid-base reaction (subject to the convention that ΔG_r° of the reaction (2.42d) is set at zero).

As will be commented on below, certain electrodes are reversible to hydrogen ions. Such electrodes, when placed in aqueous systems attain an electric potential (as measured against a reference electrode) which follows Nernst's law (that is they show a potential change of about 60 mV for each tenfold increase in the hydrogen ion activity). Comparing the potential reading in an arbitrary system with that in solutions of known hydrogen ion activity (usually buffer solutions, see section 2.7.3) then allows one to calculate directly the pH in the system.

2.7.2. Oxidation–reduction equilibria

This important group of reactions involve the transfer of electrons from donor to acceptor. These reactions are called oxidation-reduction reactions, shortly redox reactions. In general each reaction can be written as follows:

$$n_B A_{ox} + n_A B_{red} \rightleftharpoons n_B A_{red} + n_A B_{ox}, \qquad (2.49)$$

formally equivalent to the proton exchange reaction (2.44). The reaction (2.49) can be split into two reactions, one for each redox couple:

$$A_{ox} + n_A e^- \rightleftharpoons A_{red} \quad \text{with } K_{red}^\circ = \frac{(A_{red})}{(A_{ox})(e^-)^{n_A}}$$

$$B_{ox} + n_B e^- \rightleftharpoons B_{red} \quad \text{with } K_{red}^\circ = \frac{(B_{red})}{(B_{ox})(e^-)^{n_B}}$$

Such reactions are called half-reactions and are formally equivalent to the reactions involving the transfer of protons. Aqueous solutions contain only minute amounts of free electrons, but it is nevertheless possible to define a relative electron activity:

$pe = -\log(e^-)$

in a similar way as was done for the relative proton activity $pH = -\log(H^+)$ for the acido-basic reactions. Large positive values of pe (low electron activity) represent strongly oxidizing conditions, while small values of pe correspond to reducing conditions.

In analogy to the convention used above (viz. putting ΔG_r° (2.42d) equal to zero), one may now assign a zero value to ΔG_r° of the electron acceptance reaction of H^+-ions:

$$H^+ + e^- \rightleftharpoons \tfrac{1}{2}H_{2(g)} \qquad (2.49a)$$

The corresponding $K_{(2.49a)}^\circ$ then equals unity. Applying this convention to the redox reaction:

$$Fe^{3+} + \tfrac{1}{2}H_{2(g)} = Fe^{2+} + H^+ \text{, with } K^\circ = K_1$$

one finds the composing half-reactions as:

$Fe^{3+} + e^- = Fe^{2+}$, with $K^\circ = K_2$

$\tfrac{1}{2}H_{2(g)} = H^+ + e^-$, with $K^\circ \equiv 1$

Accordingly:

$$K_1 = K_2 \times 1 = K_2 = \frac{(Fe^{2+})}{(Fe^{3+})(e^-)} \,,$$

or:

$$pe = \log K_1 + \log \frac{(Fe^{3+})}{(Fe^{2+})} \qquad (2.50)$$

It is further customary to put:

$(+)\log K^\circ = pe^\circ$

or generally for exchange of n electrons:

$\dfrac{1}{n} \log K^\circ = pe^\circ$

The quantity pe° than indicates the (relative) electron activity when reacting species other than the electrons are at unit activity. Relation (2.50) may thus be written as:

$$pe = pe^\circ + \log \frac{(Fe^{3+})}{(Fe^{2+})} \tag{2.50a}$$

While the pK_D°-value of many acid-base couples is often given as such, this is usually not the case with pe°. As the pe is generally obtained with electrometric methods (cf. below) - which are in turn based on a comparison with the EMF reading obtained in a standard H_2-gas system - the literature usually specifies pe° in terms of the EMF reading in volts or millivolts. This reading is indicated as $E_h^\circ = + pe^\circ \times 60$ mV. Reintroducing this into equation (2.50) then gives the well known Nernst relation:

$$E_h = E_h^\circ + \frac{RT}{F} \log \frac{(Fe^{3+})}{(Fe^{2+})} \tag{2.50b}$$

Writing out the corresponding expression for a half-reaction of the type $Ox + ne^- \rightleftharpoons Red$, one finds:

$$E_h = E_h^\circ + \frac{RT}{nF} \log \frac{(Ox)}{(Red)} \tag{2.51}$$

Although the value of E_h° for many half-reactions is available from tables, it may also be derived from the corresponding standard free enthalpy, according to:

$$E_h^\circ = -\frac{\Delta G_r^\circ}{nF} = -\frac{\Delta G_r^\circ}{23.06 \times n}, \tag{2.52}$$

the F(araday) constant being equal to 23.06 kcal/volt. Using

$$\Delta G_r^\circ = -RT \ln K^\circ$$

one finds also:

$$\ln K^\circ = \frac{nF}{RT} E_h^\circ \tag{2.53}$$

such that $\log K^\circ = n \times 16.9 \, E_h^\circ$ (at 25°C). Alternatively equation (2.53) may be written as:

$$pe^\circ = 16.9 \times E_h^\circ \tag{2.54}$$

2.7.3. The electrometric determination of pH and pe

Foregoing any discussion on the theoretical background of electrodes, it is assumed here that electrodes reversible to a particular ion species, k, with valence z, acquire a potential with respect to the solution which follows Nernst's law, according to:

$$E = \text{constant} + 2.303 \frac{RT}{zF} \log(k)$$

where E indicates the electrode potential (which may not be measured as such). If such an electrode is built into a circuit comprising also a reference electrode (i.e. an electrode which acquires a constant potential, like e.g. the calomel electrode) the potential difference measured between reversible and reference electrode (EMF) again follows the above equation corrected for E_{ref}. Thus neglecting any diffusion potentials:

$$\text{EMF} = E - E_{ref} = \text{constant} + 2.3\,(RT/zF)\log(k) - E_{ref}.$$

Using a glass electrode - which is reversible towards H^+-ions - against a calomel electrode one may thus compare the potential difference, EMF_x, observed in an unknown solution with pH_x, with that in a standard system, EMF_s. Using to this purpose a system at 25°C containing H_3O^+-ions at unit activity, in equilibrium with H_2-gas at 1 atm pressure (again corresponding to unit activity), the pH of this standard system is zero according to the convention introduced before. Thus:

$$\text{EMF}_x = -2.303\,(RT/F) \times pH_x + \text{constant} - E_{ref} \text{ and}$$

$$\text{EMF}_s = -2.303\,(RT/F) \times 0 + \text{constant} - E_{ref}$$

Accordingly;

$$\text{EMF}_x - \text{EMF}_s = -2.303\,(RT/F) \times pH_x.$$

Using any other standard system (as nowadays commercially available), one finds that in general:

$$pH_x = pH_s - \frac{F}{2.303\,RT}(\text{EMF}_x - \text{EMF}_s)$$

Specifying the EMF in Volt, and selecting a temperature of 25°C, this yields:

$$pH_x = pH_s - (\text{EMF}_x - \text{EMF}_s)/0.059.$$

The buffer solutions from which standard reference solutions of reproducible and precisely defined pH value can be prepared, were suggested by the U.S. National Bureau of Standards. The pH values of these solutions are established from EMF measurements of cells having no significant liquid junction potential. The cells used involve chloride ions, and the accuracy of the value of the pH assigned to the reference solutions rests ultimately upon the convention used in assigning the value to f_{Cl^-} in a solution. It is believed, that the pH reference solutions of the National Bureau of Standards have theoretical significance in terms of (H^+) to about 0.01 pH unit. However, experimental measurement of pH_x in terms of pH_s, as described above is furthermore subject to the error arising from the change in the liquid junction potential when the reference solution in the cell is replaced by the test solution. The more alike in composition and concentration both solutions are, the smaller this error will be. If the pH of both solutions lies between 2.5 and 11, colloids are absent and the ionic strength is about 0.1 molar, the overall error will be about 0.02 pH units.

The oxidation potential, E_h, is measured with an electrode pair consisting of an inert electrode and a reference electrode with constant potential. The inert electrodes used commonly are bright platinum or gold. As in pH measurements, the role of the reference electrode is to supply a known EMF and to make electric connection with the system to be measured. The inert electrode acts as an electron acceptor or donor to the ions in the solution. As with the pH measurement, the uncertainty concerning the value of the liquid junction potential may be one source of error. Furthermore, the E_h-value is generally a function of the activity ratio of the oxidized and reduced forms in the solution. For the interpretation of the measured E_h in terms of these activity ratios, some assumptions are necessary. These assumptions include the following:

1. All species involved in the oxidation-reduction system are in internal equilibrium. In general, this assumption is true for simple dissolved species, with some important exceptions (some organic species, sulfates and dissolved oxygen).
2. Bright Pt- and Au-electrodes are truly inert; in contact with certain ions, however, the electrode may become coated (sulfides and chlorides are such offenders !).

2.8. GRAPHICAL PRESENTATION OF SOLUBILITY EQUILIBRIA

Logarithmic diagrams are frequently used to visualize the relations between the activities of the different components involved in a precipitation reaction at equilibrium. Applying this approach to a reaction represented by:

$$A_a B_b \rightleftharpoons aA + bB , \qquad (2.55)$$

it means that the activities of components A and B in the equilibrium solution have to be plotted. The relation between the activities of the above components in the equilibrium solution is described by the solubility product, i.e. $K_{SO}^\circ = (A)^a (B)^b$. Taking logarithms of this product gives:

$$\log(A) = (1/a)\log K_{SO}^\circ - (b/a)\log(B), \qquad (2.56)$$

which represents the expression sought. So in a logarithmic diagram with (A) and (B) as variables the above relation is represented by a straight line, having a slope $-b/a$ and an intercept $(1/a) \log K_{SO}^\circ$. Many times, however, the majority of the solubility relations are influenced by the formation of the complexes of either cation, or anion, or both of them. These complexes can be formed either between the components of the precipitate themselves, or with the ions produced by the solvent (protonization of anions, formation of hydroxo-complexes with cations if water is the solvent) or with other ligands present in the solution.

As an illustration of the mentioned reaction mechanisms, the dissolution of solid ferric-hydroxide, $Fe(OH)_{3(s)}$, in water is considered with the purpose to construct the complete solubility diagram for this system. In table 2.4 the equilibrium reactions occurring in this system with their constants are given. These reactions show that beside the precipitation reaction (reaction 1, table 2.4) also complex forming reactions between the ferric ion, Fe^{3+}, and hydroxyl ions are taking place leading to the formation of hydroxocomplexes containing one or more metal, Fe^{3+}, ions (mono- and polynuclear complexes). The concentration of such species is often higher than the concentration of the free metal ion, Fe^{3+}, and this causes a substantial increase in the solubility of the solid phase. The activities of these complexes, like the one of the ferric ion in the equilibrium solution are fixed by the presence of solid $Fe(OH)_{3(s)}$ and can be presented in a solubility diagram. Referring to the reactions presented in table 2.4 it seems obvious to use OH^- and the activity of the metal ions as variables in such a diagram. For practical reasons, however, the former variable is replaced by pH. Introduction of pH as a variable in the reactions listed in table 2.4 is possible by combining these reaction equations with the simultaneously proceeding dissociation reaction of water.

TABLE 2.4.

Equilibrium reactions with their constants for the system $Fe(OH)_{3(s)} - H_2O$.

				$\log K°$
1	$Fe(OH)_{3(s)}$	$\rightleftharpoons$	Fe^{3+} *1 $+ 3OH^-$	-37.5
2	$Fe^{3+} + OH^-$	$\rightleftharpoons$	$FeOH^{2+}$	11.0
3	$Fe^{3+} + 2OH^-$	$\rightleftharpoons$	$Fe(OH)_2^+$	21.6
4	$2Fe^{3+} + 2OH^-$	$\rightleftharpoons$	$Fe_2(OH)_2^{4+}$	24.9
5	$Fe^{3+} + 3OH^-$	$\rightleftharpoons$	$Fe(OH)_3^°$	≈ 28.5 *2
6	$Fe^{3+} + 4OH^-$	$\rightleftharpoons$	$Fe(OH)_4^-$	≈ 32.5
7	H_2O	$\rightleftharpoons$	$H^+ + OH^-$	-14.0

*1 Many metal ions usually exist in aqueous solutions as hydrated ions. Likewise the Fe^{3+}-ion is usually surrounded by 6 molecules of water, $Fe(H_2O)_6^{3+}$. In the hydration shell of the hydroxocomplexes these water molecules have dissociated successively into hydroxyl ions. Thus $Fe(H_2O)_5(OH)^{2+}$, $Fe_2(H_2O)_8(OH)_2^{4+}$, etc. In table 2.4 these H_2O molecules have been omitted in order to simplify the notation.

*2 The uncertainty connected with the constant of reaction 5 (table 2.4) is due to the experimental difficulties in determining the concentration of an electroneutral complex in the solution in contact with the solid phase of the same composition because of the possible presence of colloidal particles.

This gives for the precipitation reaction (cf. 2.6):

$$Fe(OH)_{3(s)} + 3H^+ \rightleftharpoons Fe^{3+} + 3H_2O \tag{2.57}$$

with log K° = 4.5
Taking logarithms and rearranging, leads to:

$$-\log (Fe^{3+}) = 3pH - 4.5 \tag{2.57a}$$

The activity of the hydroxocomplex $Fe(OH)^{2+}$ depends upon the activity of Fe^{3+} and pH according to:

$$Fe^{3+} + H_2O \rightleftharpoons FeOH^{2+} + H^+ \tag{2.58}$$

with log K° = -3.0   (11.0 - 14.0).
As the activity of Fe^{3+} ions is governed by the solid phase present, the same should hold for $FeOH^{2+}$. Realizing that all reactions concerned (formation of the complex, dissociation of water, solubilizing of solid) are proceeding simultaneously, it follows for the equilibrium of the hydroxocomplex ($FeOH^{2+}$) with its solid phase:

$$Fe(OH)_{3(s)} + 2H^+ = Fe(OH)^{2+} + 2H_2O \tag{2.59}$$

Equation (2.59) leads to:

$$-\log (FeOH^{2+}) \rightleftharpoons 2pH - 1.5 \tag{2.59a}$$

The expressions for the other complexes are derived in a similar way, giving:

$$-\log (Fe(OH)_2^+) = pH + 1.9 \tag{2.60}$$
$$-\log (Fe_2(OH)_2^{4+}) = 4pH - 5.9 \tag{2.61}$$
$$-\log (Fe(OH)_3^\circ) = + 9.0 \tag{2.62}$$
$$-\log (Fe(OH)_4^-) = 19 - pH \tag{2.63}$$

The above logarithmic relationships are represented by straight lines in fig. 2.1 with slopes varying from -1 to +4. These latter values correspond with the number of OH⁻ groups added to or dissociated from the $Fe(OH)_3$ molecule (or in the case of polynuclear complexes group of molecules). The slope is thus equal to the charge of the complex formed and has the opposite sign.

According to fig. 2.1 it follows for the $Fe(OH)_{3(s)}-H_2O$ system that ferric ions are dominant in the solution of pH < 3 (at pH < 1 the dinuclear complex $Fe_2(OH)_2^{4+}$ dominates as in this pH range the activity of the Fe^{3+} ion is sufficient to permit the formation of such a species). In solution with a pH value ranging from 3 to 7.5 the dihydroxocomplex, $Fe(OH)_2^+$, is the most abundant one. The minimum solubility in the pH range 7.5 to 10 is represented by the undissociated but soluble portion of $Fe(OH)_3^0$ which has a concentration of about 10^{-9} mole l^{-1}. At higher pH values (i.e. pH > 10) the tetrahydroxocomplexes $(Fe(OH)_4^-)$ becomes the most important one, causing the solubility of $Fe(OH)_{3(s)}$ to increase again.

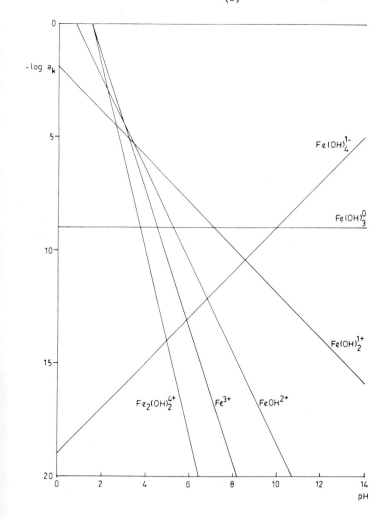

Fig. 2.1. The activity of Fe(III) species in equilibrium with $Fe(OH)_{3(s)}$ (log $K_{SO}^0 = -37.5$).

2.9. SURFACE STRUCTURE AND SOLUBILITY

The considerations leading to the calculation of the solubility of a given compound were based on the existence of equilibrium between 'pure' substances characterized by a definite value of their free enthalpy of formation. For a solid the free enthalpy of formation is determined by the interaction between the composing atoms or ions and thus depends on the detailed arrangement between these ions in the solid. A 'pure' solid in thermodynamic sense thus implies a singular arrangement of the composing ions in the solid as is found only in the interior of a large, perfect crystal. It must then be admitted immediately that, strictly speaking, such pure solids cannot exist, since the surface layer of a perfect crystal differs structurally from the situation found in its interior (e.g. incomplete coordination of surface atoms or ions). The specific free enthalpy of an ensemble of atoms arranged in such a less perfect arrangement is logically higher than corresponds to the perfect arrangement. As in practice the exact extent (in depth) of a surface layer is not known, the presence of such a layer may be taken care of by assigning to a crystal of finite size an extra amount of free enthalpy per unit surface area. This then implies that the (total) free enthalpy of a solid will increase with its surface area. If specified per unit amount of solid, the free enthalpy of formation of a solid thus increases with its specific surface area. Accordingly the solubility of a solid phase constituent increases with decreasing particle size. Also in mixed systems, the larger particles will tend to grow at the expense of the smaller ones, a phenomenon often referred to as ageing.

The above reasoning about the influence of crystal structure on solubility may be extended along the line: macro-crystal → micro-crystalline solid → amorphous solid. In this context amorphous forms of chemical compounds which are known to exist also in crystalline form (e.g. gibbsite and a-morphous $Al(OH)_3$) may be regarded as substances containing a very high fraction of imperfectly ordered atoms (such that X-ray diffraction gives no indication of the pattern characteristic for the perfect crystal). Obviously there exist degrees of imperfection in amorphous precipitates of solids and it is thus no surprise that $\bar{G}_f^\circ$ and log $K°$ of such substances show considerable variation depending on the conditions during formation of the compound (cf. table 6.1 reactions nr. 1 through 8). Particularly in case of soil systems this variability should be taken into account when applying values of $\bar{G}_f^\circ$ as found in the chemical literature, as conditions in soil may deviate from those in laboratory experiments.

The dependence of the solubility of crystalline substances on crystal size may be estimated as follows. Assuming one 'mean' type of surface one finds for a single crystal at constant T and P:

$$dG = \mu^° dn + \sigma dA \text{, or} \tag{2.64}$$

$$\mu = \mu^° + \sigma dA/dn \tag{2.65}$$

in which σ indicates the (mean) free enthalpy per unit surface area (surface tension), A is the area of the crystal and n gives the number of moles in the crystal. Expressing the volume of the crystal as $V = nM/\rho$, with M = formula weight and ρ = density, the above equation becomes:

$$\mu = \mu^° + \frac{M\sigma}{\rho} \cdot \frac{dA}{dV} \tag{2.66}$$

As $A \sim d^2$, while $V \sim d^3$ one finds:

$$\frac{dA}{dV} = \frac{2}{3} \frac{A}{V} = \frac{2}{3} \frac{g}{d} \tag{2.67}$$

in which d indicates a mean diameter of the crystal and g is a geometry factor characterizing the shape ($g = 6$ for isodiametric particles). Substitution in (2.66) thus gives:

$$\mu = \mu^° + \frac{2gM\sigma}{3\rho d} \tag{2.68}$$

The solubility product of particles with a diameter d (and a molar surface $Mg/\rho d$), $K^°_{SO,d}$, may thus be related to that of (infinitely) large crystals, $K^°_{SO}$, via the expression:

$$RT \ln \frac{K^°_{SO,d}}{K^°_{SO}} = \mu - \mu^° = \frac{2gM\sigma}{\rho d} \tag{2.69}$$

With this equation the value of σ may be determined from solubility experiments.

Conversely one may estimate that for a value of σ in the order of 1 Joule per m^2, $\mu - \mu^°$ of one micrometer particles is of the order of 8×10^5 J/kmole = 0.2 kcal/mole. This indicates that the effect of particle size on solubility becomes significant once the particle diameter is less than 1 μm.

LITERATURE CONSULTED

Butler, J.N., 1964. *Ionic equilibrium*. Addison-Wesley, Reading
Davies, C.W., 1962. *Ion Association*. Butterworths, London.
Garrels, R.M. and Christ, C.L., 1965. *Solutions, Minerals and Equilibria*. Harper and Row, Publ., New York.
Guggenheim, E.A., 1950. *Thermodynamics*. North Holland, Amsterdam.

Kielland, J., 1937 Individual activity coefficients of ions in aqueous solutions. J.Am.Chem. Soc. 59. 1675-1678.

Pourbaix, M., 1963. *Atlas d'Equilibres Electrochimiques*. Gauthiers-Villars, Paris.

Stumm, W. and Morgan, J.J., 1970. *Aquatic Chemistry. An introduction Emphasizing Chemical Equilibria in Natural Waters*. Wiley-Interscience. New York.

CHAPTER 3

SURFACE INTERACTION BETWEEN THE SOIL SOLID PHASE AND THE SOIL SOLUTION

G.H. Bolt

In the preceding chapters the solid phase and the liquid phase of soil were treated as separate entities. In reality, however, these phases interact with each other along the plane of contact between the phases. This interaction between liquid and solid phase is a very common phenomenon which is related to the difference in atomic structure of both phases on either side of the plane of contact. In the particular case of the plane of contact between the ionic solid phase and the aqueous liquid phase of the soil the high dielectric constant of the soil solution is highly significant: the polar water molecules possess a strong tendency to cause dissociation of surface groups of the solid.

In this respect it is important to recognize that the radial symmetry of interatomic forces inside a crystalline solid is necessarily absent at the terminal layer of atoms along its surface. Confronted with the polar water molecules this terminal layer is bound to develop a specific 'surface' structure, which is often more 'open' than the internal structure of the crystal (cf. section 2.9).

On the other hand also the liquid layer adjacent to the solid phase is modified in comparison to the 'bulk' solution, both with regard to structure and ionic composition. In fact one might state that the interaction between solid and liquid phase gives rise to the formation of a 'surface phase', situated between solid and liquid. The properties of this surface phase are then the domain of surface chemistry. In the following sections both 'sides' of this surface phase will be discussed briefly.

3.1. THE SURFACE CHARGE OF THE SOLID PHASE

In ionic solids the bonds between the different atoms in the lattice are primarily electrostatic in nature. When brought in contact with water, the polar water molecules tend to penetrate between the surface atoms, which weakens the binding force between these atoms and may lead to dissociation of surface ions.

In the particular case of clay minerals one finds a situation which is very vulnerable to this attack by water molecules. As was pointed out in chapter 1, the internal structure of the clay mineral has an imbalance of charges due to substitution of Si^{4+} and Al^{3+} ions by other cations of lower valence. The electroneutrality of the clay particle is then provided for by the presence of certain cations (Na, Ca, K, Mg, etc.) on the exterior surface. It is not surprising that these exterior cations (which are hardly part of the crystal lattice) dissociate completely from the surface if the latter is in contact with water. As a result one might conclude that clay minerals in water have a constant surface charge, equal in magnitude to the 'substitution charge' of the lattice.

A somewhat different situation exists at the surface of oxides. Whereas in the interior electric neutrality is maintained by the particular three dimensional coordination of metal ions (e.g. Si, Al, Fe, etc.) with O-ions, at the surface the coordination is necessarily incomplete, as this is a terminal layer. Electroneutrality is then maintained (in the dry crystal) by an appropriate number of protons, attached to a terminal layer of O-ions. Thus the surface of an oxide crystal could be presented schematically as (MOH), in which M is the interior lattice, O a surface O-ion and H the neutralizing proton. In contact with water dissociation of these protons may take place, but the degree to which this occurs will now depend on the concentration of protons in the liquid phase. In conclusion it may be stated that oxide surfaces are partly hydroxylized and in contact with water will attain a surface charge which is strongly dependent upon the pH of the solution. In fact it may be shown that actually the surface potential of the oxide colloid is fully determined by the pH of the solution (cf. part B). The H-ion is, therefore, often referred to as the potential-determining ion of oxides. As will be discussed in section 3.2.1 this implies that the surface charge of oxides in water varies with pH, the electrolyte level of the bulk solution and the valence of the counter--ions.

A (logical) extension of the above reasoning implies that if the H-ion concentration in solution is high enough, the surface may adsorb a higher number of protons than needed for neutralization, thus attaining a positive charge. In that case the situation amounts to association of surface OH-groups with excess protons to form $M\text{-}O_H^{H^+}$ groups.

The particular concentration of H-ions in solution needed to acquire a positive surface charge, depends very much on the valence of the metal cation in the oxide lattice and on the number of coordinated O-atoms. Comparing e.g. the tetravalent Si-ion coordinated with (only) four O-ions in Si-oxides, with the trivalent Al-ion coordinated with six O-ions in gibbsite, it is not surprising that while the gibbsite surface is positively charged at H-ion concentrations of 10^{-7} or even lower, the Si-oxides are hardly, if ever, positively charged.

Thus oxide surfaces may be characterized by the particular pH-value at which the surface is just electrically neutral. This pH-value is referred to as the zero-point of charge, pH_0. At pH-values above pH_0 the surface is negatively charged, whereas below pH_0 it is positively charged (cf. also section 4.6.1.4.).

It should finally be pointed out that the above mechanism operates wherever an oxide lattice is terminated. Thus the edges of clay minerals contain 'broken bonds' of the Si-O-Si structure of the tetrahedral layer. So, in addition to the constant 'substitution' charge, clay minerals will possess a pH-dependent charge along the edges. Where pH_0 for gibbsite is rather high the exposed 'gibbsite' layer along the edges are likely to be positively charged at low pH-values. In contrast, the Si-OH groups of the tetrahedral layer are negatively charged or near neutral even at low pH-values.

A third category of solid surfaces in soil are the internal and external surfaces of organic matter constituents. Referring to the polymeric structure of these compounds as discussed in chapter 1, the situation is characterized by the presence of R-OH groups, in which R symbolizes the polymer chain. These groups are moderately to weakly acidic and the degree of dissociation of the H-ions will again depend on pH and electrolyte level of the bulk solution. The 'surface' charge of organic matter constituents is thus negative and variable (cf. section 4.6.1.3.).

3.2. PROPERTIES OF THE LIQUID LAYER ADJACENT TO THE SOLID PHASE

In the previous section it was concluded that in aqueous environment usually a (partial) dissociation of surface ions takes place, imparting an electric charge to the surface of the solid phase. Much in contrast to e.g. partly dissociated molecular acids, where the dissociated H-ions spread themselves rather homogeneously throughout the solution, the ions dissociated from a solid surface tend to remain in the neighborhood of the charged surface. The cause of this phenomenon is the fixation of the surface charges to the solid phase, preventing them from spreading themselves throughout the system (like the anions of a soluble acid). Depending on the proximity of the charges on the surface, the electric fields of the individual charges reinforce each other, giving rise to a combined electric field in the liquid layer adjacent to the solid surface. This field will attract the dissociated ions (and other ions of like sign) which will thus accumulate close to the charged surface. The resulting spatial distribution of charges (e.g. a negatively charged surface with an accumulation of positively charged ions close to it) is in principle analogous to that of a plate condensor. It is given the name 'electric double layer'.

In contrast to the plate condensor, however, the accumulated 'counter' ions of the surface can move freely through the solution phase. This implies that they are subject to two opposing tendencies, viz.: a) they are attracted towards the surface by the electric field (adsorption tendency) and b) they tend to distribute themselves evenly throughout the solution phase by diffusion (diffusion tendency). In general terms one might say that the distribution strives towards minimum Energy (condition a) but also towards maximum Entropy of the sytem (condition b), as is shown schematically in figure 3.1.

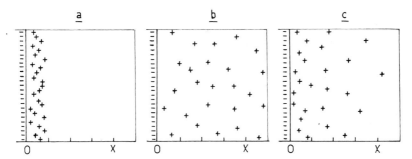

Fig. 3.1. Distribution of counterions
a: condition of minimum Energy; b: condition of maximum Entropy; c: condition of minimum Free Energy, i.e. actual distribution

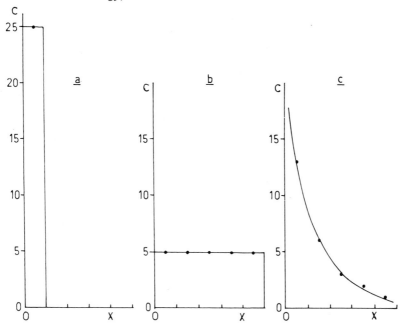

Fig. 3.2. Concentration of counterions in the double layer
a, b, c as in 2.1; c corresponds to the DDL

Maximum entropy in this context implies the most probable distribution of the ions, i.e. disappearance of concentration gradients of the counterions and thus disappearance of any accumulation.

The resulting (equilibrium) distribution corresponds to a minimum of the 'Free Energy' of the system and amounts to a 'diffuse' accumulation zone (fig. 3.1c) much alike the distribution of gas molecules in the earth's atmosphere. The concentration distribution in this zone follows the Boltzmann equation:

$$c_1/c_2 = \exp(-\Delta E/kT), \tag{3.1}$$

in which ΔE (= $E_1 - E_2$) is the difference in potential energy of the ions in the attractive field, comparing the arbitrary positions 1 and 2, while kT is the (thermal) kinetic energy of the ion.

The arrangement of charges (in particular of the counterions) described above is termed a Diffuse (electric) Double Layer (DDL). The corresponding concentration distribution is given in figure 3.2c, in which one recognizes a certain extent, u, of the DDL. This parameter could be defined as the width of the zone in which the electric field of the charged surface is still noticeable, or where a noticeable accumulation of counterions is found (cf. the Boltzmann equation). Granted that u is not sharply defined, it suffices for the moment to establish that the extent of the DDL may vary in magnitude, depending on conditions.

It may also be noted that the total electric charge of the counterions in the DDL just equals the surface charge (thus satisfying the condition of overall electric neutrality of the system). If now the concentration of the counterions is specified in keq per m^3 and the distance coordinate, x, in m, then the surface area underneath the concentration curve yields a number of keq of counterions per unit surface area of the solid phase. This number equals the surface density of charge of the solid surface, Γ, according to:

$$\Gamma \equiv \int_0^{d_1} c\,dx\,(keq/m^2) \tag{3.2}$$

in which d_1 is the thickness of the liquid layer on the solid surface.

3.2.1. The extent of the diffuse double layer at high water content

The situation created above is somewhat artificial, as only the dissociated counterions of the solid surfaces were considered. In reality (in soils) solid surfaces are in contact with a liquid phase containing dissolved salts. Accepting that the double layer has a finite extent, there will exist (at least at high liquid content of the system) a region in the solution phase which is not influenced by the solid surface. In this region the solution is simply a salt solution containing e.g. c_o keq/m^3 of Na-ions and of Cl-ions. Applying now the Boltzmann equation to this system, remembering that the attractive field of the surface is an electrostatic field, one finds for the concentration of Na-ions in the double layer:

$$c_{Na} = c_{o,Na} \cdot \exp(-e\psi/kT) \tag{3.3}$$

in which $e\psi$ is the electric energy of a monovalent cation with charge e, when present in an electric field with a potential ψ, relative to the bulk solution. Putting the electric potential in the bulk solution at zero, equation (3.3) relates the concentration of Na-ions somewhere in the DDL to the local value of ψ. If the surface charge is negative (e.g. for clays), ψ is also negative and $\exp(-e\psi/kT)$ is a number larger than unity. Accordingly one finds that Na-ions are accumulated in the region of negative electric potentials.

In contrast to the Na-ions the Cl-ions (referred to as co-ions of the surface as they carry a negative charge, i.e. of like sign as the surface), tend to be expelled from the double layer. Limiting the discussion to a system containing only Na-ions and Cl-ions (e.g. a so-called Na-clay dispersed in a solution of NaCl) one may now sketch the ionic distribution in the double layer as is given in figure 3.3.

The actual distribution of the ions may only be drawn if the electric potential, ψ, is known as a function of the position coordinate in the double layer (cf. part B).

The total counter charge in the double layer now equals the surface area between the curves for the counterion (i.e. the accumulated cation) and the co-ion (i.e. the expelled anion), so:

$$\Gamma \equiv \int_0^{d_1}(c_+ - c_-)dx, \tag{3.4}$$

where c_+ and c_- are the concentrations of Na- and Cl-ions, respectively.

The extent of the DDL now corresponds to the width of the zone where the ionic concentrations, c, differ significantly from the concentration of these ions in the bulk (equilibrium) solution, c_o. Remembering the cause for the formation of the DDL, - viz. the tendency of the counterions to diffuse away from the accumulation zone towards the region of lower concentration in the bulk solution - one may conclude that the addition of salt to the bulk solution suppresses the tendency of the counterions to diffuse away from the accumulation zone and thus decreases the extent, u, of the DDL.

An equally important factor in determining the extent of the diffuse layer is the valence of the counterion. Divalent cations (Ca!) are attracted more than monovalent cations, so u must decrease with increasing valence of the counterion (if the equilibrium concentration remains the same).

This increased accumulation of e.g. Ca-ions is found back in the Boltzmann-equation for this case:

$$c_{Ca} = c_{o,Ca} \exp(-2e\psi/kT), \qquad (3.5)$$

where 2e equals the charge of the divalent cation.

In practice this implies that u is very large in Na-soils at low electrolyte level (e.g. around 200 Å at 10^{-3} normal NaCl). As will be discussed in chapter 9 such conditions may occur in 'sodic' soils. In Ca-soil, at higher salt levels, the extent of the DDL is very much smaller (a few tens of Å only). It is worthwhile mentioning here that once the surface density of charge is of the order of magnitude as found for clay minerals, the extent of the DDL is hardly dependent upon the precise value of the charge density (cf. part B).

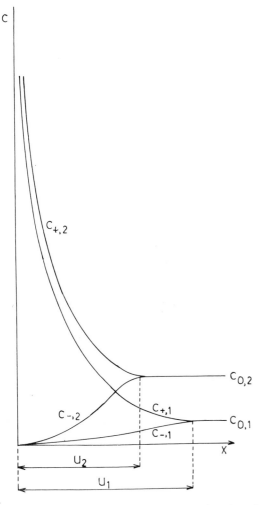

Fig. 3.3. Concentration of counter- and co-ions in a DDL on a negatively charged surface, for two values of the equilibrium concentration, $c_{o,1}$ and $c_{o,2}$.

A final remark is made here about the electric capacity of the DDL. Just as in the case of the plate condensor the capacity, C, increases when the distance between the plates becomes smaller, the electric capacity of the DDL will increase with diminishing u. Accordingly C_{DL} increases with increasing electrolyte and/or increasing valence of the counterion. This conclusion will be used when discussing the adsorption of H-ions by soil constituents in section 4.6.

For a plate condensor one defines the electric capacity, C, according to:
$$Q = C.V$$
in which Q is the charge and V is the potential difference between the plates. If the medium between the condensor plates is given, the capacity is inversely proportional to the distance between the plates. In analogy one might put for the double layer:
$$q = C_{DL}\psi_s,$$
where q is equal to Γ expressed in esu, and ψ_s is the potential difference between the charged surface (one plate) and the equilibrium solution ('end' of the other plate). But then the electric capacity of the double layer, C_{DL}, will be inversely proportioned to u, the extent of the diffuse ion swarm. Increase of salt and/or increase of the valence of the counterions then causes an increase of C_{DL}. For the above case of a surface with fixed charge Γ (or q) this implies a decrease of ψ_s, and consequently also of ψ at any place in the double layer. If, however, the potential of the surface, ψ_s, is fixed (oxide surfaces!), then an increase of c_o and/or the valence of the counterion will cause an increase of q (and of Γ!).

3.2.2. The diffuse double layer at low liquid content of the system

As ions are forced to stay in the liquid phase, the extent of the double layer may not exceed the thickness of the liquid layer, d_l, in contact with the charged surface. Starting from a system as discussed in section 3.2.1, with a given value of u, one may now gradually remove water, until $d_l \approx u$. Further removal of water then forces both the counterions and the co-ions into a liquid layer with decreasing thickness. This causes a readjustment of the ionic concentrations, leading to a concentration distribution as is pictured in figure 3.4. Such a double layer may be termed a truncated diffuse double layer.

The important feature of a truncated double layer is its tendency to reabsorb (forcefully) more water until the double layer has developed its full extent, corresponding to the composition of its equilibrium solution. In fact the situation may be compared to the compulsory uptake of water by an osmometer: the overall amount of ions (per unit volume) in the truncated double layer is too high in comparison to the situation in the fully extended double layer in equilibrium with an excess of its own equilibrium solution. Truncated double layers will, therefore, develop a swelling pressure when brought in contact with water, comparable to the osmotic pressure developed by the solution in an osmometer if in contact with water. The magnitude of this swelling pressure may amount to several tens of bars.

The situation may be summarized as follows. At low moisture content the double layer formed on charged surfaces may be truncated and the system develops a swelling pressure if water becomes accessible. The magnitude of this swelling pressure depends on the degree of truncation of the double layer. In other words, it depends on the relative value of the potential extent of the double layer (i.e. the extent which the double-layer would have if completely developed), and the actual thickness of the liquid layer. Accordingly the swelling pressure at a given moisture content will again depend on the valence of the counterions and the salt concentration in the soil solution. For orientation purpose it is mentioned that the potential extent of the double layer for Na-clay in 10^{-3} normal NaCl has a value of about 400 Å as against 40 Å at 10^{-1} normal NaCl. For divalent counterions these values should be halved.

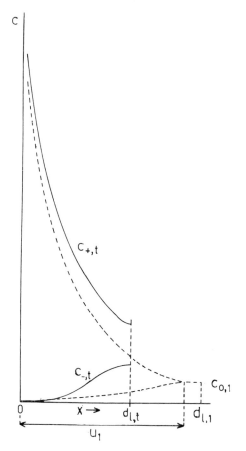

Fig. 3.4. Concentration distribution in a truncated double layer formed by removing part of the water from a system with an initial thickness of the water layer $d_{1,i} > u_i$ while in equilibrium with $c_{o,i}$. Note that $\int_0^{d_1} c_- dx$ remains constant as no salt is removed.

3.3. THE INFLUENCE OF THE INTERACTION BETWEEN SOLID AND LIQUID PHASE ON SOIL PROPERTIES

The soil is a medium in which the plant root must exercise its functions. The parts of the root which play an active role in ion uptake are covered with a water layer. This liquid layer is in fact part of the soil liquid phase; as furthermore the liquid phase in the soil consists of liquid layers on the soil particles which are interconnected, the root is in contact with a continuous liquid phase of often rather irregular shape. The uptake of water and nutrient ions is the result of the flow of water with its dissolved ions, plus diffusion of the ions through the water, towards the root. Other conditions being constant the geometry of the liquid phase is determining for the magnitude of this transport.

An important conclusion from the material discussed in section 3.2 is that the ionic composition of the liquid layers will depend on their position with respect to the solid phase. Thus it was indicated in the preceding part that a layer of a thickness of 20-200 Å has a composition which differs from that of the equilibrium solution. For a 'heavy' clay soil with a specific surface area of 100 m^2/g (cf. section 1.1.1) one finds at a double layer extent of 50 Å, that the total volume of solution with a composition differing from that of the equilibrium solution, equals 0.5 cm^3/g. This implies that even at a moisture content (by weight) of 50 % the total liquid phase must be counted towards this zone of deviating composition. Although this is a rather extreme case, it must be kept in mind that for the interpretation of the above-mentioned transport processes towards the root, the composition of the soil solution (i.e. the solution outside the range of influence of the solid phase) may not always be the reference point. This limitation will be more severe, the lower the moisture content, the greater the surface area of the solid phase, the lower the salt concentration, and the greater the percentage of monovalent counterions.

A comparable argumentation is valid for the uptake of water by the plant. Just like an osmotic pressure of a solution may be expressed as a decrease of the potential of the water ('osmotic potential', negative compared to the potential in pure water), the swelling pressure mentioned in section 3.2.2 corresponds to a decreased potential of the soil water and should be recognized as a component of the matrix potential of soil water. This decrease of the water potential inhibits the uptake of water by plant; the degree of inhibition is again dependent upon the factors moisture content, surface area, salt concentration and valence of counterions.

Another aspect of the influence of the double layer on soil properties is the physical behavior of soil. Upon addition of water the developed swelling pressure causes an increase of the thickness of the water layers on clay particles, forcing the individual particles away from each other. The resulting swelling of soil aggregates now causes closure of the inter-aggregate pores, which pores are mainly responsible for the transport of water. The 'hydraulic conductivity' of the soil will thus decrease strongly upon swelling of the aggregates. Accordingly it appears that the drainage of water is likely to be impeded if water with a low salt content (rainy season!) enters into soils with a high percentage of monovalent counter-ions (sodic-soils).

Related to the above is the dispersion of individual clay particles following the breakdown of aggregates upon swelling. This phenomenon is recognized in the field by the presence of a high turbidity in ponded water and, after drying out, by the formation of a crust with fissures. Conversely, the dispersion of soil particles is needed for mechanical analysis. For that purpose the soil is thus purposely saturated with sodium-ions, after which the salt concentration is decreased. Usually spontaneous dispersion will then occur (necessary in order to determine the grain size of the individual particles).

Finally the expulsion of anions from the double layer (which depends again on the extent of the double layer) gives rise to several typical phenomena, as e.g. the variation of the composition of the solution which one obtains by pressing out a soil, the 'abnormal' behavior of the soil solution upon dilution and the selective retention of salts upon the passage of an electrolyte solution through a soil column (the 'salt-sieving'-effect). For a further discussion of these phenomena reference is made to part B.

A number of physical and chemical properties of soil are thus related to, or are explained by, the presence of a zone in the liquid phase - adjacent to the solid phase - in which the composition differs considerably from that of the equilibrium soil solution. The greater the proportion of the soil moisture which is present in this zone, the greater is its influence on the behavior of the soil. In general, this influence is increased as the soil is heavier textured (greater surface area of the solid phase). A high percentage of monovalent counterions and a low salt concentration increase the effects described.

RECOMMENDED LITERATURE

van Olphen, H., 1963. *An introduction to Clay Colloid Chemistry*, Wiley Interscience, New York. Chapter 3, p. 30-40, chapter 7, p. 89-108, chapter 8, p. 109-119.

CHAPTER 4

ADSORPTION OF CATIONS BY SOIL

G.H. Bolt, M.G.M. Bruggenwert, A. Kamphorst

In earlier chapters it was mentioned that solid phase particles of the soil often carry a negative surface charge. The overall electroneutrality of the system is then maintained by the presence of an excess of cations (in comparison to the anions present) in close proximity of the solid surface. As follows from the above this excess of cations cannot be separated from the surface (i.e. taken out of the system). It is possible, however, to exchange these cations against others, while maintaining the electroneutrality of the system by means of the replacing cations. The cations adsorbed by the soil solid phase are thus available to the plant, e.g. by exchange against H-ions liberated by the plant root.

Alternatively, cation exchange makes it possible that ions, that are brought into the system via the solution phase, are retained by the soil (e.g. K-ions from fertilizer salts, NH_4-ions from fertilizers or other sources, Na-ions from seawater following inundation, heavy metal ions from waste water, etc.). In such a case the cations which were held originally by the solid phase are exchanged and released into the soil solution. As a rule the exchange-adsorption of cations is a beneficial characteristic of the soil system: retention of fertilizer cations will decrease leaching losses of these cations, whereas unwanted polluting cations may be retained in the soil instead of being passed on to the groundwater. Excessive adsorption of unwanted cations could, however, disturb the soil system as a biotope. A special case here is the adsorption of Na-ions, which, if present in sizable amounts, often cause undesirable characteristics of the soil (cf. chapter 9).

The totality of the solid phase components that are responsible for the adsorption of cations in soil is often referred to as adsorption complex (for cations). This adsorption complex gives to the soil a buffering capacity with respect to changes in the cationic composition of the soil solution (caused by addition of fertilizer, leaching, ion uptake by plants, pollutants entering the soil, etc.). It also serves as a storage for nutrient ions.

4.1. QUALITATIVE DESCRIPTION OF THE EXCHANGE REACTION

Schematically, cation exchange in soil may be indicated according to:

Soil $-A_+$ + B_+ ⇌ Soil $-B_+$ + A_+

The double arrow signifies the reversibility of the exchange. One often uses in this respect the word 'reaction', although in case the adsorbed ions are present in the diffuse double layer the exchange process is hardly a reaction in the sense as used in chemistry. From the condition of electroneutrality, mentioned previously, it follows that the exchange must be equivalent, e.g. two monovalent cations are replaced by one divalent ion, etc.

With respect to the rate of the above reaction it may be stated that, at least under favorable conditions, this rate is very high (half-time of the reaction is a matter of minutes or even less). These favorable conditions refer to the accessibility of the exchanger surface from the solution side. Shaking experiments in the laboratory with dilute suspensions usually fulfill this condition. Inaccessibility may be caused by aggregation and/or condensation of clay plates, implying diffusion through thin liquid films towards the exchange sites.

The total amount of cations held exchangeably by a unit mass (weight) of soil is termed the cation exchange capacity of the soil, CEC. The name 'adsorption' capacity is somewhat ambiguous in this respect, as sometimes a fraction of the adsorbed cations is virtually non-exchangeable (cf. section 4.5). In this text the CEC will be indicated with the symbol γ and will be expressed in meq per 100 grams of dry material in accordance with established practice. Referring to chapter 3, the presence of these exchangeable cations may be associated with the surface charge of the solid surface, as exchangeability against other common cations implies that the resident cation is dissociated from the surface in aqueous environment. Thus γ may be related to the surface density of charge, Γ, via the specific surface area of the solid phase, S, according to:

$$\gamma = 10^5 \ S \ \text{x} \ \Gamma \ (\text{meq}/100 \ \text{g}) \tag{4.1}$$

in which the factor 10^5 is necessary because S and Γ are expressed in m^2/kg and keq/m^2, respectively. With this equation the surface density of charge of soil colloids may be estimated from their CEC, provided information is available about the specific surface area. For the clays montmorillonite, kaolinite and illite Γ is about 10^{-9}, 2×10^{-9} and 3×10^{-9} keq/m^2, resp. The surface area is usually determined from B.E.T. analysis (N_2-adsorption at low temperature) and/or negative adsorption of anions (cf. chapter 5). For completely dispersed Na-montmorillonite one has found $S = (60-80) \times 10^4 \ m^2/kg$, for Na-illite $S = (5-20) \times 10^4 \ m^2/kg$, while for kaolinites S varies between values less than $0.1 \times 10^4 \ m^2/kg$ up to about $(2-4) \times 10^4 \ m^2/kg$.

The CEC of a soil may sometimes be estimated from its clay and humus contents, using e.g. for illitic clays about 40 meq/100 g and for the humus 200-300 meq/100 g. Obviously such estimates are applicable only in a limited area in which one knows the types of clay and organic contstituents.

The specification of the relative amounts of the different cations exchangeably adsorbed by the complex (expressed as percentages of the CEC) is usually referred to as (cationic) composition of the adsorption complex. In section 4.2.2 some examples of compositions as occur in nature are given (table 4.1). If a clay, or soil, is (almost) completely saturated with one particular cation, the name of that cation is often attached to that of the material, e.g. a (homoionic) Na-clay, a Ca-soil, etc.

4.2. EXPERIMENTAL APPROACH

Already in the middle of the 19th century Thompson and Way discovered that cations are adsorbed by soil, by observing that upon percolation through a layer of soil a solution of ammonium sulphate lost its NH_4-ions while Ca--ions appeared. Leaving aside certain complications, a quantitative determination of the amounts of different cations adsorbed by a given soil sample may be carried out in much the same manner by means of percolation. To this purpose a known amount of soil (at known moisture content) is percolated (on a Buchner funnel or in a percolation tube) with a solution of a salt of a cation that is virtually absent in the soil. The percolate of the soil is collected until its composition becomes identical with the percolating solution. Analysis of this percolate then yields the total amounts of the (displaceable) cations present in the soil before percolation. As will be discussed later, these amounts are then corrected for the amounts present in the soil solution to yield the amounts adsorbed. To this purpose the composition of the original equilibrium solution must also be determined, e.g. by analysis of a (pressure) filtrate of the original soil.

A variant on the above method is a shaking procedure. The sample is then repeatedly shaken with an excess of a replacing solution. After each shaking period the suspension is centrifuged and the supernatant is collected, again until its composition differs negligibly from the solution added to the sample. Analysis of the combined supernatants yields the total amounts present of the different cations. Finally the 'modern' method of isotopic dilution is mentioned here. For details reference is made to part B.

4.2.1. Interpretation of the analysis-data

The treatment of the data obtained from the above determination is best illustrated by means of an example. The accuracy of the analytical data is

TABLE 4.1

Examples of cationic compositions of different soils

	CaCO$_3$ (wt.%)	clay (%)	humus (%)	CEC (me/100g)	cationic composition (% of CEC)					pH (H$_2$O)	pH (KCl)
					H+Al	Ca	Mg	K	Na		
1. Holland, recent marine clay (periodically submerged)	0.3	29.3	23.3	60.7	0	26.2	32.8	6.9	34.1	7.3	7.1
2. Holland, young seaclay soil	5.2	48.9	3.0	18.5	0	81.8	10.8	6.5	0.9	7.4	7.0
3. Holland, young lake deposit (IJsselmeerpolders)	10.1	19.9	2.7	36.5	0	90.3	7.5	2.0	0.2	8.2	7.4
4. Holland, river basin soil	1.3	40.3	5.3	78.1	0	90.6	8.5	0.6	0.3	7.9	7.2
5. Surinam, acid sulphate soil (reclaimed coastal plain)	0	52.0	4.3	38.0	20	39	27	3	0.5	6.1	5.0
6. U.S.A., gray brown podsolic (alfisol)	0	12.8	1.2	13.1	33.6	51.1	12.2	3.1	0.0	6.7	-
7. Holland, humus podsol (spodosol)	0	2	4.2	8.8	53.7	35.6	6.0	3.3	1.4	5.7	4.6
8. Puerto Rico, Latosol (oxisol)	0	50.3	7.4	30.4	94.9	1.6	2.6	0.3	0.6	4.8	4.1
9. U.S.A.,reddish brown lateritic soil (ultisol)	0	42.0	7.7	50.6	56.9	28.6	9.3	5.0	0.2	5.8	-
10. Rumenia, chernozem (mollisol)	0.2	39.5	2.6	33.9	10.2	66.1	19.6	1.8	2.3	7.1	-
11. U.S.A.,self-mulching clay soil (vertisol)	2	43.0	1.3	50.8	2.2	78.3	16.2	3.1	0.2	7.5	-
12. U.S.A., saline soil	8	?	?	25.6	0	67	22	3	8	7.2	-
13. Turkey, saline alkali soil	5	?	?	10.4	0	27	24	2	47	7.4	-
14. India, saline alkali soil	2	20.0	?	9.0	0	2	1	2	95	10.3	-
15. Turkey, alkali soil	0	?	?	11.2	0	30	29	2	39	8.5	-
16. Rumenia, solodized solonetz	0	27.4	2.4	19.7	17.9	25.9	25.4	1.4	29.4	7.4	-

left out of consideration. Starting with a 100 gram sample (dry weight) which is brought to water saturation one finds a moisture content, W, of 60 cm³ water per 100 gram dry soil. The concentration of the ions present in the soil solution at this moisture content is determined in a small amount of filtrate obtained from a sub-sample and is found as:

$c_{o,Na}$ = 0.0060 normal (meq/ml)

$c_{o,Ca}$ = 0.0010 normal

$c_{o,Cl}$ = 0.0070 normal

Other ions appear to be present only in negligible amounts. The sample is percolated with a sufficient amount of a 1 normal NH_4NO_3 solution to ensure complete exchange. Analysis of the percolate yields the following total amounts of the ions in the system (per 100 gram dry soil):

T_{Na} = 3.40 meq/100 g

T_{Ca} = 22.12 ”

T_{Cl} = 0.28 ”

All experimental data necessary for the calculation of the CEC and the cationic composition of the soil are now available.

4.2.1.1. Cation exchange capacity

According to the definition in section 4.1 the CEC equals the total amount of cations exchangeably adsorbed by the exchange complex. As the sample contains also dissolved salts, the cations present as salt must be subtracted from the total amounts found in the percolate. Naturally one cannot decide whether one particular cation is adsorbed (i.e. is located in the double layer) or whether it is present as a 'free' cation, accompanied by an anion, in the soil solution. Nevertheless, the total amount of cations associated with the anions in solution may be determined unambiguously (assuming that the anions, as co-ions, are expelled from the region close to the solid surface). Thus:

$$\gamma = \Sigma_{cat} T_{cat} - \Sigma_{an} T_{an} \tag{4.2}$$

in which cat refers to different cations present, and an to the anions. Substituting the above experimental data this gives:

$\Sigma_{cat} T_{cat}$ = 3.40 + 22.12 = 25.52 meq/100 g

$\Sigma_{an} T_{an}$ = = 0.28 ”

γ = 25.24 meq/100 g

4.2.1.2. The cationic composition of the exchange complex

Whereas in the above treatment the CEC, or the total amount of exchangeably adsorbed cations, was found by subtracting $\Sigma_{an} T_{an}$ (i.e. total amount of anions in the system = total amount of cations present in the form of salts) from $\Sigma_{cat} T_{cat}$ (i.e. total amount of cations in the system), it would appear logical to define the amount adsorbed of cation k as the difference between T_k and the amount of k present as salt. There remains, however, an arbitrary decision to be made, viz. which part of the total anions present should be considered to be associated with cation k as k-salts. As these salts are completely dissociated, while both cations and anions exhibit particular distribution curves in the neighborhood of the charged solid surface (cf. part B) there appear to be two 'reasonable' (though arbitrary) ways of going about this partition of the adsorbed cations.

a. The first approach is to subtract from T_k a proportional fraction of $\Sigma_{an} T_{an}$, that is a fraction equal to the relative equilibrium concentration of ion k. This gives the amount adsorbed of cation species k, which is then conveniently expressed in terms of the product of the equivalent fraction adsorbed, N_k, and the cation exchange capacity γ. Applying this to the above experimental data one finds:

$$\gamma N_{Na} = T_{Na} - \frac{c_{o,Na}}{c_{o,Na} + c_{o,Ca}} \cdot T_{Cl}$$

$$= 3.40 - 0.006 \times 0{,}28/0.007 \quad = 3.16 \text{ meq}/100 \text{ g}$$

$$\gamma N_{Ca} = T_{Ca} - \frac{c_{o,Ca}}{c_{o,Na} + c_{o,Ca}} \cdot T_{Cl}$$

$$= 22.12 - 0.001 \times 0{,}28/0.007 \quad = 22.08 \text{ meq}/100 \text{ g}$$

$$\gamma N_{Na} + \gamma N_{Ca} = \gamma = 3.16 + 22.08 \quad = 25.24 \text{ meq}/100 \text{ g}$$

In a generalized form this implies:

$$\gamma N_k = T_k - \frac{c_{o,k}}{\Sigma_k c_{o,k}} \cdot \Sigma_{an} T_{an} \tag{4.3}$$

and

$$\Sigma_k \gamma N_k = \Sigma_{cat} T_{cat} - \Sigma_{an} T_{an} = \gamma \tag{4.3a}$$

b. The second approach is to introduce the excess of ion k, $\overset{+}{\gamma}_k$, according to:

$$\overset{+}{\gamma}_k = T_k - Wc_{o,k}, \tag{4.4}$$

in which the excess is defined in comparison to the situation where the exchange complex would be absent. In that situation the concentration of k would be constant everywhere in the system. Accordingly the amount of k-salt would be equal to the product of the moisture content (ml per 100 grams of soil), W, and the equilibrium concentration of ion k, $c_{o,k}$. Applying these definitions to the above experimental data, this gives:

$\overset{+}{\gamma}_{Na} = T_{Na} - Wc_{o,Na} = 3.40 - 0.36 = 3.04$ meq/100 g

$\overset{+}{\gamma}_{Ca} = T_{Ca} - Wc_{o,Ca} = 22.12 - 0.06 = 22.06$ meq/100 g

For the anion this leads to:

$\overset{+}{\gamma}_{Cl} = T_{Cl} - Wc_{o,Cl} = 0.28 - 0.42 = -0.14$ meq/100 g

Thus, defined in this manner, the excess of anions (as compared to the product of moisture content and equilibrium concentration) is negative, i.e. there is a deficit of Cl-ions in the double layer. This checks with the conclusion of chapter 3: anions are expelled from the liquid layer adjacent to negatively charged surfaces.

This deficit is termed the anion exclusion of the adsorber (or negative adsorbtion of the anion) and indicated with $\bar{\gamma}_{an}$, defined as:

$$\bar{\gamma}_{an} = -\overset{+}{\gamma}_{an} = Wc_{o,an} - T_{an} \tag{4.5}$$

Summing up over the excesses of the cations and the deficit(s) of anions then gives again γ, according to:

$$\Sigma_k \overset{+}{\gamma}_k + \Sigma_l \bar{\gamma}_l = \Sigma_k T_k - W\Sigma_k c_{o,k} + W\Sigma_l c_{o,l} - \Sigma_l T_l$$
$$= \Sigma_k T_k - \Sigma_l T_l = \gamma, \tag{4.6}$$

because in the equilibrium solution $\Sigma_k c_{o,k} = \Sigma_l c_{o,l}$.

There are then two ways of specifying the composition of the exchange complex:

a: as the relative amounts of cations adsorbed, with $\Sigma_k \gamma N_k = \gamma$

b: as the relative amounts of excesses of cations and deficits of anions, with
$\Sigma_k \overset{+}{\gamma}_k + \Sigma_l \bar{\gamma}_l = \gamma$

The choice under b gives a more complete picture, as also the situation with regard to the anions is explicitly given. Thus for the experiment described the specifications read:

a: $\gamma N_{Na} = 3.16$ meq/100 g   b: $\overset{+}{\gamma}_{Na} = 3.04$ meq/100 g

$\gamma N_{Ca} = 22.08$,, $\overset{+}{\gamma}_{Ca} = 22.06$,,

$\bar{\gamma}_{Cl} = 0.14$,,

In the present case the differences between γN_k and $\overset{+}{\gamma}_k$ are very small, and in practice negligible in comparison to experimental error.

As may be shown the differences are equal to proportional parts of the anion exclusion, i.e. 6/7 and 1/7 of 0.14, respectively.

In general one may say that the two definitions give almost the same result in soils with low electrolyte levels (cf. chapter 9). In that case one may as well specify the cationic composition in terms of the relative amounts of $\overset{+}{\gamma}_k$ only, which then implies that only cationic concentrations have to be determined. In the present treatise no distinction will be made between both quantities, while serving a warning that in saline soils this neglect may be unwarranted. As may be shown $\bar{\gamma}_{an}$ may then amount to 10-15% of the CEC. When attempting to interpret published data on cationic compositions of saline soils one must, therefore, ascertain in which manner the author concerned has treated his experimental data.

4.2.1.3. The distribution ratio of ions in the system

A very important number with respect to ionic behavior in soil is the distribution ratio, R_D, defined as:

$$R_{D,k} \equiv \overset{+}{\gamma}_k / Wc_{o,k} \tag{4.7}$$

Obviously, this number signifies the relative storage for each ion, that is the excess amount adsorbed compared to the amount in solution, both quantities being specified in meq per 100 g of soil. This number may be interpreted as a buffer capacity of the soil (for a given ion), with respect to changes in the solution concentration as a result of e.g. ion uptake and fertilizer application. As will be discussed in chapter 7 the distribution ratio also determines the effective rate of transport of an ion species in comparison to the velocity of the soil solution.

According to the previous section one finds:

$$R_{D,k} = T_k / Wc_{o,k} - 1$$

Furthermore, the overall distribution ratio for all cations could be taken as:

$$\bar{R}_{D,cat} = \gamma/W\Sigma_k c_{o,k} \qquad (4.8)$$

which for a given soil depends solely on the (variable) total amount of salts present.

4.2.2. Some experimental data

The CEC and cationic composition of different soil types vary widely. The CEC ranges from 1 to 100 meq per 100 grams of soil. Below 1 meq/100 g the CEC has little practical significance, as then the amount of the adsorbed ions hardly exceeds that in solution: the mentioned buffering effect then disappears. 'Soils' with a CEC-value below 1 meq/100 g are only certain sandy soils with very low organic matter contents (e.g. dune sands). The other extreme is found in soils with a very high content of certain clay minerals (e.g. montmorillonite, vermiculite) or high in organic matter (cf. tabel 4.1). Examples of the former are Vertisols (black cotton soils) and also sediments consisting of fairly pure clay minerals (Wyoming bentonite). In the Netherlands the heavy clay soils (e.g. 50 % illites and 5% organic matter) exhibit a CEC-value of up to 40 meq/100 g, whereas the sandy soils low in organic matter (1-2%) have CEC-values of about 5 meq/100 g.

As was commented on in chapter 3, the surface density of charge (and thus the CEC, cf. equation (4.1)) of certain materials may vary with pH and with electrolyte level. Therefore specification of these parameters is required when listing CEC (and also cationic composition) for such materials (cf. also part 4.6.2 of this chapter).

For later usage it is of interest here to look at the magnitude of the overall distribution ratio. For temperate region soils the electrolyte level is usually around about 0.01 normal at a high moisture content of e.g. 20% by weight. Thus Wc_o is of the order of 0.2 meq per 100 g of soil. If the CEC equals, for example, 20 meq/100 g, $\bar{R}_D$ is about one hundred!

The cationic composition of 'normal' arable soils consists for about 90% of the ions Ca, Mg and H/Al, of which the latter pair comprises less than 20%. While referring to section 4.6 for a discussion on the role of Al-ions in soil, it suffices for the moment to establish that both ions contribute in a comparable manner to soil acidity. Beyond the above limit of 20% H/Al one often uses the term 'acid' soil, which then covers the range from that limit up to near full saturation with H/Al, occurring occasionally under exceptional circumstances like those prevailing in so-called acid-sulfate soils (cf. chapter 8). The remaining 10% of the CEC in the above 'normal' soils is made up by the ions of the common alkali metals Na and K, occasionally also by NH_4^+.

In 'acid' soils these alkali ions often occupy even less than 10%, with the exception of solodized soils (cf. chapter 8). In contrast the percentage of alkali ions, particularly Na may reach values up to about 80% in so-called (saline)-alkali- or (saline)-sodic-soils (cf. chapter 9). Finally it is mentioned that the percentage Mg-ions in many arable soils in temperate regions is around one tenth to one fifth of the percentage of Ca-ions on the exchange complex. A number of examples of the cationic composition of representative soils from different locations are given in table 4.1, whereas in fig. 4.1 a schematic picture is presented of the whole range of cationic compositions encountered, omitting exceptional cases.

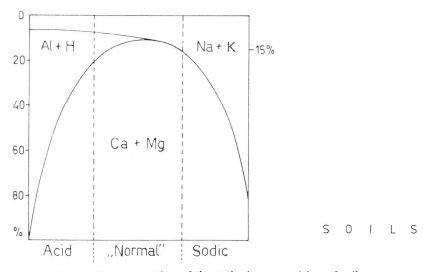

Fig. 4.1. Schematic presentation of the cationic composition of soils.

4.3. MODEL CONSIDERATIONS

The significance of the experimental definitions given in section 4.2. may be illustrated by comparing the quantities defined with the surface areas underneath the concentration-distribution curves of the ions in the neighborhood of a charged surface. As was pointed out in chapter 3 the electric field emanating from a negatively charged surface causes an increase in concentration of positive counterions and a decrease of the anion concentration in the direction of the charged surface. Granted that the precise distribution is influenced by both the electrostatic field and other attractive (or repulsive) forces, the situation may be schematized as pictured in fig. 4.2.

In this figure one recognizes the equilibrium concentrations:

$c_{o,-} \equiv c_{o,+} + c_{o,2+}$

occuring at large values of x, the distance from the charged surface. The mean thickness of the liquid layer on the solid surface, in m, is indicated with $d_l = W/(10^5 S) \equiv W/S^*$. The actual concentrations of the cations increase towards the surface, the rate of increase for the divalent cations, c_{2+}, exceeding that for the monovalent cations, c_+, according to Boltzmann's equation (cf. chapter 3). The anionic concentration c_-, decreases towards the surface. The excess of monovalent cations held by the charged surface then equals:

$$F_+ = \int_0^{W/S^*} (c_+ - c_{o,+}) dx,$$ i.e. the cross-hatched surface

area underneath the c_+-curve, in keq per m² charged surface. Writing-out:

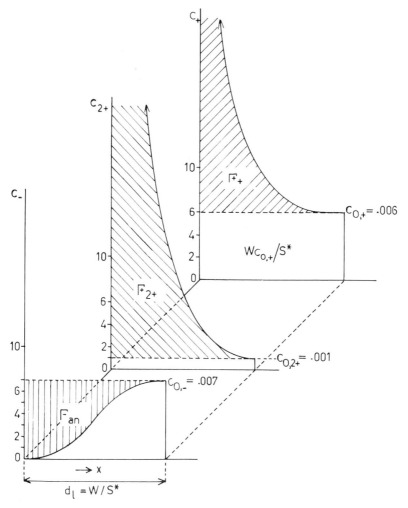

Fig. 4.2. Schematic (perspective) presentation of the concentration distributions in the DDL in a system in equilibrium with a solution of 0.006 normal Na, 0.001 normal Ca and 0.007 normal Cl, and with a water content of W cm³ water per 100 g. The cross-hatched areas define F for the cations, Γ for the anions; (+) and (2+) refer to mono- and divalent cations, respectively.

$$F_+ = \int_0^{W/S^*} c_+ dx - Wc_{o,+}/S^* \tag{4.8}$$

In the same manner one finds for the excess of divalent cations and the deficit of the anions, respectively:

$$F_{2+} = \int_0^{W/S^*} c_{2+} dx - Wc_{o,2+}/S^* \tag{4.8a}$$

$$\Gamma_{an} = Wc_{o,an}/S^* - \int_0^{W/S^*} c_{an} dx \tag{4.8b}$$

Summing up one finds again:
$F_+ + F_{2+} + \Gamma_{an}$ = sum of the three cross-hatched areas = Γ, the surface density of charge of the colloid, in keq/m^2

Generalizing the above to a system containing different cationic species, k, and anionic species, l, one finds the counterparts of equations (4.4), (4.5) and (4.6) as:

$$F_k = \int_0^{W/S^*} c_k dx - Wc_{o,k}/S^* = \frac{T_k - Wc_{o,k}}{S^*} = \overset{+}{\gamma}_k/S^*$$

$$\Gamma_l = Wc_{o,l}/S^* - \int_0^{W/S^*} c_l dx = \frac{Wc_{o,l} - T_l}{S^*} = \overset{-}{\gamma}_l/S^*$$

$$\Gamma = \Sigma_k F_k + \Sigma_l \Gamma_l = [\Sigma_k \overset{+}{\gamma}_k + \Sigma_l \overset{-}{\gamma}_l]/S^* = \gamma/S^*$$

Thus the experimental quantities defined in the previous section, and expressed in meq per 100 g, have been identified with surface areas underneath concentration-distance curves, expressed in keq per m^2 surface area of the solid surface. A model theory predicting these concentration-distance curves may thus be applied to predict the ionic composition of an exchanger (cf. part B).

4.4. THE EXCHANGE EQUILIBRIUM

As was pointed out before, the rate of exchange reactions on well-accessible surfaces is very high. Once equilibrium has been reached there exists a relationship between the composition of the exchange complex and that of the soil solution. From way back in soil science history one has attempted to generalize this relationship in the form of an 'exchange equation'. Such exchange equations are in fact indispensable in practice, as they must give the basis for predictions of changes in the soil system as a result of external influences, such as the application of fertilizers, addition of irrigation- and waste-water, ion uptake by plants, etc. In developing such an equation one must strive towards a compromise between ease of application and accuracy of prediction. It should be pointed out here that one has not found an exchange equation valid for all different exchange materials in the soil At the same time it appears that often a reasonable accuracy is found with equations containing only one empirical parameter, provided that the conditions for application are not too extreme.

4.4.1. Exchange equations

Referring to the previous paragraph, in this section attention will be paid mainly to situations commonly occurring in soils. Thus if the exchange capacity of the soil is due mainly to clay minerals with a reasonably constant surface charge in the pH-range from e.g. pH 5 to pH 7, it is found that the distribution over solution and exchange complex of ions of equal valence follows the equation:

$$\frac{\overset{+}{\gamma}_a}{\overset{+}{\gamma}_b} = K_{a/b} \cdot \frac{c_{o,a}}{c_{o,b}} \tag{4.9}$$

In this equation (often referred to as the Kerr-equation) $K_{a/b}$ is a selectivity 'coefficient' which is reasonably constant over a fairly wide range of compositions. $K_{a/b}$ is specific for any ion pair and may differ for different exchangers.

Referring to the earlier observation that the electric field of the charged exchanger surface is the main cause for the adsorption of cations, it is obvious that such a field will not distinguish between cations of equal valence. Thus $K_{a/b}$ should be equal to unity if only electrostatic forces play a role and if geometry factors are absent. The latter condition is, however, not fulfilled, as the 'minimum distance of approach' of different cations towards the surface of the adsorber varies with ion size. From the very steep gradients of the electric potential close to clay surfaces ($\sim$ 1 million Volt per cm in 10^{-2} normal electrolyte) one may deduce that 1Å difference in ionic radius could easily lead to a selectivity coefficient of about two in favor of the smaller cation.

With respect to the constancy of $K_{a/b}$ for a given pair of cations, it may be shown that this selectivity coefficient bears a relation to a thermodynamic exchange constant comparable to the thermodynamic reaction constants used in chapter 2. Such a thermodynamic 'exchange constant', if used in equation (4.9), would require that both the concentrations in solution and the amounts adsorbed are replaced by activities. As the relation between amount adsorbed of a certain cation and its activity in the adsorbed phase is far from obvious, practice requires the use of equation (4.9) as given, granted that the use of solution phase activities on the right hand side might lead to a better constancy of the selectivity coefficient for a wider range of total electrolyte levels. Reference is made to part B of this text for a detailed discussion of the thermodynamics of exchange reactions.

Experimental evidence shows that the relative preference of clays for the monovalent cations follows in this respect the lyotropic series, i.e.

$Cs > Rb > K \cong NH_4 > Na > Li$,

indicating that the radius of the hydrated cation is decisive for the magnitude of $K_{a/b}$ for clays.

Another factor that may enter is the effect of the surface structure of the adsorber. If the latter is porous (Fe- and Al-hydroxides, the 'hexagonal holes' in clay plates, exchange spots between clay plates, internal adsorption in zeolites) the binding preference may be determined by the radius of the 'naked' cation, giving a reverse of the above order.

For the common cations Na and K it is found that in most clays K is preferred with a selectivity coefficient of about five (cf. also section 4.5). A similar effect, though of much smaller magnitude is found for the selectivity between divalent cations: Ca tends to be preferred over Mg with a factor of about 1.2, whereas its selectivity with respect to other divalent cations (e.g. Ba, Sr, Zn, Co, Ni, Cu) is close to unity. It is noted here that for organic adsorbers (soil organic matter) the situation with respect to divalent ions may be quite different due to chelate formation (cf. section 4.5).

Of special interest is the exchange equation describing heterovalent exchange, especially the exchange equilibrium between the dominant cations Na (+K) and Ca (+Mg). Obviously (Boltzmann equation, cf. chapter 3) the relative preference for mono- and divalent cations now depends on the magnitude of the electric potential in the zone of accumulation, particularly very close to the surface where the majority of the adsorbed cations are located.

In fact, if the exchangeable cations were concentrated in a monolayer on the surface one would find:

$$\overset{+}{\gamma}_+ / \overset{+}{\gamma}_{2+} = c_{s,+}/c_{s,2+} = (c_{o,+}/c_{o,2+}) \cdot e^{-e\psi/kT}$$

in which s refers to this surface layer. The selectivity coefficient $K_{1/2}$ would then equal the Boltzmann factor for this layer. In that case one would also find that the 'reduced' ratio of the concentration in the surface layer equals the same for the equilibrium solution, so:

$$c_{s,+}/\sqrt{c_{s,2+}} = c_{o,+}/\sqrt{c_{o,2+}}$$

It may now be shown (cf. volume B) that for highly charged surfaces the total molar concentration of cations in the surface layer of a colloid with given charge density is roughly constant, i.e. $c_{s,+} + c_{s,2+}$ = constant. Combination of these two conditions then leads to the conclusion that $c_{s,+}$ and $c_{s,2+}$ are fully determined by the value of the reduced ratio in the equilibrium solution. Assuming finally that the amount of cations in the surface layer is dominant with respect to the composition of the exchange complex, one would expect that for a particular highly charged surface the value of $\overset{+}{\gamma}_+/\overset{+}{\gamma}_{2+}$ is determined by the reduced ratio of these ions in solution, or:

$$\frac{\overset{+}{\gamma}_+}{\overset{+}{\gamma}_{2+}} = F\,(c_{o,+}/\sqrt{c_{o,2+}}),$$

in which F is a function which could be calculated for any model of the ionic distribution in the double layer. The above conclusion is in line with Schofield's so-called ratio law which states that if a (negatively charged) surface is in equilibrium with a solution containing concentrations $c_{o,+}$, $c_{o,2+}$, $c_{o,3+}$ of, respectively, mono-, di- and trivalent cations, this surface is also in equilibrium with a solution containing $p.c_{o,+}$, $p^2.c_{o,2+}$, $p^3.c_{o,3+}$ of the same ions (p being an arbitrary multiplier).

Experimental data have shown that for most soils the mono-divalent exchange equilibrium follows roughly the expression:

$$\frac{\overset{+}{\gamma}_+}{\overset{+}{\gamma}_{2+}} = K_G \cdot c_{o,+}/\sqrt{c_{o,2+}/2} \qquad (4.10)$$

in which + and 2+ refer to mono- and divalent ions, and K_G is the empirically determined Gapon exchange constant (after Gapon who proposed this equation at an early date).

In contrast to the dimensionless selectivity constant $K_{a/b}$, K_G has the dimension of [concentration]$^{-\frac{1}{2}}$. Following the convention introduced by Gapon in specifying $c_{o,+}$ and $c_{o,2+}/2$ in moles/liter, one finds in practice that for the Na-Ca exchange equilibrium many soils exhibit a value of K_G equal to about ½ (mol/l)$^{-\frac{1}{2}}$. Notwithstanding its approximate nature the Gapon equation is the simplest, reasonably reliable mono-divalent exchange equation, which may be used in all those cases where no information is available as to the particular conditions locally. It forms the basis of the assessment of alkalinity problems (cf. chapter 9). Some of its limitations are: underestimation of the exchangeable Na-percentage in the high range (> 40% Na), and in montmorillonitic soils (where K_G tends to be closer to unity). Where the selectivity coefficient between different divalent cations is close to unity, one may use in the equation the sum of the amounts of these ions. In contrast $K_{K/Na}$ is quite different from unity (about 5), so in this case the summing up of amounts should perhaps be limited to systems containing a high excess of Na-ions in comparison to K-ions (in the equilibrium solution). For further comments on the exchange equilibrium reference is made to part B.

4.4.2. Application of the exchange equations in estimating changes in composition of solution and complex

It follows from the foregoing that if the situation in the soil solution changes as a result of external factors, the equilibrium between complex and solution is disturbed. A new equilibrium will then be established in which complex-composition and solution-composition undergo mutual adjustment. These adjustments are effectuated by equivalent exchange between both.

Although the exchange is equivalent, the resulting percentage changes in complex and solution will usually differ greatly. This follows from the difference in magnitude of the amounts of cations present initially on the complex, $\overset{+}{\gamma}$, and in solution, $W \cdot c_o$. In estimating the expected changes it is then necessary to establish the amounts of ions present in the different parts of the system.

4.4.2.1. Generalized scheme of calculation

All changes occurring in the field that influence the exchange equilibrium may be summarized as (positive or negative) additions of ions and/or water to the solution phase. Taking as an example the fairly involved case where both mono- and divalent cations are present, one may specify the original equilibrium situation according to the values of $\overset{+}{\gamma}_+$, $\overset{+}{\gamma}_{2+}$, W, $c_{o,+}$, $c_{o,2+}$, where (+) and (2+) refer to the mono- and divalent cations. These quantities are connected by means of a Gapon-type exchange equation which is written here as:

$$\frac{\overset{+}{\gamma}_+}{\overset{+}{\gamma}_{2+}} = K_G \frac{Wc_{o,+}/W}{\sqrt{Wc_{o,2+}/2W}} \qquad (4.11)$$

In this presentation one easily identifies the amounts present in solution, Wc_o (in meq per 100 g of soil), though these amounts must be divided by the moisture content to give the relevant concentrations again (in mol/l or mmole/ml), as these must be entered into the Gapon equation.

Next arbitrary changes are introduced, to be specified here as the addition of p meq of monovalent cations (per 100 g of soil), q meq of divalent cations and r ml of water, all added to the solution phase. If these quantities are entered into equation (4.11), in general an inequality results, according to:

$$\frac{\overset{+}{\gamma}_+}{\overset{+}{\gamma}_{2+}} \neq K_G \frac{(Wc_{o,+} + p)/(W+r)}{\sqrt{(Wc_{o,2+} + q)/2(W+r)}}$$

In order to re-establish equilibrium, exchange must take place, which can be characterized by a shift of x meq of monovalent cations from solution to complex, accompanied by a reverse shift of x meq of divalent ions from complex to solution. The quantity x (which may be either positive or negative) is then found from the equality:

$$\frac{\overset{+}{\gamma}_+ + x}{\overset{+}{\gamma}_{2+} - x} = K_G \frac{(Wc_{o,+} + p - x)/(W+r)}{\sqrt{(Wc_{o,2+} + q + x)/2(W+r)}} \qquad (4.12)$$

The above procedure is valid for all situations, involving a given cation pair obeying a given exchange equation. It may be used to calculate the new exchange equilibrium, e.g. after simultaneous extraction of cations and water by plant roots or after the addition of irrigation water containing salts. Although the solution of an equation like (4.12) looks cumbersome, in practice it is easily obtained graphically by plotting both the left hand side (LHS) and the right hand side (RHS) against arbitrarily chosen values of x. In selecting these values of x it should be remembered that denominator and numerator of both sides must remain positive, so in the above case:

$(Wc_{o,+} + p) > x > - (Wc_{o,2+} + q)$, while also $\overset{+}{\gamma}_{2+} > x > - \overset{+}{\gamma}_+$.

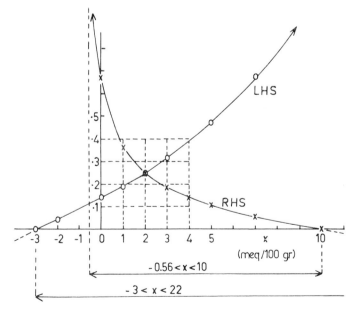

Fig. 4.3. Graphical solution of eq. (4.12a).

Using as an example the system discussed in 4.2.1.2. one finds for the initial situation:

$$\frac{3.0}{22} = K_G \frac{0.36/60}{\sqrt{0.06/120}}, \text{ or } K_G = 0.5.$$

If to a sample of this system, containing 100 g of soil (and 60 ml of solution), one adds 140 ml (r) of a solution containing 9.6 meq NaCl (p) and 0.50 meq $CaCl_2$ (q), equation (4.12) reads:

$$\frac{3.0 + x}{22 - x} = \frac{1}{2} \cdot \frac{(0.36 + 9.6 - x)/200}{\sqrt{(0.06 + 0.50 + x)/400}} \tag{4.12a}$$

Plotting LHS and RHS for x = 0.0, 1.0, 2.0, 3.0 and 5.0 (meq/100 g), resp., one finds x = 2.0 (cf. figure 4.3). The new composition of the system is thus:

$$\overset{+}{\gamma}_{Na} = 5.0, \quad \overset{+}{\gamma}_{Ca} = 20, \quad c'_{o,Na} = 0.040 \text{ normal}, \quad c'_{o,Ca} = 0.0128 \text{ normal}$$

4.4.2.2. Special cases

While the above procedure yields the correct answer in all situations (as long as the exchange equation is valid) there are several situations where the shift in equilibrium may be found in a simpler manner. The basis for these simplifications is often the fact that the distribution ratio is usually very much different from unity, e.g. around 100.

a. For homovalent exchange the square root disappears from the RHS, and thus the system is not influenced by r (i.e. dilution).
b. For heterovalent systems simple dilution (p=q=0) implies that divalent ions move towards the complex (RHS decreases in magnitude). Thus x as used above is negative but larger, i.e. lesser than $-Wc_{o,2+}$. If now the R_D-values of the original system are large (e.g. in non-saline soils with high CEC), the maximum relative change of $\overset{+}{\gamma}_k$ is limited to $1/R_{D,k}$ (for both ions) and usually negligible. This then implies that the LHS remains virtually unchanged. Consequently the RHS maintains a constant value of the reduced concentration ratio. As the new total electrolyte concentration is given as $(c_{o,+} + c_{o,2+}).W/(W + r)$, the problem amounts to dividing this new total concentration into the two components $c'_{o,+}$ and $c'_{o,2+}$ such that

$$c'_{o,+}/\sqrt{c'_{o,2+}} = c_{o,+}/\sqrt{c_{o,2+}}.$$

A similar reasoning holds for withdrawal of water (e.g. by evaporation) and for the addition of trace amounts of p and/or q (in comparison to the original amounts adsorbed). This situation could be referred to as 'complex-dominated' exchange adjustment.

c. Addition of trace amounts of an ion originally not present in the system (e.g. in case of addition of waste water to the soil). Now again the composition of the exchange complex is hardly influenced, but in this case one is mainly interested to find out what fraction of the trace component remains in solution. Assuming that the original system was dominated by e.g. Na and Ca, with distribution ratios equal to $R_{D,Na}$ and $R_{D,Ca}$, respectively, it is obvious that these numbers remain practically unchanged. The exchange equilibrium of the added cation A will now follow a Kerr type equation with the dominant ion of the same valence. Accordingly:

$$\frac{\overset{+}{\gamma}_{A+}}{\overset{+}{\gamma}_{Na}} = K_{A/Na} \cdot \frac{c_{o,A+}}{c_{o,Na}} \cdot \frac{W}{W}, \text{ or}$$

$$R_{D,A+} = K_{A/Na} \cdot R_{D,Na}, \text{ whereas} \tag{4.13}$$

$$R_{D,A_{2+}} = K_{A/Ca} \cdot R_{D,Ca}. \tag{4.13a}$$

If the relevant Kerr-constants are known the concentration of the trace compound remaining in solution may be calculated as:

$$c'_{o,A} = \frac{c_{o,A}}{R_{D,A} + 1} \tag{4.14}$$

in which $c_{o,A}$ is the concentration of A in the added waste water.

d. Another extreme situation arises if excessive amounts (in comparison to the original amounts adsorbed) are added to the system. This situation could arise if a soil is leached with large amounts of e.g. irrigation water. The solution phase is now the dominant one, and the system will eventually adjust to the composition of the leaching solution, the exchange composition being found by applying the appropriate exchange equation.

4.5. HIGHLY SELECTIVE ADSORPTION OF CATIONS BY SOIL

In the previous sections cation adsorption was treated as governed by electrostatic attraction of countercharges, modified to some degree by geometry factors like the distance of closest approach. Under those circumstances the adsorption is completely reversible and selectivity coefficients fall within the range of prediction of rather simple models.

Under certain conditions, however, one finds very high selectivity coefficients for certain ions, sometimes giving rise to partly irreversible reactions. In general one might attribute these phenomena to extreme compatibility of a certain ion with certain individual exchange sites, bearing in mind that after all the solid phase exchangers of the soil exhibit discreteness of exchange sites on a micro-scale. In some cases a reasonable explanation for these phenomena may be given. In this respect it appears logical to mention separately the 'irregular' behavior of certain cations when present in minute quantities. Depending on the mechanism of formation of solid phase exchangers (either in nature or synthetically) there will often exist failures in the crystal lattice, leaving very small crevices which could accomodate an ion of a particular size (dehydrated!) and valence much better than any other ion. As the irregular imperfections of the exchanger surface are not known, predictions seem hardly possible here.

Of more significance are 'regular' imperfections in the surface layer of e.g. Fe- and Al-oxides but presumably also present in SiO_2-surfaces, especially in particles formed during rather rapid precipitation. These surface layers are then imperfectly crystallized giving a limited number of M-OH instead of M-O-M structures. Thus a surface layer of finite depth becomes spongy, and it appears that certain cations, if small enough in the dehydrated form, can then replace some protons from the surface layer. They might then become rather tightly embedded, and this could even lead to reversal of charge of the colloidal surface, at a pH corresponding to or slightly above the zero point of charge in the absence of these ions. The above situation should be distinguished from the replacement of H-ions attached to the exterior O-ions of the surface, or OH-groups on the edges of tetrahedral and octahedral layers of clay minerals. For these groups, which stick out more or less freely in the solution, no extreme selectivity is to be expected.

4.5.1. Fixation of cations in clay lattices

As was shown in chapter 1, the 2:1 clay minerals have a tetrahedral surface layer containing 'hexagonal' holes between the O-atoms. These holes provide a possibility for cations to penetrate rather close to the seats of charge of the clay, provided the ion is small enough. The smaller the cation, however, the higher is the energy needed to dehydrate the ion, and practical experience shows that apparently the cations K, Rb, Cs (and also NH_4) will be able 'to make use of this possibility', whereas Na and Li will not do so. Thus K (and the other mentioned ions) tend to be adsorbed with a somewhat higher selectivity than could be expected from their hydrated size (i.e. a selectivity factor of about 5 with respect to Na, for high percentages of K present).

An extension of this phenomenon of much greater significance is found in clays of the mica type (e.g. illites). In nature, these multilayered clays (5-10 platelets per particle) contain considerable amounts of K-ions embedded in the hexagonal holes of adjacent tetrahedral layers of the individual platelets, thus cementing these together. Assuming five platelets per particle, the total amount of K thus 'locked in' may correspond to twice to four times the CEC associated with the exterior surfaces of the entire particle. These interlattice K-ions are held extremely preferentially with respect to other cations, and are, therefore, largely non-exchangeable (in dilute electrolyte solutions). Along the edges, however, some K-ions may be exchanged against e.g. H or Ca, if the clay is brought in contact with a solution of very low K-concentration. Upon prolonged exposure of these clays to the above conditions, the edge-situated interlattice sites thus become occupied by ions other than K, presumably leading to partial opening of the interlattice space. If now these clays come in contact with K-ions added to the solution, the latter will rapidly disappear into the edge-situated interlattice sites. The lattice then closes again and the adsorbed K becomes once more very inaccessible. This phenomenon is termed K-fixation by 'open' illites. The irreversibility of the exchange reaction is thus attributed to a change of geometry upon adsorption of K. It should be mentioned that NH_4, Rb and Cs are fixed roughly to the same extent as K-ions in these clays.

In addition to illites also the vermiculites are highly selective for K-ions, but also for Mg ions and presumably certain other divalent cations. Again in this case these ions become preferentially adsorbed inside the 'expanding lattice' of vermiculites. Though highly selective this adsorption remains reversible to a fairly high degree.

4.5.1.1. Practical aspects

The illitic clays are often an important constituent of the solid phase in alluvial soils. In certain areas these soils may have been used for agricultural purposes for many centuries, without application of K fertilizers. Considerable amounts of K could thus be removed, at a concentration sufficient for a rather low production level. As a result these illites have reached a low level of readily available K, which is accompanied by the 'open' structure described above. Upon intensification of the agricultural production the K-supply becomes insufficient. Application of K-fertilizer, however, gives only a meager response, because the increased concentration of K in the solution leads to entry of K in the vacated interlattice positions and subsequently to closing of the lattice. The K applied thus becomes unavailable for the crops. The term 'potassium fixation' indeed arose in agricultural practice. It appears to be promoted by liming, because the increase in pH facilitates the removal of H-ions which often have taken the place of the K-ions in the 'open' illite.

K-fixation in soils is hardly ever remedied, although its effects may be moderated. Thus filling up all vacant positions in the clay lattice with K-ions is economically not feasible. On the other hand one may make use to some extent of the strong competition of NH_4-ions for the interlattice sites. Application of NH_3- or NH_4-salts thus gives (temporarily) NH_4-fixation, suppressing the K-fixation to some degree. The fixed NH_4 is then gradually liberated throughout the growing season and oxidized to nitrate. Also it appears to be of help to use low grade K-fertilizers, which limit the concentration of K in solution to a lower value relative to e.g. Na-ions.

Another practical implication of the above is the fixation of Cs in these soils. Inasfar as radio-active Cs is a much feared pollutant derived from atomic explosions, it may safely be stated that this ion does not move at all in K--fixing soils. As in this case only trace amounts are concerned, even high potassium illites (e.g. all river clays) are very effective in retaining radio-cesium.

4.5.2. Complex formation of cations by organic matter ligands

In section 1.1.2 the existence of ligands in the polymer chains of humus was pointed out. These ligands will form complexes with certain cations (particularly transition metals). Such complexes should be seen as quite different from the regular Coulombic adsorption of cations in a double layer. In principle the cation present in a complex cannot be 'exchanged' against e.g. Na-ions, that is, a Na-ion will never occupy the position of a Cu^{2+}-ion on an organic matter ligand.

Reasoning from the standpoint taken in chapter 3, the complexed cation should be considered as part of the solid phase, and its presence will thus cause a decrease of the (negative) surface charge of the organic matter. In other words the CEC of soil organic matter will decrease if certain ions of the transition metals are introduced into the system. As was discussed in chapter 3, a similar distinction could be made with regard to the 'position' of H-ions in the acidic OH-groups in comparison to the regular (alkali and earth-alkali) cations. Mentioning again the observed connection between the acidic OH-groups and the complexing ligands one might thus argue that the CEC of organic matter will strongly depend on the concentrations of H-ions and transition metal ions present in the system.

The binding of the above ions by organic matter ligands is thus taken out of the range of cation exchange phenomena, discussed in section 4.4.1, and the exchange equations should be considered as inapplicable. Rather should one attempt to specify the relative preference of organic matter in forming complexes with ions of especially the transition metals in terms of stability constants of the complexes (cf. also chapter 2). It seems improbable that particular values of the stability constants of organic matter complexes with different metals would be generally valid for different types of organic matter. Indeed different authors have listed stability constants which differ even in following order of magnitude. Consensus of opinion appears to exist on the fact that Cu^{2+} is very tightly bonded over a wide range of pH values. Other metal complexes appearing often high in stability are Fe^{3+}, Pb^{2+} and to a lesser degree Ni, Co, Mn. The stability 'constants' are, as could be expected, very sensitive to pH: at pH > 7 these cations are complexed much more strongly than at low pH-values.

For practical purposes the above observations are of great importance with regard to the retention by soil of heavy metal ions present as pollutants in waste water. Much in constrast to the clays (and also the oxide surfaces) the organic components in soil may be very effective in this respect.

4.6. THE ADSORPTION OF H- AND AL-IONS BY SOIL CONSTITUENTS

The adsorption of H-ions deserves separate treatment, as its role in exchange equilibria is often complicated. At the same time a considerable amount of information has been collected in the course of the years, partly because of the presumed significance of soil pH with regard to the growth of higher plants and microbes but also because of the relative ease of the determination of H-ion activities in the soil solution with the help of glass-electrodes. Thus titration curves of soils and soil constituents have been determined in abundance. Unfortunately, part of these data have not contributed much to the understanding of the adsorption of H-ions by soils, because of lack of information about the composition of the solution in which the soil was titrated (cf. below).

4.6.1. Analysis of the different types of adsorption mechanisms

A fairly complete story on the adsorption of H-ions by different soil constituents is considered to be beyond the scope of this part of the text; accordingly, only the main aspects will be outlined here, leaving details for part B.

4.6.1.1. Non-selective adsorption of H-ions

Starting at one end of the range of adsorption mechanisms (for H-ions) occurring in soil one might consider the non-selective adsorption of H-ions by surfaces with a fixed charge density. The best example of such a surface is the strongly-acidic exchange resin. Here complete dissociation of the H-ions from the surface OH-groups prevails, and thus the charge density is fixed. The H-ions of the system will then serve to neutralize the surface charge together with all other cations present in the system. They will be accumulated (with the other cations) in the neighborhood of the charged surface as a swarm of counter-ions. The relative proportion of H-ions in this swarm should then follow an exchange equation of the Kerr- or Gapon-type. As the (hydrated) size of H_3O^+ is comparable to that of K^+, the selectivity coefficient for the H-K-exchange appears to be close to unity. If a surface of the above type contains a given amount of adsorbed H-ions, the pH of the equilibrium solution will clearly depend on the concentration of the competing ion: at $\overset{+}{\gamma}_H = 0.5\ \gamma$ one would expect a pH value of about 2 at a K-concentration in the equilibrium solution of 10^{-2} normal, while in 10^{-1} normal KCl the pH should be around 1.

This competition effect of other cations should be seen as the result of the proximity of the surface charges, which leads to the formation of a double layer as was discussed in chapter 3. This double layer is the result of the interaction between the electrostatic fields of the individual surface charges. As the resultant electrostatic field is non-selective for ions of equal valence, one might regard the competition of other monovalent cations as a non--selective competition.

In more general terms, the competition effect may be coupled to the increase in electric capacity of the double layer upon addition of neutral electrolyte. For a given (i.e. fixed) surface charge, this increased capacity causes a drop in the potentials in the counterion layer. Addition of electrolyte to a surface with e.g. $\overset{+}{\gamma}_H = 0.5\ \gamma$ will thus cause the H-ions present in the counterion layer to move into the equilibrium solution, until (at each point of the double layer) the 'new' ratio of the (reduced) concentration in the double layer to the (increased) concentration in the equilibrium solution corresponds to this reduced electric potential in the counterion layer (cf. Bolzmann equation, chapter 3). In other words, addition of electrolyte to the above system causes a reduction of the distribution ratio of the H-ions (and of all other cations). Depending on the actual distance between the surface charges this effect of the addition of electrolyte will eventually vanish. That is, once the electrostatic field of the double layer has been reduced to negligible potentials, further addition of electrolyte will no longer influence the pH of the equilibrium solution (aside from a small effect on the activity coefficients in solution). In practice this means that the lower the surface density of charge of the colloid, the lower is the electrolyte level above which no effect of the pH is noticed. To restate the above argument in other words: an individual surface charge gives rise to a local electrostatic field. The extent of these individual fields depends on the electrolyte level (cf. the Debye--Hückel interaction length $2/\kappa$), and depending on the particular conditions, i.e. the electrolyte level and the distance between neighboring individual charges, these individual fields may or may not interact with each other to form an electric double layer. If so, the above pH-effect will be found; if not, the effect is absent. 'Proximity' of surface charges is thus a relative notion. The higher the electrolyte level, the closer they must be spaced to exhibit the proximity effect.

4.6.1.2. The role of Al-ions in acidified clay systems

Looking now at the behavior of H-ions on clay minerals, one finds that the above non-selective competition effect due to proximity of the surface charges is present, but only in freshly prepared H-clays. In contrast, H-clays that have been allowed to 'age', exhibit the proximity effect allright (as evidenced by the drop in pH upon addition of neutral electrolyte, e.g. KCl), but the actual value of the pH is far above the one predicted from a Kerr-equation using $K_{H/K} \approx 1$. Instead one finds that a selectivity coefficient in the order of one thousand would be needed to explain a pH value of e.g. 5 in a system with $\dot\gamma_H = 0.5\ \gamma$ and $c_{o,K} = 10^{-2}$ normal.

Analysis of the equilibrium solution of such an 'aged' clay system shows, however, that in addition to H-ions and K-ions there is also a finite concentration of Al-ions in the solution. Detailed studies on these systems have now shown that the above observations may be explained as follows. Freshly prepared H-clays will spontaneously 'decompose', that is, protons will penetrate into the octahedral layer, replacing the Al-ions (and sometimes also Mg-ions) present along the edges. The trivalent Al-ions thus freed are then adsorbed preferentially in the counterion layer on the planar sides of the clay mineral. Thus an H-clay will soon become a $H-Al^{3+}$-clay (within hours, at room temperature).

At the same time such an Al-clay (aside from its strong tendency towards flocculation) behaves like a weakly acidic exchanger, because in aqueous systems the hydrated Al^{3+}-ions behave like a weak acid, according to:

$$Al(H_2O)_6^{3+} \rightleftharpoons Al(H_2O)_5 OH^{2+} + H^+$$

Indeed, addition of excess electrolyte to an aged H-clay yields a system that has a titration curve similar to that of a solution of $AlCl_3$ (the surface Al-ions having been exchanged by the cations of the added electrolyte), which is discussed in section 4.6.2 (see figure 4.4e). At low or moderate electrolyte levels such systems exhibit a behavior which may be interpreted as the superposition of the proximity effect upon the weakly acidic Al^{3+}-ions (cf. also section 4.6.2). In conclusion: the proximity effect explains the shift in pH upon addition of neutral electrolyte while the presence of the weakly acidic Al^{3+}-ions on the clay explains the high 'selectivity' for H-ions as deduced from the relatively high pH-value for a given composition of the exchange complex and the equilibrium solution.

4.6.1.3. Preferential adsorption of H-ions by organic matter

In contrast to the non-selective adsorption of H-ions by the completely 'dissociated' clay surface, organic matter chains contain acidic groups which behave like moderately weak to very weak acids. As was pointed out in sections 1.1.2 and 4.5.2 these acidic groups usually serve at the same time as ligands for certain metal ions. So, depending on the pH, part of these groups remain associated with a proton. Alike the situation with a molecular weak acid (e.g. acetic acid), these associated protons are now considered as part of the group and are, as a rule, not exchangeable against cations. Whereas in the molecular acid this distinction between H-ions and e.g. K-ions present in the system is clear-cut, here the proximity of acidic groups at the surface makes the situation slightly more complicated. Thus if, under a given set of conditions, a high proportion of the acidic groups is dissociated, the surface charge of the polymer chain becomes considerable and again the interaction of the electric fields of neighbouring charged groups causes the build-up of a regular counterion layer. In this layer one finds cations (and in principle also H-ions) equal in number to the surface charges. As these counterions are situated at a comparatively large distance from the surface groups, one finds again a non-selective accumulation. Using the Kerr-equation again one might expect that the counterion layer contains at pH 5 in 10^{-2} normal KCl about 0.1% H-ions and almost 100% K-ions. The system thus contains three groups of H-ions: 'free' H-ions in the equilibrium solution, non-selectivily adsorbed (and exchangeable) H-ions and H-ions associated with the polymer chain. Using as an example the above system, assuming that under these conditions e.g. 50% dissociation of the surface groups has taken place one finds for e.g. 1 gram of organic material, containing 2 meq of acidic groups, that the system contains 1 meq associated H-ions, a negligible amount of 10^{-3} meq of adsorbed H-ions and, if the solution phase is 100 ml, also 10^{-3} meq of free H-ions. The system contains also 1 meq of adsorbed K-ions and 1 meq of free K-ions.

Further addition of KCl to this system (say to 10^{-1} normal) will now lead again to an increase of the electric capacity of the double layer, resulting in a reduction of the electric potentials in the counterion layer and also at the surface of the solid phase (corresponding, in fact, to a decreased interaction between the surface groups). Accordingly the degree of dissociation of the surface groups will tend to rise, thus freeing protons which then move towards the equilibrium solution. Thus the pH decreases just as in the case of the non-selective surface treated in 4.6.1.1, but in this case the decrease of pH is due to surface-protons and is not caused by exchange of counterions.

This behavior of the organic polymer may be contrasted with that of the molecular weak acid (e.g. acetic acid). In that case a change in the concentration of neutral electrolyte (KCl) will hardly affect the pH of the system (aside from small changes due to activity coefficients). The conclusion is then again that organic polymers containing acidic groups partly associated with H-ions show a pH-sensitivity towards the electrolyte level due to the proximity of the surface groups. In this case the pH-change is connected with a change of the surface charge and thus with the CEC. Obviously the magnitude of this effect depends on the actual distance between neighboring groups. In practice one finds usually only a weak effect (as compared to that for the highly charged clays).

An exception to the above mechanism is the effect on the pH of certain special cations (notably some transition metals, like e.g. Cu). The pH change is then profound even for fairly small additions. The explanation for this phenomenon was already indicated in section 4.5.2: these ions form chelates with the surface groups and are then competing directly with the surface-associated H-ions. Inasfar as these chelated ions are again treated as non-exchangeable (viz. they are located at, or in, the surface and not in the counterion swarm) one would thus reason that the addition of a small amount of e.g. Cu^{++}-ions leads to a decrease of the CEC (while freeing H-ions) while addition of KCl leads to an increase of the CEC (while freeing H-ions).

4.6.1.4. H-ion adsorption by oxide surfaces

The oxides of certain metals (in soil e.g. Al and Fe-oxides) may be treated in a similar fashion as the organic polymers with their acidic groups. Often these surfaces are amphoteric, i.e. depending on the pH of the system the oxide surface is either negatively charged or positively charged, while becoming electrically neutral at a certain characteristic pH value (zero point of charge, pH_o). As an example it is mentioned that Hematite (Fe_2O_3, a common compound in many tropical soils) exhibits a pH_o of around 7; particles of this material may thus be either positively charged or negatively charged depending on the soil pH. If the pH exceeds pH_o one finds a situation very similar to that of the partly dissociated polymers discussed above. Applying the same reasoning as above, one may conclude that at $pH > pH_o$, the pH will drop upon the addition of neutral electrolyte, the freed H-ions coming from the surface layer and not from the counterions, thus giving rise to a (slight) increase of the CEC.

If the pH is below pH_o an 'opposite' effect arises. The increased electrical capacity of the double layer (at increased electrolyte level), causes an increased association of the surface with protons, giving an increase in pH. At the same time the positive charge of the surfaces tends to increase, giving a (slight) increase of the AEC (Anion Exchange Capacity). In contrast to the organic polymer, neighboring surface O-atoms (which are able to associate with one or more H-ions) are situated very close to each other, so that the proximity effect is always considerable. An exception is the zero point of charge, where no effect is found (because the surface has no charge and does not emanate an electric field).

A final remark may be made about the Al-clay discussed in 4.6.1.2. A dense 'coating' of the clay surface with $Al(H_2O)_6^{3+}$ ions, presents a situation in some ways analogous to that of a Gibbsite surface at a $pH < pH_o$. Thus the 'coating' is charged positively, but as it is deposited on a highly (negatively) charged clay surface, the overall charge of clay surface-plus-coating is still negative. Addition of electrolyte then gives a decrease of pH. In fact one might notice that the very high positive charge of the coating (3 positive charges per Al-atom) is much beyond what one could possibly obtain with Gibbsite particles. If the latter particles are charged positively by lowering the pH gradually one finds that the structure becomes unstable long before the positive charge is three units per Al-atom. The structure then simply dissolves to form Al-salts. If, however, clay surfaces are present in the system, these will serve as a template to keep the trivalent Al-ions close together under the binding influence of the negatively charged clay. The present story may serve as an indication that if Gibbsite is present in soil this material will also serve to supply the exchange complex with Al-ions if the pH is lowered enough.

4.6.1.5. Summary of the mechanisms of H-adsorption

All components of the soil solid phase interact with H-ions. For freshly prepared H-clays the interaction amounts to non-selective accumulation in the counterion swarm. For organic ligands and oxide surfaces the interaction is a highly selective association between protons and surface groups. In that case the associated H-ions are considered as part of the surface and are in principle not exchangeable against other cations.

Due to the proximity of surface charges (arising from substitution in the clay and from dissociation of surface OH-groups of the other compounds discussed) addition of neutral electrolyte to a soil system containing the above materials will in general lead to a decrease of pH. This is caused either by exchange of H-ions from the counterion swarm (freshly prepared H-clays) or by release of surface associated H-ions accompanied by an increase of the CEC of the material (provided the original pH of the system exceeded the value corresponding to the zero point of charge of the material).

'Aged' acidified clays are in reality Al-M-clays and the decrease of pH upon addition of electrolyte amounts to a release of Al^{3+}-ions from the surface, which in turn leads to a decrease in pH due to the partial dissociation of these ions into $Al(OH)^{2+}$ and H^+-ions. The above principles are illustrated schematically in figure 4.4.

4.6.1.6. Base saturation; lime potential

Above it was inferred that in field soils (which are aged in principle) the total amount of protons associated in some way with the solid phase may form a substantial part of the exchange capacity even at near neutral pH-values. For organic material and oxide surfaces these protons are then directly associated with the surface; in clay soils they are essentially part of the Al-hydroxide coating. As will be discussed below this gives rise to an extended buffering range of soils up to near neutral pH. The complement of the fraction of the CEC associated directly or indirectly with the above protons is customarily designated as the 'base saturation' of the soil.

If furthermore the dominant complementary cation on the exchange complex is Ca^{2+}, one may conclude that slightly acidic soils (i.e. pH 5-7) tend to be buffered with respect to both Ca-ions and H-ions. The ratio law now states that if a soil is in equilibrium with a solution of given composition in Ca and H, it is also in equilibrium with all other solution compositions for which $c_H/\sqrt{c_{Ca}}$, or actually $a_H/\sqrt{a_{Ca}}$, has the same value as in the original equilibrium solution. Accordingly the composition of the equilibrium solution of the above slightly acidic soil is characterized by a particular value of $a_H/\sqrt{a_{Ca}}$, which thus remains invariant with small additions of salt and/or dilution with water. The same then applies to the negative logarithm of the above 'reduced' activity ratio, i.e. pH-½pCa. The latter quantity has been termed the 'lime potential' as it corresponds also to a particular value of the ion product $\sqrt{a_{Ca} \times a_{OH}^2}$. In contrast to the pH of a soil, which for a given composition of the exchange complex with regard to H- and Ca-ions is still highly variable with the total electrolyte level (or with moisture content), the above lime potential is thus rather constant and therefore a better characteristic of the composition of the exchange complex with regard to H/Ca, i.e. in practice the base saturation.

4.6.2. The titration curves of soil constituents

The discussion given in section 4.6.1 is illustrated by the titration curves of soil constituents at different electrolyte levels. For comparison, the titration curves of 1 meq of HCl and 1 meq HAc (acetic acid) in 100 ml H_2O are pictured in figure 4.5a (cf. chapter 2). Neglecting activity corrections these curves are independent of the level of neutral electrolyte (e.g. KCl) also present. In addition the titration curve of a mixture of ½ meq HCl and ½ meq HAc is dotted in. It shows that the strong acid (which is completely dissociated) is titrated first, whereafter the pH rises until the weak acid starts dissociating.

In figure 4.5b the titration curves of 1 meq of a 'strongly acidic' exchanger (H-resin or freshly prepared H-clay) are given. Using KOH as base and KCl as neutral electrolyte these curves may be estimated with the help of an exchange equation of the form:

$$\frac{\overset{+}{\gamma}_H}{\overset{+}{\gamma}_K} = \frac{c_{o,H}}{c_{o,K}}$$

i.e. assuming that the selectivity coefficient $K_{H/K}$ equals unity. As expected the proximity effect of the surface charges shows up in a downward shift of the curves with increasing electrolyte level.

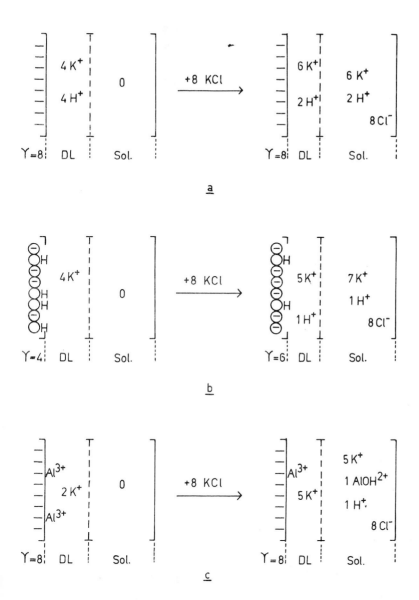

Fig. 4.4. Decrease of pH upon addition of KCl to negatively charged adsorbers (note also decrease of extent of diffuse layer of counterions (DL)
a. Non-selective adsorption of H-ions (e.g. freshly prepared H-clay)
   H-ions are released from diffuse layer; γ remains constant
b. Highly selective oxide surface. H-ions are released from surface OH-groups, thus increasing γ
c. Al-clay. Al^{3+}-ions are released into the solution, which then hydrolyse partly to Al^{2+} and H^+

It is noted in passing that in the figures presented the effect of the electrolyte level is somewhat suppressed due to the excessive amount of solution present (100 ml per meq of clay). As may be verified the curves for 10^{-2} and 10^{-1} normal salt are shifted downward, starting at pH 2 and pH 1.3, respectively, if only 10 ml of solution are present.

The titration curve for organic matter OH-groups (figure 4.5c) may now be seen as a combination of figures a and b. These are a mixture of many acidic groups with different acid strengths, so a curve is obtained resembling those of the acid mixtures as shown by the dotted line in figure 4.5a. However, one finds a more gradual transition from stronger to weaker acidic groups, leading to somewhat irregular titration curves that often show no distinctive buffer-regions. At the same time the proximity of these groups shows up in an upward shift of the titration curve for the very low electrolyte levels.

In figure 4.5d the titration curve of an oxide surface is pictured (hematite). At pH-values above the zero point of charge the proximity of the surface groups gives rise to a downward shift upon increase in electrolyte level, below pH_o the opposite is seen.

Finally the 'aged' clay (i.e. an Al-M-clay with about 75% Al on the exchange complex) is shown in figure 4.5e, together with the curve of freshly prepared H-clay (both at 10^{-2} and 1 normal KCl). It is noted here that the Al-clay in 1 normal salt resembles very closely the titration curve of $AlCl_3$. Thus the Al-clay shows the superposition of the salt effect (due to proximity of surface charges) on the weak acid behavior of $AlCl_3$.

The titration curve of a soil should be interpreted as a combination of figures 4.5c, d and e. A few examples are shown in figure 4.5f. In deviation of the titration curves of common acids one may distinguish three peculiar characteristics of the titration curve of soils, viz:

a. A rather broad range of pH over which the slope is fairly constant, without distinct inflection points, (cf. the use of a liming factor as discussed in the following section).
b. A great sensitivity towards the total electrolyte level at which the curve is determined (cf. the need for a standardized salt level when using the curve for the prediction of the lime requirement as discussed in the following section).
c. A long time period needed for complete equilibration with added base (cf. the role of Al-ions and their polymers as discussed above).

In conjunction with b above it is clear that the titration curve and also the original pH-value of a soil if measured in a solution of unspecified composition, yield little information as to the base saturation of that soil. The pH--value measured in that way at best characterizes a prevailing condition with regard to e.g. root growth and microbial life and then only if the soil is at the same moisture content as used during the determination of the pH.

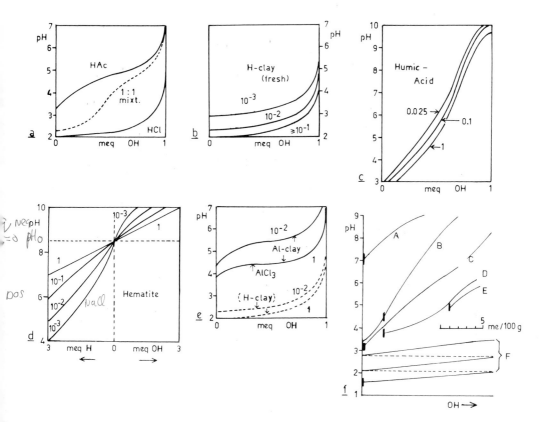

Fig. 4.5. Titration curves of soil constituents, etc.
a. 1 meq of HCl, HAc and a 1:1 mixture, respectively, in 100 ml solution.
b. 1 meq freshly prepared H-clay in 100 ml KCl solutions of different normalities.
c. 1 meq of a humic acid in 60 ml NaCl solutions of different normalities (after van Dijk).
d. 100 grams Hematite with a specific surface area of 30 m²/g (after Breeuwsma).
e. 1.3 meq Na-Al-clay containing 1 meq of Al, in 100 ml NaCl solutions of different normalities. Note that the curve in 1 normal electrolyte corresponds to the curve of 1 meq AlCl₃-solution (after: Bruggenwert).
f. 100 grams of different soils in 250 ml 1 normal KCl;
   A. Calcareous river clay (river Rhine, Netherlands)
   B. Humic sandy soil (Wageningen, Netherlands)
   C. Silt loam derived from loess (Middachten, Netherlands)
   D. Loamy Soil (Serawak, Malaysia)
   E. Non-calcareous river clay (river Rhine, Netherlands)
   F. Tropical Peat (Sumatra, Indonesia) (indicates original pH in 1 normal KCl)

Broady speaking the titration curve of a soil may also be divided into a number of partly overlapping pH-ranges characteristic for certain deprotonization reactions. Thus up to pH 4 neutralization takes place of free and exchangeably adsorbed H_3O^+ ions. Between pH 4 and pH 5.5 the adsorbed Al^{3+}-monomers become gradually deprotonized (the first pK-value of Al^{3+} is about 5). From pH 5 to pH 7 carbonic acid, Al-polymers and organic carboxyl groups are being-titrated. Above pH 7 then follow phenolic OH--groups, bicarbonates, large Al-polymers and Gibbsite surfaces.

4.6.3. Correction of the soil pH

It may be shown that ion uptake by plants is influenced by the soil pH, both in a direct manner and in an indirect manner via solubility of certain nutrient ions. In this text only the soil chemical aspects of the pH-correction will be considered. Three aspects must be mentioned here: 1) The magnitude of the correction, 2) the compound which is used for the correction, 3) the amount applied of the latter. As regards the first point, i.e. the difference in pH (ΔpH) between the original situation and the required one, it is important to reconsider the effect of salt concentration on soil pH. As this concentration varies with moisture content during the growing season it is necessary to formulate the required pH value at a normalized composition of the soil solution. It would be possible to take a solution of 0.01 molar $CaCl_2$ for this purpose; in temperate regions this is a satisfactory average composition under field conditions. A drawback of this choice may be that in certain cases the original salt level is higher than this value. In the Netherlands one has chosen to specify the required pH at 1 normal KCl, taking into account that the required value in the field is usually higher than the pH (KCl). If the latter is given as e.g. pH(KCl) = 5.5 the field value would often be around pH 6.5. Obviously the original pH of the soil is then also determined in 1 normal KCl.

In the Netherlands the required pH for most soil types has been established on the basis of extensive experimentation by the Institute of Soil Fertility at Groningen.

The material to be used for pH correction must contain acid-neutralizing components. These could be hydroxides (e.g. NaOH, $Ca(OH)_2$), or carbonates. The latter react with H-ions under formation of CO_2. Upon neutralization of the soil the adsorbed H-ions will, however, be replaced by the cation of the hydroxide or carbonate. As was indicated in chapter 3, the Na-ion is not acceptable for this purpose (swelling, peptization etc.). In contrast Ca-ions are quite satisfactory and accordingly correction of the pH is effected by 'liming', i.e. application of $CaCO_3$ (and $MgCO_3$) or sometimes $Ca(OH)_2$. The carbonates have the advantage that local surplusses will not lead to excessive pH-values. This must be avoided since at high pH the organic matter will oxidize rather easily, whereas certain trace elements become insoluble. For the following considerations the material used for pH-correction will be referred to as 'liming material'. The amount applied will be specified in terms of its neutralizing power, expressed as kilograms of CaO (i.e. 28 kg per kg equivalent).

The lime requirement for a given value of ΔpH may be found directly from the titration curve of the soil (in 1 normal KCl). In principle this titration curve must thus be determined. It is often found that certain soils (e.g. acid clay soils and sandy soils with some organic matter, i.e. soils with weak acid character) exhibit a fairly linear curve in the range of pH(KCl) between 4 and 5.5. The lime requirement is then proportional to ΔpH. The 'mean slope' of the titration curve is then sometimes indicated as the liming factor. It is often expressed in tons of CaO per ha plow layer of e.g. 10 cm thickness, per 0.1 unit of pH. Similarly, the lime requirement is defined as the amount of CaO (tons per ha) needed to raise the pH-value of 10 cm plough layer from the actual to the required value.

Obviously the liming factor (or the mean buffer capacity) is determined to a large extent by the CEC of the soil. For soils containing the same type of cation adsorbing materials the liming factor becomes proportional to the amount of adsorber per 100 g soil. This renders it possible to express the liming factor in terms of clay content and organic matter percentage.

In the Netherlands such relationships have been determined by statistical analysis of a large number of data.

If such data are available the lime requirement may be computed directly from the observed value of pH(KCl) and humus and clay content of the soil. In all other cases the titration curve must be determined in order to calculate the lime requirement.

4.6.4. Measurement of pH in soil; the suspension effect

Referring to the discussion on the electrometric determination of the pH in aqueous solutions, some cautioning remarks must be added with respect to the common practice to determine in a similar fashion the pH in soil suspensions. In the latter case the electrodes are bathed in a heterogeneous system comprising a solution plus suspended solid particles. Unless the suspension is vigorously shaken during the measurement (which tends to produce unstable readings), the suspension exhibits a considerable density gradient in the vertical direction. It is then found that if the electrode-pair (i.e. glass- and calomel-electrode) is first placed in the clearest upper part of the supernatant liquid and subsequently lowered into the sediment that the pH-reading changes - usually from a higher value to a lower one. This pH--change has been termed the suspension effect.

In the early days the suspension effect was explained rather superficially as the result of the increased 'concentration'(i.e. amount per unit volume of sediment) of H-ions in the sediment as compared to that in the (clear) supernatant liquid ($R_{D,H} \gg 1$!). Granted that indeed the H-ion concentration increases sharply towards negatively charged surfaces, it must nevertheless be ascertained that the value of the partial molar free enthalpy of the H-ions, $\bar{G}_{H^+}$, is invariant with the position in the suspension (at equilibrium), as any increase in the local concentration is in balance with a decrease of the local value of the electric potential. To put it more precisely, in regions where an electrostatic field is operative, the electrochemical potential, $\bar{\mu}_k$, comprising both concentration and electric field effects, is constant (at equilibrium). So in this case:

$$\bar{G}_{H^+} \equiv \bar{\mu}_{H^+} = \mu_{H^+} + F\psi = \text{constant}.$$

Accordingly the glass electrode, being a reversible electrode with regard to H-ions, will register the same potential difference with respect to a fixed reference electrode, regardless of the position of the glass electrode in a suspension or its supernatant solution.

Returning now to the suspension effect as found by lowering the electrode pair into the sediment, it may be verified experimentally that it is necessary and sufficient to shift the calomel reference electrode from supernatant to sediment to observe this effect, while the position of the glass electrode is immaterial (if equilibrium prevails). The behavior of the calomel electrode in this respect is then explained in terms of a diffusion potential originating at the liquid junction between a concentrated KCl solution and a sediment containing charged surfaces. Foregoing details it suffices here to point out that in case the solid surface is negatively charged, the transference number of the cations may grossly exceed that of the anions. As a result a diffusion potential will arise at the orifice of the calomel electrode to the effect that the glass electrode appears more positive with respect to the calomel electrode when the latter is placed in the sediment instead of in the supernatant liquid. Obviously this results in a lower value of the pH if the calomel electrode is placed in the sediment. For details see chapter 11 of part B.

Aside from proving that the sediment contains negatively charged surfaces, the suspension effect does not lend itself easily to quantitative conclusions. Accordingly it should be avoided when measuring the pH of a soil suspension. This leads to the conclusion that care must be taken to let a suspension settle out before measuring its pH. The electrodes must then be inserted such that the calomel electrode is positioned in the supernatant liquid. As to the glass electrode, its position is immaterial if indeed equilibrium prevails. As in practice the supernatant liquid is often rather sensitive to external influences (e.g. CO_2 affecting strongly the pH of a near neutral dilute salt solution) it appears safest to insert the glass electrode into the well buffered sediment for most dependable readings (For details see chapter 11 in part B).

ILLUSTRATIVE PROBLEMS

1. 200 gram of a water-saturated soil sample (W = 100 ml/100 g dry soil) is percolated with an excess of a 1 normal NH_4NO_3 solution. The percolate contains: 2.10 meq Na, 24.60 meq Ca, 5.90 meq Mg, 8.30 meq K, 0.13 meq Cl and 0.13 meq SO_4. Calculate the CEC of the soil. (Answer: 40.64 meq/100 g. dry soil).

1a. From another portion of the above sample a small amount of the solution phase is extracted. This extract contains in eq per liter: 0.0020 Na, 0.0050 Ca, 0.0050 K, 0.0010 Mg, 0.0080 Cl and 0.0050 SO_4. Calculate: 1) the amounts of cations adsorbed and anions excluded and the CEC (in meq/100 g dry soil); 2) the cationic composition of the exchange complex expressed in percentages of the CEC; 3) the distribution ratios of the individual cations.

(Answer:

	Na	K	Ca	Mg	Cl	SO_4	CEC
$\overset{+}{\gamma}, \overset{-}{\gamma}$	1.90	7.80	24.10	5.80	0.67	0.37	40.64 meq/100g
γN_k	5.1	20.2	60.2	14.5			%
$R_{D,k}$	9.5	15.6	48.2	58.0			

2. Calculate the cationic composition (in %) of the exchange complex of an illitic soil sample ($K_{Na/Ca}$ = 0.5) in equilibrium with a solution containing: a) 0.1 normal NaCl and 0.1 normal $CaCl_2$. (Answer: 18% Na and 82 % Ca). b) 0.01 normal NaCl and 0.01 normal $CaCl_2$. (Answer: 7 % Na and 93 % Ca).

3. Estimate the composition of the equilibrium solution at 0.02 normal total electrolyte, for an illitic soil containing about 2 % K, 5 % Na and 93 % Ca on the exchange complex. (Answer: ≈ 0.0008 normal K, 0.0097 normal Na and 0.0095 normal Ca).

4. The concentration of the soil solution of an illitic soil sample (γ = 30 meq/100 g, $K_{Na/Ca}$ = 0.5, W = 25 ml per 100 g dry soil) is 0.12 normal NaCl and 0.08 normal $CaCl_2$. Calculate the cationic composition (in meq/100 g dry soil) of the exchange complex. (Answer: 6.9 meq Na and 23.1 meq Ca per 100 g dry soil).

4a. 60 ml of a solution containing 0.18 normal NaCl and 0.02 normal $CaCl_2$ is added to 150 g of the above moist sample. Calculate the new compositions of the exchange complex and the equilibrium solution. (Answer: $\gamma.N_{Na}$ = 8.5 meq/100 g. and $\gamma.N_{Ca}$ = 21.5 meq per 100 gram dry soil; $c_{o,Na}$ = 0.139 normal and $c_{o,Ca}$ = 0.062 normal).

4b. The moist sample of question 4 is diluted to a suspension containing 500% moisture by adding distilled water. Calculate the composition of the exchange complex and of the equilibrium solution of this suspension. (Answer: $\gamma.N_{Na}$ = 5.4 meq/100 g and $\gamma.N_{Ca}$ = 24.6 meq/100 g dry soil; $c_{o,Na}$ = 0.0091 normal and $c_{o,Ca}$ = 0.0009 normal).

5. 1.2 meq NaCl is added to a system containing 100 grams salt free H-resin ($\gamma = 60$ meq/100 g) in 120 ml (distilled) water. Calculate the electrolyte concentration and the pH of the solution after the exchange equilibrium has been established ($K_{H/Na} = 5$). (Answer: $c_o = 0.01$ normal, pH = 2.03).

6. Line C in figure 4.5f (page 85) shows the titration curve in 1 normal KCl of a loess soil. The pH (KCl) of the original soil is 3.1. Calculate the liming factor of this soil and the lime requirement (ton per ha) to bring the pH (KCl) to 6, if given that the bulk density = 1500 kg per m^3. (Answer: the liming factor = 0.12 ton/ha; the lime requirement = 3.5 ton/ha).

6a. How many tons of liming material (containing 20% silicates and 80% (by weight) $CaCO_3$) have to be added to the above soil in order to raise the pH of a 20 cm deep plough layer from the original pH-value to the required value ? (Answer: 15.6 ton liming material per ha).

RECOMMENDED LITERATURE

Russell, E.W., 1973. *Soil Conditions and Plant Growth*. Longman, London. Chapter 7, p. 89 - 128.

Scheffer, F. and Schachtschabel, P., 1970. *Lehrbuch der Bodenkunde*. Enke Verlag, Stuttgart. Chapters B I, p.105-140, B III p. 143-166.

CHAPTER 5

ADSORPTION OF ANIONS BY SOIL

G.H. Bolt

The interaction between anions and the surface of the solid phase is somewhat involved. It was indicated already in chapter 3 that soil particles, though predominantly negatively charged, may also carry some positive charges. Both the oxide surfaces (notably Fe- and Al-oxides/hydroxides) and the edges of clay minerals are likely to be positively charged at pH values below seven. Accordingly positive adsorption of anions is to be expected at these sites. In contrast, the anions are expelled from the diffuse double layer formed on negatively charged surfaces, which leads to a negative adsorption (exclusion) of anions from these regions. The net adsorption, as assessed experimentally, must thus be interpreted as the sum of a positive and a negative quantity.

Although it would appear that a similar reasoning applies to the adsorption of cations (if both positively and negatively charged sites are present in the system) it may be shown that this is in practice of no concern (cf. section 5.2).

5.1. ANION EXCLUSION AT NEGATIVELY CHARGED SURFACES

Referring to figure 3.3 in chapter 3 and to section 4.2.1.2 it appears that the deficit of anions, $\bar{\gamma}_{an}$, constitutes only a small fraction of the CEC. This follows also from the Boltzmann equation, which implies that if the concentration of e.g. monovalent cations in the double layer is found by multiplying the equilibrium concentration by a certain 'Boltzmann factor', $\exp(-e\psi/kT) > 1$, that then the corresponding concentration of monovalent anions is found by multiplying its equilibrium concentration with the reciprocal of the factor $\exp(-e\psi/kT)$. As the Boltzmann factor increases towards the surface to values of, say, one hundred, it is clear that almost everywhere in the double layer $(c_+ - c_{o,+})$ by far exceeds $(c_{o,-} - c_-)$. Accordingly one should expect that $\bar{\gamma}_{cat} \gg \bar{\gamma}_{an}$. Indeed, calculations based on double layer theory (cf. part B) show that under conditions usually prevailing in soil, $\bar{\gamma}_{an}$ amounts to about 1-5% of the CEC. For saline soils this percentage may be somewhat higher (e.g. 15%).

The mentioned calculations show that the anion exclusion may be estimated with an equation of the type:

$$\bar{\gamma}_{an} = S \cdot d_{ex,an} c_{o,an} \qquad (5.1)$$

in which S is the specific surface area of the soil and $d_{ex,an}$ the 'effective' distance of exclusion of the anion. In turn $d_{ex,an}$ is inversely proportional to the square root of the product of the total electrolyte concentration in the system and the valence of the dominant cation in the system, z_+, according to:

$$d_{ex,an} \approx q/z_+ \sqrt{C_o} \tag{5.2}$$

Specifying C_o in keq/m^3 and the $d_{ex,an}$ in Å, the proportionality constant q may be taken equal to 6 for monovalent anions. Applying these equations to the system specified in section 4.2.1, where Ca is the dominant cation, eq (5.2) gives:

$$d_{ex,Cl} \approx 3/\sqrt{0.007} \approx 35 \text{ Å}$$

The system thus behaves as if the solid particles are surrounded by a salt free zone of 35 Å thickness (which could be taken as a measure of the extent of the double layer). The anion exclusion, $\bar{\gamma}_{Cl}$, having been determined at 0.0014 meq/g for this system (section 4.2.1.1) one may back-calculate the value of S at about 55 m^2/g.

The anion exclusion of soils, though of limited magnitude as compared to the CEC, may nevertheless constitute a considerable fraction of the product Wc_{o^-}, the first estimate of the amount of anions in solution. Obviously the value of $\bar{\gamma}_k/Wc_{o,k} = d_{ex,k}/d_l$ - with $d_l \equiv W/S$ - increases with increasing S or decreasing moisture content, W. There are at least four practical aspects involved if $d_{ex,an}$ is a non-negligible fraction of the liquid layer d_l.

a. Upon dilution the concentration of the soil solution will decrease more than proportionally, because part of the liquid phase is deficient in anions.
b. For the same reason the isolation of the soil solution by means of pressure filtration may lead to a concentration below the equilibrium value.
c. Upon passage through soil anions may move faster than the mean velocity of the liquid phase calculated from the filter flux divided by the moisture content, because of the interference between the non-uniform concentration and a Poiseuille distribution of the liquid velocity across a liquid film.
d. Alternatively $d_{ex,an}$ acts as a 'forbidden zone' for the passage of salts through soil pores, giving rise to 'salt sieving' effects (reverse osmosis).

It should also be mentioned that the determination of anion exclusion is an important method for the estimation of the specific surface area of soil or its components, following the application of equations (5.1) and (5.2). Derivations and a further discussion of the phenomena described above are given in part B.

5.2. THE POSITIVE ADSORPTION OF ANIONS

As was pointed out before, positively charged sites are limited to clay mineral edges and to oxides with usually a limited surface area. Accordingly the adsorption of anions at these sites is considerably smaller than the adsorption capacity of the soil for cations. Using a value of 1-5% of the CEC for the adsorption capacity for anions, AEC, one finds a situation as pictured below (putting the CEC of the soil arbitrarily at 100 units):

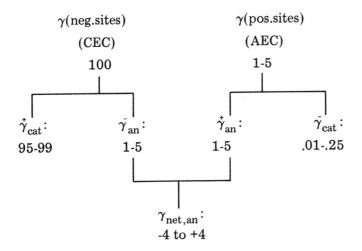

The net adsorption of anions, $\gamma_{net,\,an}$ could thus vary from positive to negative depending on conditions, while the negative adsorption of cations on positively charged sites, $\bar{\gamma}_{cat}$, is always negligible in comparison to the positive adsorption on the negatively charged sites. This rough calculation points to the fact that experimentally determined values of $\gamma_{net,an}$ must be interpreted carefully.

Limiting the discussion to the positive adsorption of anions one should notice the differences with the positive adsorption of cations. Aside from the usually rather limited capacity to adsorb anions, this capacity is very sensitive towards pH and electrolyte level (cf. section 3.1). Also the selectivity of the positive sites with respect to different anions is as a rule much larger than is the case for cation adsorption. This last observation is related to the composite nature of many anions, and the lower degree of hydration due to the larger size (in comparison to cations). For common anions found in soil experimental evidence indicates the following order of preference:

$SiO_4 > PO_4 \gg SO_4 > NO_3 \approx Cl$

As a result SO_4- and Cl-ions are often not adsorbed even if only very low concentrations of PO_4-ions are present. In systems free of PO_4-ions, sometimes a positive adsorption of Cl and/or SO_4 may be shown to exist, at least when the pH does not exceed a value of 6. Again the Cl-adsorption is suppressed by SO_4-ions; as in soils usually PO_4-ions and also SiO_4-ions are present, the positive adsorption of Cl and SO_4 can not be detected in many field soils of the temperate region. In certain tropical or subtropical soils with a high content of Al- and Fe-hydroxides, positive adsorption of SO_4 could be shown to exist.

The 'overall' adsorption of anions may thus be either positive or negative, depending on whether one or the other mechanism dominates the situation.

For Cl the overall effect is usually negative, as is the case for SO_4. For PO_4 the situation is different. When a rather low concentration of PO_4 solution is combined with a much higher concentration of Cl (as often occurs in the field), the negative adsorption of PO_4 (which is not specific) is very small, whereas the positive adsorption of anions is limited to the PO_4-ions due to the high preference for this ion. Under such conditions the 'overall' adsorption of PO_4 is certainly positive.

5.3. PHOSPHATE 'FIXATION'

Clay minerals and the sesqui-oxides/hydroxides exert a rather strong preference for PO_4-ions (in comparison to other anions), indicating the existence of chemical bonds between the solid phase and phosphate. The presence of these chemical bonds is supported by the observation that at a fairly high concentration of PO_4-ions in solution the edges of the clay minerals become also negatively charged. This would obviously be impossible if the binding mechanism between phosphate and clay were electrostatic in nature. The presence of chemical bonds between edge Al-groups of clay plates and PO_4-ions is also supported by the low solubility of Al-phosphates in general.

It could be shown that the maximal amount of PO_4 that was adsorbed by certain illites and montmorillonites corresponds with about one PO_4-group per two edge-situated Al-groups. The reversal of charge of the edges of clay minerals with excess PO_4-ions is brought to use when it is attempted to peptize a clay suspension (e.g. for mechanical analysis). Flocculation of clay suspension is caused in first instance by the simultaneous existence of positively charged edges and negatively charged planar sides of the clay plates, giving rise to a card-house arrangement of the clay plates. Peptizing agents for clay suspensions thus usually contain phosphates.

As a result phosphate ions are bonded rather strongly by clay minerals, although the adsorption-capacity for phosphate is not very large (comparatively small edge-surface!). For soils containing sesqui-oxides in abundant amounts this adsorption capacity for phosphate is much greater. In these soils the concentration of PO_4-ions in solution is, therefore, very low. Inasfar as in these (often acidic) soils there exists a fairly high concentration of Al-ions in solution, precipitation of Al-phosphates may also occur (cf. chapter 6).

Addition of fertilizer phosphate to these soils gives usually low response. In analogy with K-fixation, this phenomenon is termed phosphate-fixation. It should be recognized that in the case of phosphate, irreversibility due to changes in geometry of the system (lattice collapse) does not occur. Although different in nature, in practice 'PO_4-fixation' has a similar appearance as K-fixation, i.e. low response to high gifts of fertilizer. Here again fixation cannot be avoided entirely, but may be moderated by the simultaneous application of competitors for binding-sites. In this case silicate appears to be useful, as are also organic anions (manure!).

Another aspect of the strong bonding of PO_4-ions by Fe- and Al-oxides/hydroxides is the rather efficient retention of phosphates by certain soils when used to filter waste waters containing phosphates. Probably coatings of Fe- and Al-oxide/hydroxide on sand grains, as are often found in sandy soils, contribute to observed retention of phosphate in these soils. Referring to the following chapter it might be stated that due to the combined action of adsorption and precipitation reactions, inorganic phosphate is rather immobile in soil. In view of eutrophication problems in surface water this may be considered a fortunate fact.

ILLUSTRATIVE PROBLEMS

1. 10 gram of a Na-clay is dispersed in 50 ml of an acidified NaCl-solution. Two small dialysis bags containing 5 ml each of distilled water are brought into the suspension (A). After shaking overnight one bag is removed and its contents are analyzed to contain 5 ml of 0.0100 normal NaCl. Next one adds 0.050 mmol Na_3PO_4 to the remaining system, which is shaken again (B). Analysis of the second bag gives 5 ml of 0.0110 normal NaCl with a trace of phosphate (10^{-4} molar). Finally the remaining system is extracted repeatedly with 0.5 normal $Ca(NO_3)_2$. The extract indicates that in system (B) the total Cl-content was 0.55 meq while the total Na-content was 3.20 meq. Calculate the amounts adsorbed of Cl and PO_4 in system (B) and (A), estimate the specific surface area of the clay, its exchange capacity and its surface density of charge.
(Answers: in (B): $\bar{\gamma}_{Cl} = 0.55$ meq/100 g, $\overset{+}{\gamma}_{PO_4} = 0.44$ mmol/100 g;
in (A): $\bar{\gamma}_{Cl} + \overset{+}{\gamma}_{Cl} = 0.00$ meq/100 g implying $\bar{\gamma}_{Cl} \approx \overset{+}{\gamma}_{Cl} = 0.55$ meq/100 g; $S \approx 90$ m^2/g (cf. eq. 5.1, 5.2), $\gamma = 25$ meq/100 g, $\Gamma \approx 2.8 \times 10^{-7}$ meq/cm^2

2. One adds 45 ml of distilled water to a duplicate of system B (consisting of 50 ml of suspension and 5 ml of solution in the second bag, with a composition as specified above). Calculate the chloride concentration in the bag after equilibrium has been established. (Answer: 0.0059 normal, to be compared with $0.011 \times 55/100 = 0.0061$ normal for straight dilution of a solution).

RECOMMENDED LITERATURE

Russel, E.W., 1973. *Soil Conditions and Plant Growth*. Longman, London. Chapter 7, p. 128-133.
Scheffer, F. and Schachtschabel, P., 1970. *Lehrbuch der Bodenkunde*. Enke Verlag, Stuttgart. Chapter B II, p. 141-142.

CHAPTER 6

COMMON SOLUBILITY EQUILIBRIA IN SOILS

I. Novozamsky and J. Beek

6.1. CARBONATE EQUILIBRIA

The omnipresence of CO_2-gas in soil systems delimits the maximun possible concentration of many cations in the soil solution. The most abundant one of the cations forming carbonates with low solubility is the Ca^{2+}-ion and accordingly many soils contain solid calcium carbonates.

The crystalline calcium carbonates found in soils are calcite and aragonite, the latter being metastable with respect to calcite as is indicated by the solubility constants listed in table 6.1. In addition one may find several other carbonates, such as dolomite, $CaMg(CO_3)_2$, or magnesite, $MgCO_3$. Siderite, $FeCO_3$, occurs in certain anaerobic soils. Solubility constants of the above carbonates are to be found in the literature or can be calculated from thermodynamic data.

While the local pressure of CO_2-gas in soil, together with the pH of the system (cf. below), controls the maximum concentration of the relevant cations, conversely the presence of solid phase carbonates in soil will stabilize the actual concentration of the relevant cations in the soil solution (again as a function of pH and the CO_2-pressure). Accordingly one finds that under conditions of excess leaching with rain water (and thus low total electrolyte level in the soil solution) the exchange complex of soils containing $CaCO_3$ is nearly saturated with Ca^{2+}-ions. This is in turn of importance with respect to the physical properties of the soil system (structural stability, hydraulic conductivity).

In the following sections the equilibrium conditions pertaining to the $CaCO_3$-system will be derived, starting with the pure carbonic acid solution.

6.1.1. The CO_2-H_2O system

The solubility of $CO_{2(g)}$ in water is governed by Henry's law for gases at low (partial) pressure. The dissolved CO_2 is partly hydrated (i.e. it occurs both as CO_2 and as H_2CO_3 molecules). Expressing the total concentration of dissolved CO_2 as $H_2CO_3^*$ one finds the 'solubility constant' as:

$$CO_{2(g)} + H_2O \rightleftharpoons H_2CO_3^*, \quad \log K^\circ = -1.46 \qquad (6.1)$$

The activity of $CO_{2(g)}$ is here expressed in terms of the gas pressure, P, in bar [*1]. Where the CO_2-pressure in the atmosphere is about 0.3 mbar, the activity of $H_2CO_3^*$ in solution in equilibrium with the atmosphere equals roughly $3 \cdot 10^{-5.46} \approx 10^{-5}$ mol/l.

In general terms one finds: ← activity of CO_2 gas.

$$-\log(H_2CO_3^*) = 1.46 - \log P_{CO_2} \tag{6.2}$$

The carbonic acid molecules entertain protolysis reactions according to:

$$H_2CO_3^* \rightleftharpoons HCO_3^- + H^+ \qquad \log K° = -6.35 \tag{6.3}$$

$$HCO_3^- \rightleftharpoons CO_3^{2-} + H^+ \qquad \log K° = -10.33 \tag{6.4}$$

Combining the reactions (6.1), (6.3) and (6.4) one finds, respectively:

$$-\log(HCO_3^-) = 7.81 - \log P_{CO_2} - pH \tag{6.5}$$

$$-\log(CO_3^{2-}) = 18.14 - \log P_{CO_2} - 2pH \tag{6.6}$$

In figure 6.1 the relationships (6.2), (6.5) and (6.6) are plotted for the above value of P_{CO_2} equal to 0.3 mbar. One may notice the increase in total carbonate in solution with increasing pH, and the difference in slope for the various carbonate species. Where in soil the value of P_{CO_2} tends to vary between the atmospheric value of 0.3 mbar and an upper limit of 0,1 bar (in anaerobic soils) one should realize that the lines drawn in figure 6.1 will shift upward with one unit for each tenfold increase of P_{CO_2}. The 'bands' corresponding to $P_{CO_2} = 0.01 - 0.1$ bar have therefore been added to figure 6.1.

The above system lends itself for a demonstration of the calculation of the actual concentration of the different ionic species at equilibrium for a specified set of conditions.

Referring to chapter 2 (section 2.6.2) it is recalled that in addition to a) the equilibrium conditions for different species involved (e.g. 6.2, 6.5, 6.6 above) one must b) always fulfill the condition of electroneutrality, while c) furthermore often certain mass balance conditions are specified. Combining the three sets of conditions one may in principle solve for the concentrations of all species involved. The difficulty with the actual execution of such a calcu-

---

[*1] Troughout this text gas pressures will be specified in bars. It should be noted, however, that the values of $\bar{G}_f°$ as tabulated in the literature are actually referring to a standard pressure of 1 atm. While foregoing the question whether such data were indeed back-calculated to a pressure of precisely 1 atm. or simply refer to experimental findings under 'atmospheric pressure', the effect of the roughly one percent difference between 1 atm. and 1 bar falls within the range of accuracy foreseen in this text.

lation scheme is that the relationships a) are expressed in terms of products and ratios of activities, while b) and c) are in terms of sums and differences of concentrations. Referring to section 2.6.3 it is obvious that the introduction of the activity coefficients, if necessary, then often implies the application of iteration procedures. In the present case this last complication may be avoided if the situation is limited to low concentrations where activity corrections are not necessary. Listing the dissolved species present in this system one finds, respectively, $H_2CO_3^*$, HCO_3^-, CO_3^{2-}, H^+, OH^-, H_2O. Making use of the different dissociation constants of the carbonic acid and water and the solubility constant for CO_2 in water (reaction 6.1) one finds the following relations between the above species:

$$(H_2CO_3^*) = 10^{-1.46} \times P_{CO_2} \tag{6.7a}$$

$$(HCO_3^-) = 10^{-7.81} \times P_{CO_2} / (H^+) \tag{6.7b}$$

$$(CO_3^{2-}) = 10^{-18.14} \times P_{CO_2} / (H^+)^2 \tag{6.7c}$$

$$(OH^-) = 10^{-14} / (H^+) \tag{6.7d}$$

The electroneutrality condition for this system gives:

$$[H^+] = [HCO_3^-] + 2[CO_3^{2-}] + [OH^-] \tag{6.8}$$

and the mass balance condition:

$$c_{CO_3} = [H_2CO_3^*] + [HCO_3^-] + [CO_3^{2-}] \tag{6.9}$$

where c_{CO_3} represents the total dissolved carbonate concentration. The pH of this system (i.e. at equilibrium) is found by substituting equation (6.7b, c, d) into equation (6.8). As in the present case no activity corrections are applied this gives directly a relation in which pH and P_{CO_2} are the only variables. Selecting P_{CO_2} at 0.3 mbar, it than follows that the corresponding pH of the aqueous system attains a value of 5.65. The actual calculation of the pH is frequently simplified by making suitable approximations by which certain terms in additive equations may be neglected. In this situation the last two terms in relation (6.8) can be neglected. giving $[H^+] = [HCO_3^-]$, thus allowing the pH to be calculated directly from equation (6.7b).

Relation (6.9) is for the chosen system H_2O-$CO_{2(g)}$ of little help, because no information is given about the value c_{CO_3} (total carbonate species concentration in solution), and also not needed for the description of the system in question, where the mass-balance condition is replaced by the relation (6.7a). If, however, no gaseous phase would be present (i.e. water saturated with $CO_{2(g)}$ and kept out of contact with the atmosphere), the knowledge of c_{CO_3} and the use of relation (6.9) would be indispensable.

The above relations are also used to calculate the relative distribution of the different carbonate species in solution as a function of pH. In this respect equation (6.9) is written as:

$$\frac{[H_2CO_3^*]}{c_{CO_3}} + \frac{[HCO_3^-]}{c_{CO_3}} + \frac{[CO_3^{2-}]}{c_{CO_3}} = 1 \tag{6.9a}$$

After substitution of relations (6.7) in the above equation expressions are derived for the terms $H_2CO_3^*/c_{CO_3}$, HCO_3^-/c_{CO_3} and CO_3^{2-}/c_{CO_3}, with pH as the only variable. Using these expressions figure 6.2 was constructed, showing the relative distribution of the different carbonate species as a function of pH.

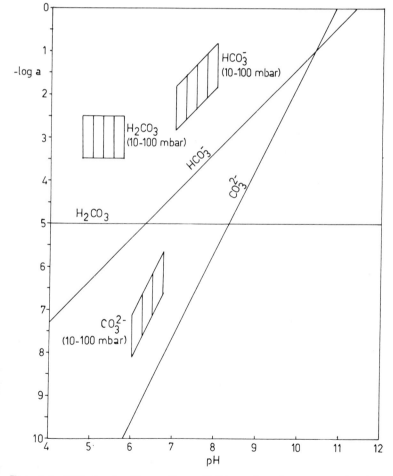

Fig. 6.1. Effect of pH on the activities of carbonate species in equilibrium with 0.0003 bar CO_2 (solid lines). The 'bands' correspond to $P_{CO_2} = 10\text{-}100$ mbar.

6.1.2. Systems containing $CaCO_{3(s)}$

In the following discussion it is assumed that only pure calcite is present as the solid carbonate phase. Such a condition is of course seldom met in a real soil system where e.g. aragonite or dolomite could be present. Moreover soil solid carbonates have regularly small amounts of impurities in their crystal structure causing a corresponding deviation of their activity from that at a pure state (cf. section 2.1). So a system where only pure calcite is present should be considered as a very simplified approximation of the situation in a calcareous soil and therefore one should be careful in applying the conclusions derived in this discussion directly to a soil system.

Starting with a system where pure water, i.e. originally free of carbonate species, is saturated with an excess of calcite while no gaseous phase is present, it follows that the solution composition will be governed by the solubility product of $CaCO_{3(s)}$ and - because of the protonization of the anion - the first and second dissociation constant of carbonic acid, i.e. relations (6.3) and (6.4), respectively, are of importance. Representing the dissolution reaction of calcite by:

$$CaCO_{3(s)} \rightleftharpoons Ca^{2+} + CO_3^{2-} \qquad \log K_{SO}^\circ = -8.35, \qquad (6.10)$$

the electroneutrality condition for this system reads:

$$2[Ca^{2+}] + [H^+] = [HCO_3^-] + 2[CO_3^{2-}] + [OH^-] \qquad (6.11)$$

As calcite is the only source that can deliver calcium ions and carbonate species, a mass balance relation may be formulated as:

$$[Ca^{2+}] = [CO_3^{2-}] + [HCO_3^-] + [H_2CO_3^*] \qquad (6.12)$$

With the relations mentioned above the equilibrium solution composition can be calculated. For this particular set of relations it appears useful to derive the proton balance equation substituting relation (6.12) into (6.11), yielding:

$$[H^+] + [HCO_3^-] + 2[H_2CO_3^*] = [OH^-] \qquad (6.13)$$

Fortunately, this relation can be further simplified for the present system, because the final solution should be alkaline. Now it follows that for solutions with pH > 8 the (H^+) becomes negligible with respect to (OH^-) as does the $(H_2CO_3^*)$ with respect to (HCO_3^-) $((H_2CO_3^*) < 10^{-1.65} \times (HCO_3^-))$. Assuming activities to be equal to concentrations this results in the simple relation:

$$[HCO_3^-] = [OH^-] \qquad (6.13a)$$

Substituting the above expression into the mass-balance relation (6.12), neglecting the last term on the RHS, expressing the activities of the other species in terms of (H^+) and again assuming that activities are equal to concentrations, gives:

$$10^{15.98} \times (H^+)^2 = \frac{10^{-24.33}}{(H^+)^2} + \frac{10^{-14}}{(H^+)} \qquad (6.12a)$$

Multiplying by $(H^+)^2$ and rearranging gives:

$$10^{15.98} \times (H^+)^4 - 10^{14}(H^+) = 10^{-24.33}$$

Solving this equation by trial and error yields:

$(H^+) = 10^{-9.94}$ i.e. pH = 9.94

$(Ca^{2+}) = 10^{-3.90}$

$(CO_3^{2-}) = 10^{-4.45}$ and

$(HCO_3^-) = (OH^-) = 10^{-4.06}$

Another system can be formulated if in addition to the solid and liquid phase also a gaseous phase is considered (i.e. a soil atmoshpere).

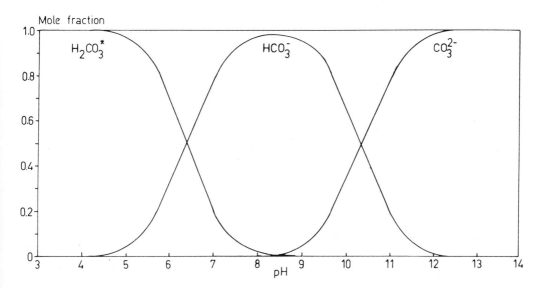

Fig. 6.2. Different carbonate species, in mole fraction of total carbonate in solution, as a function of pH.

The solution composition of this latter system can be calculated with the help of dissociation reactions of carbonic acid (6.3,4), the relation describing the solubility of $CO_{2(g)}$ in water (6.1), the solubility product of calcite (6.10) and an electroneutrality condition (similar to relation (6.11) derived for the previous system), provided one of the independent variables is held at a constant value. (The mass-balance relation (6.12) derived for the previous system is not valid for the present set of conditions as two sources of carbonate species are present, i.e. $CaCO_{3(s)}$ and $CO_{2(g)}$. If in this respect P_{CO_2} is fixed it follows that only one unique solution composition corresponds with a certain value of P_{CO_2}. Using the previous value of P_{CO_2} = 0.3 mbar and again assuming that activities are equal to concentrations, it than follows that the corresponding pH = 8.33, (Ca^{2+}) = $10^{-3.3}$, (HCO_3^-) = $10^{-3.0}$ and (CO_3^{2-}) = $10^{-5.0}$. The composition of the equilibrium solution is generally sufficiently accurate calculated by using a simplified version of (6.11), i.e. $2(Ca^{2+}) \approx (HCO_3^-)$. From this latter expression it can be derived that an increase in P_{CO_2} lowers the pH of the equilibrium solution and increases the solubility of the solid phase.

The condition mentioned above, i.e. that one of the independent variables should be fixed at a constant value, follows directly from Gibb's phase rule, stating that:

F = C - P + 2

The letter F represents here the number of degrees of freedom, C the minimum number of components necessary for the description of the composition of all phases present in the system, and P the corresponding number of phases in that system. The last term (two) accounts for the fact that temperature and pressure of the system can be varied too. At constant T and P, however, these two degrees of freedom disappear. Applying Gibb's phase rule to the latter system ($CaCO_{3(s)}$ - H_2O-$CO_{2(g)}$)indicates that at constant T and P (25°C and 1 atmosphere, respectively) 4 components are present (CaO, H_2O, $CO_{2(g)}$ and another (inert) gas), allowing the presence of 4 phases (2 solids ($CaCO_{3(s)}$, $CaO_{(s)}$, liquid and gas). If only calcite is present as a solid phase, gives:

F = 4 - 3 = 1.

If this degree of freedom is used for varying the P_{CO_2}, the solution composition should vary accordingly. On the other hand it also indicates that at fixed value of P_{CO_2} only one composition of the solution is possible, i.e. the pH of the solution is fixed. If, however, the pH is chosen as the independent variable it follows that a variation of the pH is accompanied by a change in P_{CO_2} and solution composition.

If, however, next to the calcite-H_2O-$CO_{2(g)}$ system, also the soil system is present, situations may arise that dissolved species delivered by the first system are influenced by components present in the latter system, e.g. pH is influenced by the organic matter, metal carbonates other than calcite present in the soil. Under such conditions the equilibrium solution composition may differ from the one given for the pure system calcite -H_2O-$CO_{2(g)}$.

Constructing a logarithmic solubility diagram (cf. section 2.8) may than be an useful method to represent the equilibrium solution composition. The expression describing the equilibrium composition is derived from the solubility product of calcite (6.10) and the equations (6.1), (6.3) and (6.4), giving:

$$CaCO_{3(s)} + 2H^+ \rightleftharpoons Ca^{2+} + CO_{2(g)} + H_2O \; ; \; \log K^\circ = 9.79 \tag{6.13}$$

Taking logarithms, than gives:

$$-\log(Ca^{2+}) = 2pH - 9.79 + \log P_{CO_2} \tag{6.13a}$$

The above relation is plotted in figure 6.3 together with similarly derived relations for other solid carbonates. In the same figure are also presented the dissociation products of carbonic acid and a dashed line is used to indicate the equilibrium solution composition of the pure solid carbonate $-H_2O - CO_{2(g)}$ system at a $P_{CO_2} = 0.3$ mbar. This latter composition is calculated using the simplified condition $2(M_{2+}) = (HCO_3^-)$, where M_{2+} represents the divalent cation concerned.

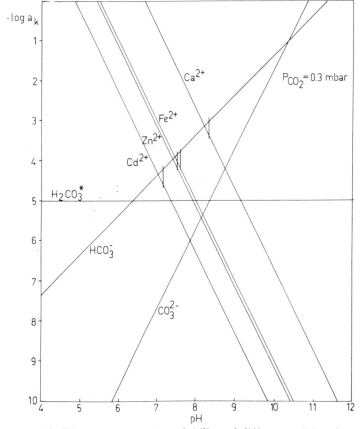

Fig. 6.3. Effect of pH on the solubility of different solid carbonates at a constant P_{CO_2}. The dashed lines present the equilibrium solution composition of the corresponding systems and have been calculated with the simplified condition $2(M_{2+}) = (HCO_3^-)$. (after Stumm and Morgan, 1970).

So far the solubility of calcite has been treated with respect to the Ca^{2+}-ion only. In actuality the ion-pairs $CaHCO_3^+$ and $CaCO_3°$ should also be considered. Whenever calcite is present in the system the ion activities of the relevant ion-pairs are ultimately determined by this solid phase. By appropriate combinations of the formation reactions of different ion-pairs with the solubility product of calcite, relations are obtained describing the equilibrium activities of the species considered. These relationships are plotted in figure 6.4. This figure indicates that the activity of $CaHCO_3^+$ increases tenfold for each unit decrease in pH, whereas the activity of the $CaCO_3°$ ion-pair is not affected by the pH. As long as calcite is present the activity of both ion-pairs is also independent of P_{CO_2}. In the present system (P_{CO_2} = = 10 mbar) the ion-pair $CaCO_3°$ becomes the dominant soluble species in solutions exceeding pH = 8.5.

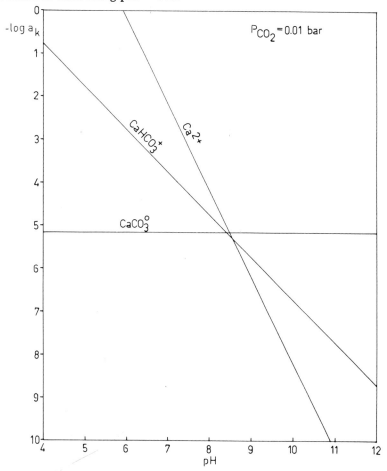

Fig. 6.4. Activity of the ion-pairs $CaHCO_3^+$ and $CaCO_3°$ in equilibrium with calcite.

A useful parameter in systems containing carbonate is the alkalinity, Alk, expressed in eq/l and defined as:

$$\text{Alk} = [HCO_3^-] + 2[CO_3^{2-}] + [OH^-] - [H^+]$$

In solutions that contain no other protic system than that of aqueous carbonate, it is a measure of the acid neutralizing capacity of the system. The alkalinity is thus a capacity parameter, in contrast to pH which is an intensity parameter. The alkalinity can also be expressed as the equivalent sum of all the cations except H^+, minus the equivalent sum of all the anions except carbonates or OH^-. From this it follows that the alkalinity is a conservative quantity which is independent of temperature or P_{CO_2}. Given a certain P_{CO_2} and Alk, the pH of the system can be calculated.

6.2. IRON OXIDES AND HYDROXIDES

Iron is present in soils in widely varying amounts; it is subject to reversible oxidation-reduction and it forms a variety of precipitates.

In aerobic soils, iron may be present in the form of $\alpha\text{-}Fe_2O_3$ (hematite), $\gamma\text{-}Fe_2O_3$ (maghemite), $\alpha\text{-}FeOOH$ (goethite), $\gamma\text{-}FeOOH$ (lepidocrocite) or $Fe(OH)_3 \cdot nH_2O$ (hydrated ferric oxide). The first four compounds represent crystalline forms whereas hydrated ferric oxide is amorphous. Hematite is the dominant form of the crystalline iron oxides in (red) tropical soils but it is less common in soils of the humid temperate climate. Goethite is formed both in tropical and temperate regions and is apparently the thermodynamically stable form at 25 °C.

Very often, however, the solubility of Fe in soils is largely controlled by the solubility of amorphous hydrous ferric oxides. Theoretically it can be expected that the ferric oxide solubility should vary between that of a fresh precipitate, and that of goethite. The choice of a particular value of the solubility product of iron oxide or hydroxide in a soil system is therefore always more or less arbitrary, unless it is based upon the experimental results of the system itself.

The composition of the aqueous solution in contact with a solid iron oxide or hydroxide can be calculated in the same way as was shown for the carbonate system. The main difference with respect to the carbonate system is that the formation of ferric hydroxocomplexes should be considered. The influence of pH upon the solubility of the solid iron compounds is best represented in a logarithmic diagram as was elucidated in section 2.8.

6.2.1. Ferrous compounds

In anaerobic soils the solubility of iron may be controlled by the solubility of Fe(II)-compounds such as ferrous hydroxide, $Fe(OH)_2$, siderite, $FeCO_3$, ferrous sulfide, FeS_2 and magnetite, Fe_3O_4, a ferric-ferrous compound. In aqueous solution the hydrated Fe^{2+}-ion, $Fe(H_2O)_6^{2+}$, may form hydroxo-complexes. The reactions with their constants are listed in table 6.1. The contribution of the ferrous hydroxo-complexes on the solubility of ferrous iron counts only for solutions with pH values above 8. Examination of formation constants of the hydroxo-complexes of Fe^{2+} and Fe^{3+} reveals that the Fe^{3+}-ion has a stronger acidic character than the Fe^{2+}-ion.

For instance the hexaaquoferric ion, $Fe(H_2O)_6^{3+}$, has an acidity constant of a similar order of magnitude as phosphoric acid, cf. section 2.7.

6.2.2. Redox reactions involving iron compounds

Discussing reactions of iron compounds in the soil system implies that oxidation-reduction reactions should be considered. As explained in section 2.7 the pe can be used as a convenient 'master variable' in equilibrium studies of redox systems. Because in most redox reactions not only electrons are involved but also protons, diagrams with pe and pH as variables (pe-pH diagrams) are particularly useful.

This will be illustrated first by considering oxidation and reduction of water, as in any environment at the earth surface the stability of water with respect to its oxidized ($O_{2(g)}$) and reduced ($H_{2(g)}$) components puts a theoretical limit on pe pH conditions that may occur. The $O_{2(g)}$ -$H_2O_{(l)}$ equilibrium can be described according to:

$$O_{2(g)} + 4H^+ + 4e^- \rightleftharpoons 2H_2O_{(l)} \tag{6.14}$$

with log K° = 83.1 (25°C and 1 atm total pressure). Conversely the reduction of $H_2O_{(l)}$ follows the equation:

$$2H^+ + 2e^- \rightleftharpoons H_{2(g)} \tag{6.15}$$

with log K° = 0. Introducing pe and pH this gives, respectively:

$$pe = 20.77 - pH + \tfrac{1}{4}\log P_{O_2} \text{, and} \tag{6.14a}$$

$$pe = -pH - \tfrac{1}{2}\log P_{H_2} \tag{6.15a}$$

Upper and lower limits of pe are obtained at $P_{O_2} = 1$ bar and $P_{H_2} = 1$ bar, respectively. The corresponding pe-pH lines are plotted in figure 6.5. Points on the line represent $H_2O_{(l)}$ in equilibrium with $O_{2(g)}$ (upper line) and with $H_{2(g)}$ for the lower line (both gases at the pressure of 1 bar). Points in between the 2 parallel lines indicate that $H_2O_{(l)}$ is the stable phase. For pe values above the upper line it follows that $H_2O_{(l)}$ will be oxidized to $O_{2(g)}$, while for pe values below the lower line water will be reduced to $H_{2(g)}$. Thus for reactions in aqueous systems only those pe values are possible (theoretically at least) that fall within the two parallel lines.

The same diagram can be used to depict the stability relationships involving hematite, magnetite and elemental Fe. The pertinent reactions and constants can be derived from table 6.1. As an illustration the half-reaction describing the reduction of hematite to magnetite will be given. This reduction reaction is described by:

$$3Fe_2O_{3,\text{hematite}} + 2H^+ + 2e^- \rightleftharpoons 2Fe_3O_{4,\text{magnetite}} + H_2O \quad \log K^\circ = 8.9 \quad (6.16)$$

The value of the equilibrium constant of reaction (6.16) can be easily calculated either from thermodynamic data or by appropriate combination of the solubility, redox and dissociation reactions of the components involved. The latter procedure may be preferable especially when reliable experimental

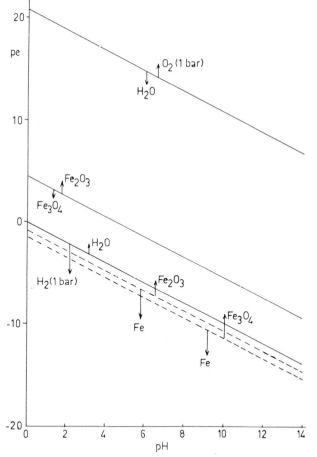

Fig. 6.5. Lines of stability of H_2O, α-Fe_2O_3, Fe_3O_4 and elemental Fe at 25 °C and total pressure of 1 bar (after Garrels and Christ, 1965).

data about the system are available. Using pe and pH as variables in half-reaction (6.16) gives:

$$p e = 4.45 - pH \qquad (6.16a)$$

Relationship (6.16a) is plotted in figure 6.5 and represents the boundary of the stability fields of hematite and magnetite. At pe-pH values on this line both phases can coexist, whereas points above this line indicate that hematite is the stable phase with respect to magnetite; accordingly the reverse is valid for points below the line.

In a comparable manner the half reactions for the reduction of magnetite to elemental Fe and the reduction of hematite to elemental Fe can be derived. The half reactions with their constants are:

$$Fe_3O_{4,magnetite} + 8H^+ + 8e^- \rightleftharpoons 3Fe_{(s)} + 4H_2O, \qquad (6.17)$$

with $\log K^\circ = -11.6$, and

$$Fe_2O_{3,hematite} + 6H^+ + 6e^- \rightleftharpoons 2Fe_{(s)} + 3H_2O, \qquad (6.18)$$

with $\log K^\circ = -4.7$. The corresponding pe-pH relationships are respectively:

$$p e = -1.45 - pH, \text{ and} \qquad (6.17a)$$

$$p e = -0.78 - pH \qquad (6.18a)$$

Plotting these pe-pH relationships in fig. 6.5 indicates that these lines are situated in the region where water is unstable (for that reason broken lines are used). From this it should be concluded that elemental Fe is always unstable in the presence of water. Moreover the line for hematite and elemental Fe is situated below the one of hematite and magnetite, i.e. hematite itself is not stable under these conditions of pe and pH. Hence this so-called 'metastable' phase boundary need not to be considered in a true equilibrium model. Furthermore it should be realized that although hematite seems to be a stable phase according to figure 6.5 (i.e. for pe-pH values on or above the stability line for α-Fe_2O_3-Fe_3O_4) it is in fact only metastable in the sense that goethite is ultimately the stable phase.

An aspect not considered sofar is the solubility of different iron oxides/ /hydroxides, with respect to varying oxidation or reduction circumstances. As revealed in the preceding discussion magnetite and hematite could be stable phases. Obviously the stability of these phases is influenced, among other factors, by the oxidation-reduction status of the soil as represented by the pe. In this respect the activity of the Fe^{2+}-ion will be calculated for magnetite and hematite. The necessary relationship for hematite is:

$$Fe_2O_{3,hematite} + 6H^+ + 2e^- \rightleftharpoons 2Fe^{2+} + 3H_2O \qquad (6.19)$$

with log $K° = 25.0$. Introducing pe and rearranging equation (6.19) gives:

$$pe = 12.5 - \log(Fe^{2+}) - 3pH \qquad (6.19a)$$

where in addition to pe and pH also $\log(Fe^{2+})$ is a variable. Selecting a pH value makes the construction possible of a diagram with pe and $\log(Fe^{2+})$ as variables.

As the pH for most soils under reduced conditions is generally close to neutral, a pH of 7 has been chosen and relationship (6.19a) has been plotted in figure 6.6 for this pH value.

A similar relationship for magnetite results in:

$$pe = 16.55 - 3/2 \log(Fe^{2+}) - 4pH \qquad (6.20)$$

Plotting this line in figure 6.6 (after substituting pH = 7) it crosses with the line for hematite. The crossing point of both lines represents the conditions where both phases could coexist. At other pe values (at least for the selected pH value) one of the phases is unstable with respect to the other one present. This is indicated in figure 6.6 with a broken line (it should be noted that although in figure 6.6 only the activity of the ferrous ion has been shown, obviously also ferric ions and the corresponding hydrolysis species are present simultaneously).

In case one prefers to use in figures 6.5 and 6.6 E_h as variable instead of pe the latter has to be multiplied by 60 (at 25 °C) to give the corresponding E_h values in millivolts.

Reducing conditions in a soil system could also be characterized, theoretically at least, by a decrease in the partial pressure of $O_{2(g)}$ of the soil air. In some cases therefore the partial pressure of $O_{2(g)}$ has been used as a variable in defining reducing conditions instead of the pe.

This approach will be elucidated for hematite being the solid phase controlling the solubility of the iron species. In this respect the (Fe^{2+}) activity will be shown for varying partial pressures of $O_{2(g)}$. The relevant relationship is found by combining equation (6.19) with equation (6.14). This gives:

$$Fe_2O_{3,hematite} + 4H^+ \rightleftharpoons 2Fe^{2+} + 2H_2O + \tfrac{1}{2}O_{2(g)} \qquad (6.21)$$

where $K° = \dfrac{(Fe^{2+})^2 (P_{O_2})^{1/2}}{(H^+)^4} = 10^{-16.5}$

Taking logarithms and rearranging gives:

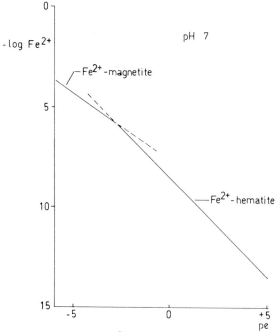

Fig. 6.6. Activity of Fe^{2+}-ion in equilibrium with hematite or magnetite respectively at a constant pH value of 7. Broken lines indicate that the compound considered is unstable with respect to other solid phase present.

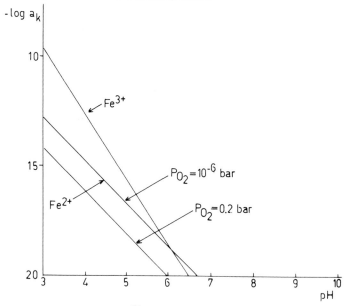

Fig. 6.7. Activity of Fe^{2+}-ion in equilibrium with hematite for two selected partial pressures of $O_{2\,(g)}$. The Fe^{3+}-ion activity is also plotted.

$$-\log(Fe^{2+}) = 2pH + \tfrac{1}{4}\log P_{O_2} + 8.25 \tag{6.21a}$$

Relationship (6.21a) has been plotted in figure 6.7 for two selected partial pressures of $O_{2(g)}$. A pressure of 0.2 bar corresponds roughly with the $O_{2(g)}$ pressure in the atmosphere whereas the lower value of 10^{-6} bar represents reducing conditions. Using partial pressures in characterizing redox reactions in aqueous systems may result in values that are very often mere calculation numbers, having no practical meaning, because equilibrium with $O_{2(g)}$ is seldom attained. In the same diagram also the activity of the Fe^{3+}-ion is shown. The activity of the latter is obviously unaffected with respect to reducing conditions as long as hematite is present.

6.2.3. pe-pH diagrams for the system hematite-magnetite-siderite-H_2O

In figure 6.5 the fields of stability were shown for hematite magnetite and water. Such a diagram, apart from only reflecting the situation at equilibrium, is only valid for the species considered. Whenever also other solid ferrous compounds, like ferrous carbonate, or sulfides (cf. section 6.2.2) might be of interest, a diagram should be constructed showing these solids too.

The relative stability of siderite, a crystalline ferrous carbonate, with respect to hematite and magnetite is derived from the equilibrium reaction of the former compound, represented by:

$$FeCO_{3(s)} \rightleftharpoons Fe^{2+} + CO_3^{2-} \quad \text{siderite} \tag{6.22}$$

with a $\log K_{SO}^\circ = -10.68$. Including the reduction reactions of the latter two compounds one finds the following two pe-pH relationships:

$$pe = 23.18 - 3pH + \log(CO_3^{2-}) \quad \text{hematite} \tag{6.23}$$

$$pe = 32.57 - 4pH + 1.5\log(CO_3^{2-}) \quad \text{magnetite} \tag{6.24}$$

describing respectively the conditions where siderite can coexist with hematite (6.23) and magnetite (6.24). The above relations can be plotted in a pe--pH diagram provided one assigns a preselected value to the activity of the component in the last term on the RHS of these relations. This latter condition is usually effectuated by either assuming that the partial pressure of $CO_{2(g)}$, P_{CO_2}, has a constant value (i.e. for an open system) or that the concentration of total carbonate species dissolved in the aqueous phase is fixed (i.e. $c_{CO_3} = [H_2CO_3] + [HCO_3^-] + [CO_3^{2-}]$ has a fixed value). This latter approach could be applied for instance in soils saturated with water containing a fixed amount of dissolved carbonate species. Starting with the situation that P_{CO_2} is a constant it follows from equation (6.6) that the equations (6.23) and (6.24) could be transformed into:

$$pe = 5.04 - pH + \log P_{CO_2} \qquad (6.23a)$$

$$pe = 5.36 - pH + 1.5 \log P_{CO_2} \qquad (6.24a)$$

These relations are plotted for a selected value of $P_{CO_2} = 0.01$ bar in figure 6.8a (showing also the previously derived stability lines for H_2O, hematite and magnetite). It is shown in figure 6.8a that under the chosen conditions only magnetite and siderite can coexist. The stability line for hematite-siderite (equation (6.23a)) falls within the stability field for magnetite and was therefore omitted from figure 6.8a.

If, however, the concentration of dissolved carbonate species is fixed, the

Fig. 6.8a pe-pH diagram for hematite, magnetite, siderite and H_2O at 25 °C and 1 bar total pressure, with P_{CO_2} fixed at 10^{-2} bar (after Garrels and Christ, 1965).

usual approach is to split the pH range in areas in which a given dissolved carbonate species is preponderant. For this purpose pH values are selected that correspond with the pK° values of the dissociation reactions of carbonic acid (equation (6.3) and (6.4)). This gives that below pH 6.35 $H_2CO_3^*$ is the dominant dissolved species, between pH 6.35 and pH 10.33 HCO_3^- is preponderant and above pH 10.33 CO_3^{2-} takes over. Moreover, it is assumed that in those pH areas the concentration of the dominant species is represented by c_{CO_3}. Returning to the equations (6.23) and (6.24) and taking into account the dissociation reactions of carbonic acid (equation (6.3) and (6.4)) it follows that the carbonate species mentioned in the former two relations can be expressed as a function of any other dissolved carbonate species and pH, leading to three pe-pH relationships representative for the indicated pH areas. In systems where no activity corrections have to be applied the selected value of c_{CO_3} is directly substituted in the derived pe-pH relationships. Each of these relationships is represented in a pe-pH diagram by a straight line in the indicated pH area although with a different slope. For a selected value of $c_{CO_3} = 10^{-4}$ molar only the lines showing the coexistance of siderite and magnetite, are presented in figure 6.8b as well as the previously derived stability lines for hematite, magnetite and water. Comparing the condition that P_{CO_2} is fixed with the situation that total dissolved carbonate is fixed it follows that for the latter situation the carbonate ion concentration $[CO_3^{2-}]$ rises to a maximum beyond a pH of 10.3 and remains constant, whereas at a constant P_{CO_2} the concentration of CO_3^{2-} rises continuously with increasing pH. This is reflected in figure 6.8b in a restricted stability field of siderite at high pH at constant total carbonate, because magnetite becomes continuously more stable as pH is increased. Note that the slope of the line changes abruptly in figure 6.8b at pH 6.35, in fact it must bend in the vicinity of this pH value.

Finally it will be stressed once again that whenever other components than the ones mentioned above have to be considered a more elaborated calculation and stability diagram will result. In this respect the highly insoluble ferrous sulfides may be of importance although these solids will not be discussed in this text.

6.3. ALUMINUM

Aluminum may be present in the soil in a number of different solids, like the crystalline silicates (e.g. feldspars, clay minerals), oxides and hydroxides, the latter often in microcrystalline or amorphous form and phosphates. Next to these solids aluminum may be present as a ligand with organic matter and in an adsorbed state. In the soil solution a mixture of complexes with inorganic or organic ligands are possible whereas under favorable conditions

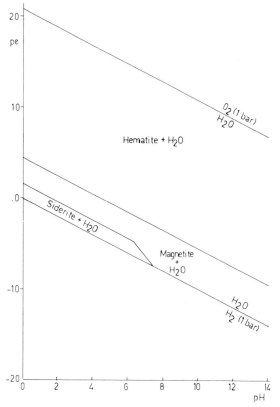

Fig. 6.8b. pe-pH diagram for hematite, magnetite, siderite and H_2O at 25 °C and 1 bar total pressure, for a fixed amount of total dissolved carbonate (c_{CO_3}) at 10^{-4} mol/l. (after Garrels and Christ, 1965).

(low pH) the hydrated $Al(H_2O)_6^{3+}$-ion may occur. In the following discussion the concentration of aluminum species in solution will be treated in equilibrium with a few solid phases. In this respect it should be mentioned that the solids might be present as a very fine or amorphous material, more soluble than corresponding with well formed and good-sized crystals (cf. section 2.9). The resulting solution composition in equilibrium with such solids is supersaturated with respect to the crystalline aluminum solids present in the soil indicating that a partial equilibrium and/or metastable situation is considered.

6.3.1. Al_2O_3 - H_2O system

The dissolution of aluminum hydroxide can be described by the reaction:

$$Al(OH)_3 \rightleftharpoons Al^{3+} + 3OH^- \tag{6.25}$$

The equilibrium constant (solubility product) K_{SO}° of this reaction may vary a factor in the order of 10^4 depending if freshly precipitated hydroxide is considered or crystalline gibbsite. Beside the Al^{3+}-ion shown in equation (6.25) also the formation of hydroxo-complexes of aluminum should be considered. The stability constants for some of these complexes are presented below:

$$Al^{3+} + H_2O \rightleftharpoons AlOH^{2+} + H^+ \qquad ; \log K^{\circ} = -5.02 \qquad (6.26a)$$

$$Al^{3+} + 4H_2O \rightleftharpoons Al(OH)_4^- + 4H^+ \qquad ; \log K^{\circ} \approx -23.57 \qquad (6.26b)$$

$$2Al^{3+} + 2H_2O \rightleftharpoons Al_2(OH)_2^{4+} + 2H^+ \qquad ; \log K^{\circ} = -6.27 \qquad (6.26c)$$

With the above relations a logarithmic solubility diagram of aluminum

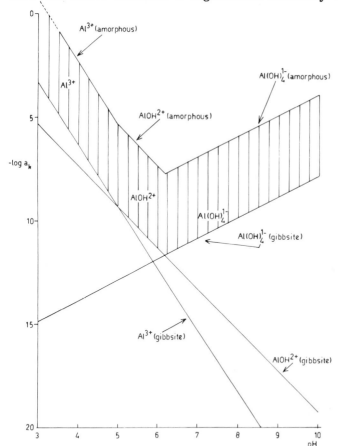

Fig. 6.9. Solubility diagram of gibbsite and $Al(OH)_3$ (amorphous) (shaded area represents possible values of aluminum solubility in soil systems).

hydroxide can be constructed. In figure 6.9 the activities of different aluminum species are shown in equilibrium with hydroxides with a $K_{SO}^{\circ} = 10^{-36.3}$ (gibbsite) and $K_{SO}^{\circ} = 10^{-32.34}$ (amorphous hydroxide). At this point it should be stressed that the constants used were arbitrarily chosen from rather varying values found in the literature and no attempt has been made to find out whether these values were correct as this is irrelevant to the purpose of this treatment. In figure 6.9 the shaded area indicates possible values of aluminum solubility in soil systems as in the soil metastable solids may control the solution composition.

6.3.2. Al_2O_3-SiO_2-H_2O system

In order to test the validity of the assumption that aluminum hydroxide is the solid phase governing the solution composition, it might be of interest to examine the conditions of the formation of $Al(OH)_{3(s)}$ from an aluminum silicate. For this purpose kaolinite was chosen. The congruent dissolution of this solid can be described by the relation:

$$Al_2Si_2O_5(OH)_4 + 6H^+ \rightleftharpoons 2Al^{3+} + 2H_4SiO_4^{\circ} + H_2O \tag{6.27}$$

with a log $K^{\circ} = 7.63$ (table 8.1, reaction 6). As long as only congruent dissolution takes place (i.e. no secondary solid phase is formed) the solution contains an equal number of moles of dissolved aluminum and $H_4SiO_4^{\circ}$. If it is furthermore assumed that no activity corrections have to be applied it follows that in the pH areas where the Al^{3+}-ion is the dominant dissolved species:

$$(Al^{3+}) = (H_4SiO_4^{\circ}) \tag{6.28}$$

Taking the logarithm of solubility product of equation (6.27) and substituting (6.28) gives:

$$-\log(Al^{3+}) = -\log(H_4SiO_4^{\circ}) = 1.5\,pH - 1.9 \tag{6.27a}$$

The above relation is represented by a straight line in the logarithmic solubility diagram of kaolinite (figure 6.10).

Here again it is necessary to take into consideration the formation of hydroxocomplexes of aluminum. The protonization of silicic acid can be neglected in the normal pH range found in soils (the first dissociation constant of silicic acid, $K^{\circ} = 10^{-9.77}$, table 6.1, reaction number 39).

Combination of the expressions for the formation of aluminum hydroxo--complexes (6.26a) through (6.26c) with equation (6.27) leads to:

$$Al_2Si_2O_5(OH)_4 + 4H^+ + H_2O = 2Al(OH)^{2+} + 2H_4SiO_4° \qquad (6.29)$$

with $\log K° = -2.41$ and

$$Al_2Si_2O_5(OH)_4 + 7H_2O = 2Al(OH)_4^- + 2H_4SiO_4° + 2H^+ \qquad (6.30)$$

with $\log K° = -39.51$

Again assuming that activity coefficients of all species concerned are equal to 1 gives respectively:

$$-\log(AlOH^{2+}) = -\log(H_4SiO_4°) = pH + 0.6 \quad \text{and} \qquad (6.29a)$$
$$-\log(Al(OH)_4^-) = -\log(H_4SiO_4°) = 9.88 - 0.5\, pH \qquad (6.30a)$$

The above relations are again represented by straight lines in the solubility diagram of kaolinite (figure 6.10).

Comparing the solubility diagram of kaolinite (figure 6.10) with that of gibbsite (figure 6.9) shows that in a wide pH-range gibbsite is more stable than kaolinite, implying that in that region incongruent dissolution of kaolinite might take place, according to:

$$Al_2Si_2O_5(OH)_4 = 2Al(OH)_{3(s)} + 2H_4SiO_4° + 7H_2O \qquad (6.31)$$

At low pH values another solid phase may be formed upon dissolution of kaolinite according to:

$$Al_2Si_2O_5(OH)_4 + 6H^+ = 2SiO_{2(s)} + 2Al^{3+} + 5H_2O \qquad (6.32)$$

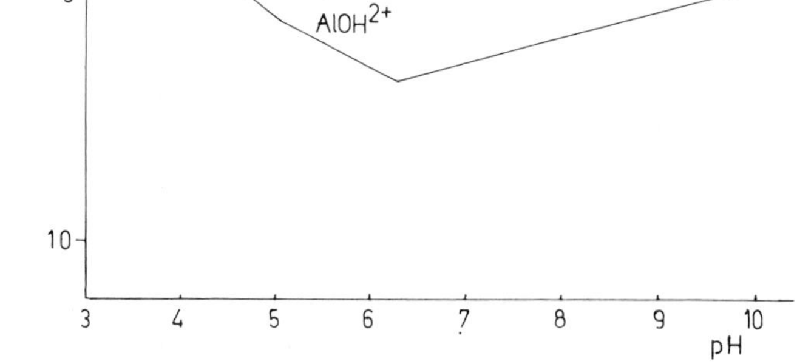

Fig. 6.10 Solubility diagram of kaolinite at congruent dissolution.

In both cases, provided a closed system is considered, the solution composition will be fixed by the presence of two solids in equilibrium (3 component system at a constant pressure and temperature is fixed by the presence of three phases). Any change in solution composition brought about by another factor, would cause one of the solid phases to disappear (e.g. drainage of dissolved $H_4SiO_4^°$, pH buffered by other soil components).

Another question is, whether one of the equations (6.31) or (6.32) will actually take place. This will depend amongst others upon the reaction kinetics. Irrespective of the previous considerations it will be clear that removal of soluble reaction products from the system will undoubtedly promote the formation of secondary solids (e.g. drainage of $H_4SiO_4^°$).

6.4. PHOSPHORUS

Phosphorus in soils occurs almost exclusively as orthophosphate in which a central phosphorus atom is surrounded by and bound to four oxygen atoms. Most of the phosphates in soils are salts of phosphoric acid, H_3PO_4. Both inorganic and organic forms are known. The relative proportions of inorganic and organic forms vary widely; generally, organic phosphorus increases and decreases with the content of organic matter and hence is comparatively low in subsoils and high in surface soils.

From the phosphate minerals which may be present in soils apatite, $Ca_{10}X_2(PO_4)_6$, where 'X' stands for F^-- or OH^--ion, strengite, $FePO_4 \cdot 2H_2O$, and variscite, $AlPO_4 \cdot 2H_2O$, should be mentioned. Under special conditions vivianite, $Fe_3(PO_4)_2 \cdot 8H_2O$, has been found. Gorceixite, $BaAl_3(PO_4)_2(OH)_5 \cdot H_2O$, and florencite, $CeAl_3(PO_4)_2(OH)_6$, have been identified in Australian soils. Wavellite, $Al_3(OH)_3(PO_4)_2 \cdot 5H_2O$, and crandallite, $CaAl_3(PO_4)_2(OH)_5$, are other possible phosphate minerals. Some of the crystalline phosphates may be present as a minor constituent of silicate minerals, where P substitutes for silicon. As a result of fertilization, more soluble calcium phosphates may be temporarily present in the soils, e.g. dicalcium phosphate, $CaHPO_4$ or $CaHPO_4 \cdot 2H_2O$, or octocalciumphosphate, $Ca_4H(PO_4)_3 \cdot 3H_2O$, as well as more soluble (amorphous, microcrystalline) forms of aluminum and iron phosphates.

In the soil solution phosphate is present, besides the soluble organic phosphates, as $H_2PO_4^-$- or HPO_4^{2-}-ions. This follows from the dissociation reactions of phosphoric acid according to:

$$H_3PO_4 \rightleftharpoons H_2PO_4^- + H^+ \qquad ; \log K° = -2.12 \qquad (6.33)$$

$$H_2PO_4^- \rightleftharpoons HPO_4^{2-} + H^+ \qquad ; \log K° = -7.20 \qquad (6.34)$$

$$HPO_4^{2-} \rightleftharpoons PO_4^{3-} + H^+ \qquad ; \log K° = -12.33 \qquad (6.35)$$

and the pH range normally found in soils. Also soluble complexes with different metal ions are possible.

6.4.1. Solubility of phosphates in soils

Various lines of evidence lead to the conclusion that the most important metal ions responsible for the binding of phosphates in soils are aluminum, iron and calcium. Other cations seem to be of minor importance in this respect, because their phosphate salts if present, are probably always in (partial) equilibrium with some of the more abundant phosphate minerals in soil. (From the practical point of view it may be worthwhile to mention precipitation of $Zn_3(PO_4)_2 \cdot 4H_2O$ as a possible explanation of phosphate induced zinc deficiency in calcareous rice soils).

Oversimplifying the system, the following phosphate salts might be considered present in soil systems:

$$AlPO_4 \cdot 2H_2O \rightleftharpoons Al^{3+} + H_2PO_4^- + 2OH^- \qquad \log K_{SO}^\circ = -30.5 \qquad (6.36)$$

$$FePO_4 \cdot 2H_2O \rightleftharpoons Fe^{3+} + H_2PO_4^- + 2OH^- \qquad \log K_{SO}^\circ = -34.9 \qquad (6.37)$$

$$CaHPO_4 \rightleftharpoons Ca^{2+} + HPO_4^{2-} \qquad \log K_{SO}^\circ = -6.66 \qquad (6.38)$$

$$Ca_4H(PO_4)_3 \cdot 3H_2O \rightleftharpoons 4Ca^{2+} + H^+ + 3PO_4^{3-} + 3H_2O \qquad \log K_{SO}^\circ = -46.91 \qquad (6.39)$$

$$Ca_{10}(OH)_2(PO_4)_6 \rightleftharpoons 10Ca^{2+} + 6PO_4^{3-} + 2OH^- \qquad \log K_{SO}^\circ = -113.7 \qquad (6.40)$$

$$Ca_{10}F_2(PO_4)_6 \rightleftharpoons 10Ca^{2+} + 6PO_4^{3-} + 2F^- \qquad \log K_{SO}^\circ = -120.86 \qquad (6.41)$$

On first sight it may be concluded that the activity of phosphate in the soil solution in contact with one or more of the solids mentioned should depend upon the activities of Al^{3+}, Fe^{3+} or Ca^{2+} and on the pH of the soil system (because of the protonization of phosphoric acid, formation of hydroxo-complexes of cations). The total phosphate concentration can be further affected by the formation of metal complexes with one or more phosphate species as a ligand.

Considering again pH as an independant variable, the question remains, what factors are controlling the activities of metals concerned in soil systems. Referring to the previous sections, $Fe(OH)_{3(s)}$ may be considered as the solid controlling Fe^{3+}-activity, $Al(OH)_{3(s)}$ or some alumino-silicate may govern Al^{3+}-activity while $CaCO_{3(s)}$ controls the activity of calcium. The metal ion activities will then further depend indirectly upon pH and redox-potential (Fe^{3+}-ions), pH and/or silicate activity (Al^{3+} ions) and the partial pressure of CO_2 and alkalinity (Ca^{2+}). The purity of solids concerned, their particle size and their degree of crystallization are some other factors influencing the total system (cf. section 2.9).

6.4.2. Phosphate solubility diagram in the system Al_2O_3 - Fe_2O_3 - CaO - P_2O_5 - H_2O

Whenever the solubilities of the cations in soils that may form a solid compound with phosphate are related to the solid phases that govern their solubility, a relatively simple phosphate solubility diagram can be developed. On this diagram any phosphates may be placed. Appropriate assumptions with respect to the solid phases that control the solubility of the cations may be respectively:
(a) that the Al^{3+}-activity in soils is determined by $Al(OH)_{3(s)}$ with $\log K^°_{SO} = -34.0$.
(b) that hydrous ferric oxide, $Fe(OH)_3$, $\log K^°_{SO} = -39.0$ controls the Fe^{3+}-activity.
(c) the Ca^{2+}-activity in neutral and acid soils is chosen at $10^{-2.5}$ molar; in calcareous soils the activity of Ca^{2+} is governed by the solubility of calcite.

For illustration purposes the equilibrium relationship will be derived for variscite in equilibrium with $Al(OH)_{3(s)}$:

		$\log K°$
$Al(OH)_{3(s)}$	$\rightleftharpoons Al^{3+} + 3OH^-$	-34.0
$Al^{3+} + 2OH^- + H_2PO_4^-$	$\rightleftharpoons AlPO_4 \cdot 2H_2O_{(s)}$	30.5
$H^+ + OH^-$	$\rightleftharpoons H_2O$	14.0
$Al(OH)_{3(s)} + H^+ + H_2PO_4^-$	$\rightleftharpoons AlPO_4 \cdot 2H_2O_{(s)} + H_2O$	10.5

Rearranging gives:

$$-\log(H_2PO_4) = 10.5 - pH \tag{6.42}$$

Similar relationships may be derived for other solid phosphates; the results are plotted in figure 6.11. Points in figure 6.11 which lie above a given isotherm represent supersaturation with respect to that phosphate compound and points below an isotherm represent undersaturation. Points of intersection indicate that both solid phases can coexist.

If one is interested in the total soluble phosphate concentration in equilibrium with a certain solid phosphate phase, also the complexes should be considered. For instance when hydroxyapatite is involved one should consider the solubility of the 'surface complex', $Ca_2(HPO_4)(OH)_2$, according to:

$$Ca_2HPO_4(OH)_2 \rightleftharpoons 2Ca^{2+} + HPO_4^{2-} + 2OH^-; \quad \log K° = -27.3 \tag{6.43}$$

The presence of such complexes will shift the lines of fig. 6.11 upward. With respect to aluminum and iron phosphates the complexes will increase total soluble P mainly in the low pH range.

From the previous discussion concerning the solubility of aluminum in the system containing silicate it may be concluded that, whenever aluminum silicates are present, the possibility exists that at the common pH values in the soils the Al^{3+}-ions activity in the solution might be governed by the silicate present. If in such a situation the phosphate bearing solid is $AlPO_4 \cdot 2H_2O$, the phosphate activities in the solution might be different from those predicted in figure 6.11.

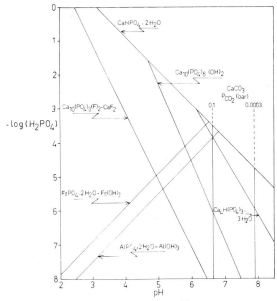

Fig. 6.11. General solubility diagram for phosphates in soils. The assumptions are that Ca^{2+} activity is $10^{-2.5}$ molar. Al^{3+} is in equilibrium with $Al(OH)_{3(s)}$ ($pK^° = 34.0$) and Fe^{3+} is in equilibrium with $Fe(OH)_{3(s)}$ ($pK^° = 39.0$), and that fluorite is present. (after Lindsay and Moreno).

6.5. RELEVANT THERMODYNAMIC DATA OF THE SYSTEMS DISCUSSED

Calculations of the type demonstrated in this chapter require the knowledge of the standard free enthalpy of formation of the compounds concerned (cf. section 2.4.). An extensive listing of $\overline{G}_f^°$-values of many different compounds may be found e.g. in 'Handbook of Chemistry and Physics' and in textbooks of thermodynamics. On the following pages some solubility dissociation and stability constants, concerning systems discussed in this chapter (which may be derived from the relevant $\overline{G}_f^°$-values) are tabulated for handy reference.

Table 6.1 Equilibrium constants for some reactions at 25 °C and 1 atmosphere total pressure

Solubility products of solids in water

No.	Solid	Reaction	log K_{SO}°	source
1	amorphous Al(OH)$_3$	$Al(OH)_{3(s)} \rightleftharpoons Al^{3+} + 3OH^-$	− 32.34 ▲	7
2	boehmite	$Al(OH)_{3(s)} \rightleftharpoons Al^{3+} + 3OH^-$	− 34.02 ▲	5
3	bayerite	$Al(OH)_{3(s)} \rightleftharpoons Al^{3+} + 3OH^-$	− 35.52 ▲	7
4	gibbsite	$Al(OH)_{3(s)} \rightleftharpoons Al^{3+} + 3OH^-$	− 36.30 ▲	7
5	amorphous Fe(OH)$_3$	$Fe(OH)_{3(s)} \rightleftharpoons Fe^{3+} + 3OH^-$	− 37.5*	7
6	amorphous Fe(OH)$_3$	$Fe(OH)_{3(s)} \rightleftharpoons Fe^{3+} + 3OH^-$	− 39.1*	8
7	hematite	$\frac{1}{2}\alpha\text{-}Fe_2O_{3(s)} + 3/2 H_2O \rightleftharpoons Fe^{3+} + 3OH^-$	− 42.5	1
8	goethite	$\alpha\text{-}FeOOH_{(s)} + H_2O \rightleftharpoons Fe^{3+} + 3OH^-$	− 44.0	7
9	ferrous hydroxide	$Fe(OH)_{2(s)} \rightleftharpoons Fe^{2+} + 2OH^-$	− 15.1	7
10	aragonite	$CaCO_{3(s)} \rightleftharpoons Ca^{2+} + CO_3^{2-}$	− 8.22	7
11	calcite	$CaCO_{3(s)} \rightleftharpoons Ca^{2+} + CO_3^{2-}$	− 8.35	7
12	dolomite	$CaMg(CO_3)_{2(s)} \rightleftharpoons Ca^{2+} + Mg^{2+} + 2CO_3^{2-}$	− 16.90	7
13	cadmium carbonate	$CdCO_{3(s)} \rightleftharpoons Cd^{2+} + CO_3^{2-}$	− 12.00	7
14	magnesite	$MgCO_{3(s)} \rightleftharpoons Mg^{2+} + CO_3^{2-}$	− 7.80	7
15	siderite	$FeCO_{3(s)} \rightleftharpoons Fe^{2+} + CO_3^{2-}$	− 10.68	7
16	zinc carbonate	$ZnCO_{3(s)} \rightleftharpoons Zn^{2+} + CO_3^{2-}$	− 10.78	7
17	monocalcium phosphate	$Ca(H_2PO_4)_2 \cdot H_2O_{(s)} \rightleftharpoons Ca^{2+} + 2H_2PO_4^- + H_2O$	− 1.14	7
18	dicalcium phosphate dihydrate	$CaHPO_4 \cdot 2H_2O_{(s)} \rightleftharpoons Ca^{2+} + HPO_4^{2-} + 2H_2O$	− 6.56	7
19	anhydrous dicalcium phosphate	$CaHPO_{4(s)} \rightleftharpoons Ca^{2+} + HPO_4^{2-}$	− 6.66	7
20	octocalcium phosphate	$Ca_4H(PO_4)_3 \cdot 3H_2O_{(s)} \rightleftharpoons 4Ca^{2+} + H^+ + 3PO_4^{3-} + 3H_2O$	− 46.91	7
21	tricalcium phosphate	$Ca_3(PO_4)_{2(s)} \rightleftharpoons 3Ca^{2+} + 2PO_4^{3-}$	− 26.00	7
22	hydroxyapatite	$Ca_{10}(PO_4)_6(OH)_{2(s)} \rightleftharpoons 10Ca^{2+} + 6PO_4^{3-} + 2OH^-$	−113.7	3
23	fluorapatite	$Ca_{10}(PO_4)_6(F)_{2(s)} \rightleftharpoons 10Ca^{2+} + 6PO_4^{3-} + 2F^-$	−120.86	7
24	variscite	$AlPO_4 \cdot 2H_2O_{(s)} \rightleftharpoons Al^{3+} + H_2PO_4^- + 2OH^-$	− 30.50	7
25	strengite	$FePO_4 \cdot 2H_2O_{(s)} \rightleftharpoons Fe^{3+} + H_2PO_4^- + 2OH^-$	− 34.9	4
26	vivianite	$Fe_3(PO_4)_2 \cdot 8H_2O_{(s)} \rightleftharpoons 3Fe^{2+} + 2PO_4^{3-} + 8H_2O$	− 36.0	4a
27	zinc phosphate	$Zn_3(PO_4)_2 \cdot 4H_2O \rightleftharpoons 3Zn^{2+} + 2PO_4^{3-} + 4H_2O$	− 32.04*	7
28	amorphous SiO$_2$	$SiO_{2(s)} + 2H_2O \rightleftharpoons H_4SiO_4$	− 2.74	1

No.		Reaction		log K°	sources
29	quartz	$SiO_{2(s)} + 2H_2O \rightleftharpoons H_4SiO_4$		-4.00	1
30	gypsum	$CaSO_4 \cdot 2H_2O_{(s)} \rightleftharpoons Ca^{2+} + SO_4^{2-} + 2H_2O$		-4.61	1
31	fluorite	$CaF_{2(s)} \rightleftharpoons Ca^{2+} + 2F^-$		-10.57	7

Solubility of gases in water

No.	Reaction			log K°	sources
32	$CO_{2(g)} + H_2O \rightleftharpoons H_2CO_3{}^*$			-1.46	7
33	$NH_{3(g)} + H_2O \rightleftharpoons NH_4OH$			$+1.75$	5

Dissociation constants of acids in water

No.				log K°	sources
34	$H_2CO_3{}^*$	$\rightleftharpoons$	$H^+ + HCO_3^-$	-6.35	7
35	HCO_3^-	$\rightleftharpoons$	$H^+ + CO_3^{2-}$	-10.33	7
36	H_3PO_4	$\rightleftharpoons$	$H^+ + H_2PO_4^-$	-2.12	7
37	$H_2PO_4^-$	$\rightleftharpoons$	$H^+ + HPO_4^{2-}$	-7.20	7
38	HPO_4^{2-}	$\rightleftharpoons$	$H^+ + PO_4^{3-}$	-12.33	7
39	H_4SiO_4	$\rightleftharpoons$	$H^+ + H_3SiO_4^-$	-9.77	7
40	H_2O	$\rightleftharpoons$	$H^+ + OH^-$	-14.00	7

Formation constants of complexes in water

No.				log K°	sources
41	$Al^{3+} + H_2O$	$\rightleftharpoons$	$AlOH^{2+} + H^+$	-5.02	7
42	$Al^{3+} + 4H_2O$	$\rightleftharpoons$	$Al(OH)_4^- + 4H^+$	-23.57	7
43	$2Al^{3+} + 2H_2O$	$\rightleftharpoons$	$Al_2(OH)_2^{4+} + 2H^+$	-6.27	7
44	$Fe^{3+} + H_2O$	$\rightleftharpoons$	$FeOH^{2+} + H^+$	-3.0^*	6
45	$Fe^{3+} + 2H_2O$	$\rightleftharpoons$	$Fe(OH)_2^+ + 2H^+$	-6.4^*	6
46	$2Fe^{3+} + 2H_2O$	$\rightleftharpoons$	$Fe_2(OH)_2^{4+} + 2H^+$	-3.1^*	6
47	$Fe^{3+} + 3H_2O$	$\rightleftharpoons$	$Fe(OH)_3^0 + 3H^+$	-13.5^*	2
48	$Fe^{3+} + 4H_2O$	$\rightleftharpoons$	$Fe(OH)_4^- + 4H^+$	-23.5^*	2
49	$Fe^{2+} + H_2O$	$\rightleftharpoons$	$FeOH^+ + H^+$	-8.3	7
50	$Fe^{2+} + 2H_2O$	$\rightleftharpoons$	$Fe(OH)_2^0 + 2H^+$	-17.2	7
51	$Fe^{2+} + 3H_2O$	$\rightleftharpoons$	$Fe(OH)_3^- + 3H^+$	-32.0	7
52	$Fe^{2+} + 4H_2O$	$\rightleftharpoons$	$Fe(OH)_4^{2-} + 4H^+$	-46.4	7
53	$Ca^{2+} + HCO_3^-$	$\rightleftharpoons$	$CaHCO_3^+$	1.26	7
54	$Ca^{2+} + CO_3^{2-}$	$\rightleftharpoons$	$CaCO_3^0$	3.20	7
55	$Mg^{2+} + HCO_3^-$	$\rightleftharpoons$	$MgHCO_3^+$	1.16	7
56	$Mg^{2+} + CO_3^{2-}$	$\rightleftharpoons$	$MgCO_3^0$	3.40	7
57	$Al^{3+} + H_2PO_4^-$	$\rightleftharpoons$	$AlH_2PO_4^{2+}$	$\sim 3^*$	7
58	$Al^{3+} + HPO_4^{2-}$	$\rightleftharpoons$	$AlHPO_4^+$	$\sim 8.1^*$	7
59	$Fe^{3+} + HPO_4^{2-}$	$\rightleftharpoons$	$FeHPO_4^+$	$\sim 9.75^*$	7
60	$Ca^{2+} + H_2PO_4^-$	$\rightleftharpoons$	$CaH_2PO_4^+$	1.08	7

61	$Ca^{2+} + HPO_4^{2-}$	$\rightleftharpoons CaHPO_4^0$		2.70	7
62	$Ca^{2+} + SO_4^{2-}$	$\rightleftharpoons CaSO_4^0$		2.31	7

Redox reactions

63	$H^+ + e^-$	$\rightleftharpoons \tfrac{1}{2}H_{2(g)}$		0.0	
64	$\tfrac{1}{4}O_{2(g)} + H^+ + e^-$	$\rightleftharpoons \tfrac{1}{2}H_2O_{(l)}$		20.77	7
65	$Fe^{3+} + e^-$	$\rightleftharpoons Fe^{2+}$		13.02	7
66	$Fe^{2+} + 2e^-$	$\rightleftharpoons Fe_{(s)}$		−14.89	7
67	$Fe_3O_{4, magnetite} + 8H^+ + 2e$	$\rightleftharpoons 3Fe^{2+} + 4H_2O$		33.1	1

▲Values for Al-species not certain, cf. table 8.1 where the corresponding value for Gibbsite would be −34.31.

*Reported values are not corresponding to 25 °C or zero ionic strength

Sources:

1. Garrels, R.M. and Christ, C.L., 1965. *Solutions, Minerals and Equilibria.* Harper & Row, Publishers, New York.
2. Lengweiler, H., Buser, W., and Feitknecht, W., 1961. *Helv. Chim. Acta* 44, p.796.
3. Lindsay, W.L., and Moreno, E.C., 1960. *Soil Sci. Soc. Amer. Proc.* 24, p.177.
4. Nriagu, J.O., 1972. *American Journal of Science* 272, p.476.
4a Nriagu, J.O., 1972. *Geochimica et Cosmochimica Acta* 36, p.459.
5. Pourbaix, M., 1963. *Atlas d'equilibres electrochimiques,* Gauthier-Villars, Paris.
6. Ringbom, A., 1963. *Complexation in Analytical Chemistry,* Interscience Publishers, New York.
7. Sillen, L.G. and Martell, A.C., 1964. *Stability constants of metal-ion complexes.* 2nd ed. Special Publication no. 17. The Chemical Society, London.
8. Schindler, P., Michaelis, W. and Feitknecht, W., 1963. *Helv. Chim. Acta,* 46, p.444.

ILLUSTRATIVE PROBLEMS

1. Consider a system $CO_{2(g)}$ - $H_2O_{(l)}$ in equilibrium with a $P_{CO_2} = 10^{-3}$ bar. Calculate the pH of the system. (Answer: pH = 5.40).
1a. Next the pH of the above system is quickly raised to pH = 7 by the addition of NaOH. Calculate the pH of this system if brought again in equilibrium with $P_{CO_2} = 10^{-3}$ bar. (Answer: pH = 6.31).
2. Calculate the solubility of $Ca_4H(PO_4)_3 \cdot 3H_2O$ in a solution having an ultimate ionic strength of 0.01 and a pH = 6.2. (Formation of complexes is neglected). (Answer: calcium = $10^{-2.57}$ mole l^{-1}; phosphorus = $10^{-2.69}$ mole l^{-1}. Could $CaHPO_4 \cdot 2H_2O$ precipitate from this solution ?; What would be the ultimate stable calcium phosphate phase.
3. A soil has been polluted with solid $PbSO_4$. Which of the following lead components, viz. $PbSO_4$, $PbCO_3$, $Pb_3(PO_4)_2$ could be a stable phase in the soil system under the following conditions: $P_{CO_{2(g)}} = 10^{-3}$ bar; pH = 6; $a_{SO_4^{2-}} = 2.10^{-3}$ mole/l; $a_{PO_4^{3-}}$ is governed by the presence of (solid) variscite and (solid) $Al(OH)_3$. Reaction constants of relevant reactions not given in table 6.1 are given below:

$PbCO_{3(s)} \rightleftharpoons Pb^{2+} + CO_3^{2-} \quad \log K°_{SO} \quad -13.14$

$Pb_3(PO_4)_{2(s)} \rightleftharpoons Pb^{2+} + 2PO_4^{3-} \quad \log K°_{SO} \quad -42.10$

$PbSO_{4(s)} \rightleftharpoons Pb^{2+} + SO_4^{2-} \quad \log K°_{SO} \quad -5.38$

$Al(OH)_{3(s)} \rightleftharpoons Al^{3+} + 3OH^- \quad \log K°_{SO} \quad -34.0$

(Answer: $Pb_3(PO_4)_{2(s)}$).

4. Consider the system $Fe(OH)_{3(s)}$ - $Fe_3(OH)_{8(s)}$ - $FeCO_{3(s)}$ - H_2O. Calculate the P_{CO_2}, pe and redox potential at the point where the three solid phases may coexist if the system is buffered at a pH = 6.
(Answer: $P_{CO_2} = 10^{-3.44}$ bar, pe = 0.92, E = 55.2 mV at 25 °C).

4a. Construct a pe - $\log P_{CO_2}$ diagram showing the stability fields of the solid phases at pH = 6. What is the lowest pe-value in the above diagram having a significance, and why. (Answer: pe = -6).
Reaction constants not specified in table 6.1 are given below:

$Fe(OH)_{3(s)} \rightleftharpoons Fe^{3+} + 3OH^- \quad \log K° \quad -37.2$

$Fe_3(OH)_{8(s)} + 8H^+ + 2e \rightleftharpoons 3Fe^{2+} + 8H_2O \quad \log K° \quad +46.54$

5. The phosphate level in a calcareous soil is governed by the presence of metastable $CaHPO_4 \cdot 2H_2O_{(s)}$. P_{CO_2} in the soil gas phase amounts to $10^{-3.5}$ bar; pH = 8.0. Calculate the activity of Zn^{2+}, supposing only zinc phosphate is formed (leaving possible complex formation out of consideration). (Answer: $(Zn^{2+}) = 10^{-5.23}$ mole/l). What would be the stable phase at a P_{CO_2} = 0.1 bar: zinc carbonate, zinc phosphate or zinc hydroxide?
Reaction constants of relevant reactions not given in table 6.1 are given below.

$Zn(OH)_{2(s)} \rightleftharpoons Zn^{2+} + 2OH^- \quad \log K° \quad -16.5$.

RECOMMENDED LITERATURE

Garrels, R.M. and Christ, C.L., 1965. *Solutions, Minerals and Equilibria*. Chapter 1,2,3,4, 5,6,7,10. Harper & Row, Publishers New York.

Lindsay, W.L. and Moreno, E.C., 1960. Phosphate Phase Equilibria in Soils. *SSSA Proc.* 24: 177-182.

Mortvedt, J.J., Giordano, P.M., Lindsay, W.L. (Ed.), 1972. *Micronutrients in Agriculture*. Chapter 3. Soil Science Society of America, Inc. Madison, Winsc., U.S.A.

Stumm, W. and Morgan, J.J., 1970. *Aquatic Chemistry. An Introduction Emphasizing Chemical Equilibria in Natural Waters*. Sections 2-4, 3-1,2,3,4,5; 7-1,2,3,4; 10-2. Wiley - Interscience. New York.

CONSULTED LITERATURE

Butler, J.N., 1964. *Ionic Equilibria. A Mathematical Approach*. Addison-Wesly Publishing Company. Inc. Reading, Massachussets.

Garrels, R.M. and Christ, C.L., 1965. *Solutions, Minerals, and Equilibria*. Harper & Row, Publishers, New York.

Lindsay, W.L., 1972. Syllabus of Lecture Series. *Chemical Equilibria in Soils*. Wageningen, Agricultural State University. Colorado State University (personal communication).

Stumm, W. and Morgan, J.J., 1970. *Aquatic Chemistry. An Introduction Emphasizing Chemical Equilibria in Natural Waters*. Wiley-Interscience. New York.

CHAPTER 7

TRANSPORT AND ACCUMULATION OF SOLUBLE SOIL COMPONENTS

G.H. Bolt

The present chapter deals with the displacement of certain soil components with respect to the soil skeleton. Accordingly the motion of the entire soil (creep, landslides) and the displacement of solid phase constituents with respect to each other (consolidation, swelling, mixing of layers by tillage operations and by burrowing animals) will not be considered. Transport of components then comprises transport in the gas phase and/or in the liquid phase. In comparing these two processes it should be noted that the density of the gas phase is only about one thousandth of that of the liquid phase. Therefore, the transport of mass (of a certain component) with the gas phase tends to be small, even at high concentration (of that component) in the gas phase. In practice the liquid phase transport of a component normally exceeds that in the gas phase, except when highly volatile compounds with a low solubility in water are involved (e.g. the common gases of the soil atmosphere, like O_2 and CO_2, but also soil fumigants). In this context the transport in the gas phase of Hg as dimethylmercury should also be mentioned. Where transport in the gas phase (often mainly by diffusion) actually belongs to the field of Soil Physics, the present discussion will be limited to transport in the liquid phase.

Due to its high mobility (at least at high moisture content of the soil) the liquid phase is indeed the main carrier for transport of different components through the soil. Referring again to Soil Physics texts, the flow of the liquid phase as a whole will be regarded here as a given quantity; the present text is then concerned only with the displacement of soluble components as determined by the liquid flow and the relevant soil characteristics.

7.1. TRANSPORT WITH AND IN THE LIQUID PHASE

The actual motion of solutes in the soil is mainly in the way of being dragged along with the liquid phase (convection). In addition, the solutes may move with respect to the liquid phase as a whole, as a result of diffusion.

Expressing the solute transport in terms of its flux, j, in e.g. keq per m^2 of soil column, per sec., one may formally write:

$$j_k \equiv J^V \cdot c_k + j_k^D \tag{7.1}$$

in which j_k is the total flux; J^V is the volume flux of the soil solution, in m³/ /m² sec (or m/sec); c_k is the concentration of k in the liquid phase, in keq/m³ (or meq/cm³). Here $J^V.c_k$ is the convective flux and j_k^D then constitutes the autonomous flux of the solute with respect to the mean motion of the liquid phase. This autonomous flux is now the combined effect of molecular diffusion and the so-called dispersion. The latter arises when a gradient of the solute concentration is combined with a distribution of the velocities in the flowing liquid around their mean value as determined by J^V (cf. part B).

If the liquid content in a soil equals θ m³ per m³ of soil, the mean velocity of the moving liquid equals J^V/θ m/sec. Obviously the liquid phase will travel faster in the wider (liquid filled) pores than in the narrow ones. Part of the solutes dragged along with the liquid phase will thus travel faster, or slower, respectively, than corresponds with the mean liquid velocity.

Fortunately, both the diffusion and the dispersion flux are proportional to the concentration gradient of the solute and may thus conveniently be combined in the term j_k^D. In practice the contribution of j_k^D to j_k in soil is often limited to 10% or less. Thus the mean concentration gradient between layers of 0.1 m or more is limited to such low values that a solution flux of a few mm per day suffices to make the convection flux the dominant component of the total flux. In contrast to this situation the solute flux in close vicinity of plant roots may depend to a considerable degree on diffusion transport.

It is interesting to compare here the transport in the liquid phase with that in the gas phase. The low density of the gas phase, while limiting the convective flux, gives rise to much higher values of the diffusion constant. In the gas phase the diffusive flux may thus exceed the convective flux by far.

7.2. SOLUTE DISPLACEMENT IN SOIL

While in the previous section it was pointed out that the solute transport is in first approximation equal to the convective flux with the liquid phase, the actual rate of displacement in the profile may be much less than the mean liquid velocity defined as J^V/θ. Such a 'delay' is caused by adsorption of the solute on the solid phase. Thus instead of moving on with the liquid phase, the solute may become 'diverted' into the adsorbed phase. The obvious example here are the cations, which are usually adsorbed in large amounts by the solid phase (cf. chapter 4). In the following sections the main aspects of one-dimensional displacement of solutes subject to exchange-adsorption, will be discussed. This proces is usually referred to as exchange-chromatography.

7.2.1. Displacement in case of complete exchange

The mean displacement in soil of a cation 'foreign' to the soil (i.e. initially not present), when fed into a soil column by means of a solution, is easily derived from the condition of integral conservation of the solute. Let the 'feed volume' be equal to V m³ per m² of column and let it contain a concentration c_f keq/m³ of the cation M_+ present as a salt MA. The soil has a cation exchange capacity of Q keq per m³ of soil (initially saturated with other cations) and upon leaching it retains a liquid content of θ m³ per m³ soil (e.g. corresponding to field capacity). The following may now be concluded after the solution has been fed into the column.

a. The liquid phase resides in the soil at a liquid content θ, and thus the feed liquid reaches to a depth V/θ m (cf. figure 7.1a).
b. Assuming that the anion A_- is not adsorbed by the soil, it will be uniformly distributed in the soil column at its original concentration $c_{A-} = c_f$. Obviously the total amount of the anion in the soil column now equals $(V/\theta)\theta c_f = Vc_f$, i.e. the amount that was present in the added feed solution (cf. figure 7.1b).
c. The cation M_+ becomes (at least partly) adsorbed by the soil and is thus distributed over the liquid phase and the adsorbed phase. Because the liquid feed passed over the 'static' solid phase, the concentration of the cation in the profile must be decreasing with depth. In fact, it is not unreasonable to assume that the top layers of the soil, which were continually confronted with a fresh solution containing only MA, have become saturated with M_+. Noting further that if local exchange equilibrium was reached during leaching there exists a particular relation between the local value of the relative amount adsorbed of M_+, q/Q, and the relative concentration in solution at the same position, c/c_f, one may expect a distribution with depth as pictured schematically in figure 7.1b.

In this figure the total amount of cations fed into the column, represented by the surface area $V.c_f$, is found back as the surface area between the lines representing θc and q as a function of depth. It is noted that beyond a certain depth the total amount of M_+ has been 'used up'. There the adsorption complex must still have its original composition.

Disregarding the precise shape of the concentration-depth curve of M_+ in the column, one may summarize the situation in terms of a 'mean depth of penetration', x_p, of the cation M_+ fed into the column. This quantity is then defined by the relation:

$$x_p.(Q + \theta c_f) \equiv V.c_f, \text{ or}$$

$$x_p = \frac{V}{\theta} \cdot \frac{1}{(1 + Q/\theta c_f)} \qquad (7.2)$$

Referring to section 4.2.1.3, one recognizes $Q/\theta c_f$ as the distribution ratio, R_D, for cation M_+ in the soil at a total electrolyte level c_f (both numerator and denominator here being specified per m^3 of soil in stead of per 100 g). Hence:

$$x_p = \frac{V}{\theta} \cdot \frac{1}{(1 + R_D)} \qquad (7.2a)$$

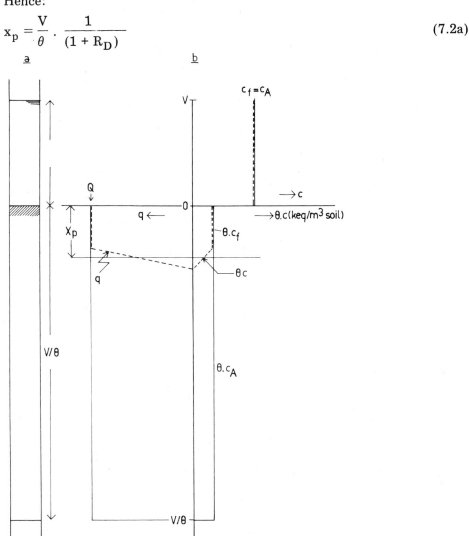

Fig. 7.1. Schematic presentation of the displacement of a cation in soil under the condition of complete exchange. The amount adsorbed, q, and the amount in solution, $\theta.c$, after addition of V m^3 of solution-with a concentration of c_f keq/m^3-per m^2 of soil surface, are given as a function of the depth, x.

Furthermore V/θ represents the depth of penetration of the liquid fed into the column, and also that of the non-adsorbed anion ($R_{D,-} = 0$!). For the present example one thus finds that a component when fed into a soil column - originally free of this component - is retained at the entrance side over a layer equal to $1/(R_D+1)$ times the depth over which the liquid penetrates into the soil. Using V as a variable one may also conclude that the depth of penetration of the adsorbed species travels with a velocity equal to $1/(R_D+1)$ times the mean liquid velocity in the column.

7.2.2. Displacement in case of incomplete exchange

The above statement may be modified slightly to obtain a wider coverage of situations. Let the feed solution be a mixture of MA with other salts, the total concentration equal to C, and the concentration of MA equal to c_f.

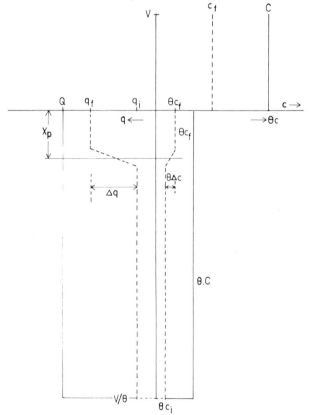

Fig. 7.2. As figure 7.1, but now for the condition of incomplete exchange, ranging from $q = q_i$ to $q = q_f$, corresponding to $c = c_i$ and $c = c_f$ at a total electrolyte level equal to C. Notice the 'leakage' of an amount equal to $(V/\theta - x_p)$. θc_i beyond $x = x_p$.

Also let the equilibrium amount adsorbed of M_+ corresponding to the composition of the feed solution be equal to q_f. Furthermore the soil column contains already an amount adsorbed of cation M_+ equal to q_i, while an equilibrium solution of this original soil contains a concentration c_i of MA when the total electrolyte level is brought to C.

In principle q_f could be determined by shaking a small amount of soil with a large excess of feed solution. The value of c_i could be determined by measuring c after adding water to the original soil until the total concentration equals C.

During passage of the feed solution its concentration in M will decrease from c_f to c_i, while the amount thus stripped from the solution is accumulated at the solid phase by raising q from q_i to q_f (cf. figure 7.2). Using the same reasoning as before one now finds:

$$x_p = \frac{V}{\theta} \cdot \frac{1}{(1+\Delta q/\theta \Delta c)} \qquad (7.3)$$

in which $\Delta q = (q_f - q_i)$ and $\Delta c = (c_f - c_i)$.

The above may be verified from the material balance, which then shows that part of the amount fed into the column, Vc_f, now resides below x_p (at a concentration c_i).

7.2.3. The influence of the exchange isotherm on solute displacement

The situation pictured above may be further elucidated by plotting the relation between q and θc. Such a plot represents the exchange isotherm in appropriate units of keq/m^3 soil both for q and θc (figure 7.3a). Constructing such a plot for a system with two cations M_1 and M_2 in which the total electrolyte level is fixed at C, while c_{M_1} varies between 0 and C, one finds $q_{M_1} = 0$ at $c_{M_1} = 0$ and $q_{M_1} = Q$ at $c_{M_1} = C$. Connecting these endpoints is the adsorption isotherm pertaining to a total concentration equal to C. The slope of the diagonal connecting the endpoints now represents R_D for any cation when the soil system (at concentration C) is fully saturated with this cation (here taken at 10 and equal to $Q/\theta C$). At a certain partial saturation with M_1, however, the value of R_{D,M_1} may differ from 10, as is shown in figure 7.3a (i.e. at 75% saturation $R_{D,M_1} = 15$). Recognizing that for a given soil (at given C and θ) the value of $\bar{R}_D$ at full saturation (with any cation) is a constant (cf. $\bar{R}_D$ of section 4.2.1.3) one may scale figure 7.3a by plotting $q_{M_1}/Q \equiv N$ against $c_{M_1}/C \equiv f$, i.e. the fractional saturation of the incoming cation M_1 in the adsorbed phase and in solution (the same ratios for the outgoing cation M_2 are then represented by (1-N) and (1-f), respectively). Such a plot (figure 7.3b) affords better accuracy of reading. The value of R_{D,M_1} for a certain fractional saturation is then found by multiplying the slope of the chord - connecting this point with the origin - with $\bar{R}_D = 10$.

In this graph one may now identify the value of $\Delta q/\theta \Delta c$ as the product of $\bar{R}_D$ with the slope of the chord connecting the original fraction adsorbed, N_i, with the final fraction adsorbed, N_f, i.e. as $\bar{R}_D \Delta N/\Delta f$. Accordingly the mean depth of penetration of an adsorbed component in soil is found as:

$$x_p = \frac{V}{\theta} \cdot \frac{1}{(1 + R_D \cdot \Delta N/\Delta f)} \qquad (7.3a)$$

7.3. THE PENETRATING SOLUTE FRONT

In the previous section it was inferred that for a given value of the feed volume, V, the concentration of the solute, c, varies with the position in the column. The region around x_p, where c changes from the initial value c_i to the final value c_f is termed the solute front (in the column). In this front the amount adsorbed, q, then changes from q_i to q_f. Considering again only the one-dimensional penetration into a column, the shape of the pair of curves relating q and c with the position in the column will be referred to as the shape of the front. One may thus distinguish sharp and diffuse fronts in a soil column.

Limiting the discussion to the situation where the concentration of the feed solution changes abruptly to the value c_f (to be termed 'step-feed'), the question arises whether and when the shape of the front penetrating into the column will become altered. Although a detailed discussion about this process of front modulation is outside the scope of this section (cf. part B), it is necessary to mention here the main conclusions that may be derived from such a discussion. Thus the applicability of equation (7.3a) in practice is critically dependent on the sharpness of the penetrating front as one is primarily interested to assess to what depth significant increases of c and q of an added solute are expected to occur. In case an originally steep front becomes very much flattened during its passage through the soil column, this could be far beyond the 'mean' depth x_p. Two factors are of prime concern in this respect, viz. the precise shape of the 'normalized' exchange isotherm and the (relative) magnitude of the diffusion/dispersion flux j^D of equation (7.1).

7.3.1. Influence of the exchange isotherm

It may be shown (cf. part B) that the rate of displacement of each point of the solute front (i.e. the travelling velocity through the column of a particular concentration) is a function of its concentration, c. In fact the depth of penetration for a particular concentration, x_c, equals in principle:

$$x_c = \frac{V}{\theta} \cdot \frac{1}{(1 + N'_c \cdot \bar{R}_D)} \tag{7.4}$$

in which N'_c is the slope of the normalized exchange isotherm at concentration c. Accordingly, the smaller the slope at a particular value of c, the faster will this concentration travel through the soil. This leads to the following important conclusions:

a. If the exchange isotherm is linear, all points of the front travel with the same velocity and the front remains unaltered during passage (figure 7.4a).

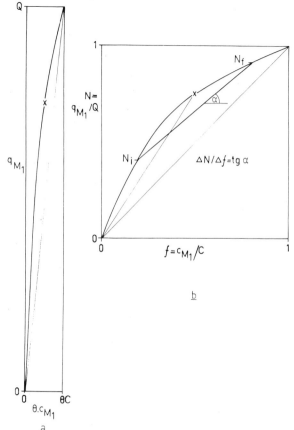

Fig. 7.3a. Exchange isotherm in units of q_{M_1} (amount of ion M_1 adsorbed, in keq per m^3 of soil) as a function of θc_{M_1} (amount in solution, in keq per m^3 of soil) for a given total electrolyte level, C.

Fig. 7.3b. The same, as the dimensionless fraction adsorbed, $N \equiv q_{M_1}/Q$, as a function of the fraction in solution, $f \equiv c_{M_1}/C$. The mean slope $\Delta N/\Delta f$ in the region $N = N_i$ to $N = N_f$ equals $\mathrm{tg}\,\alpha$.

b. If the exchange isotherm is convex upwards (figure 7.3) the slope N' at low values of c is greater than at high values of c. Accordingly the high values of c tend to travel faster than the low values, so the front will 'steepen' during passage. This situation is called favorable exchange. In the particular case here, where a step-feed was employed, this implies that the front travels as a step-front: the front cannot 'overturn' and equation (7.4) reverts to (7.3a) (cf. figure 7.4a).

c. If the exchange isotherm is concave upwards, the reverse occurs. Low values of c travel faster than high values of c, so the front flattens upon passage (non-favorable exchange, cf. figure 7.4b).

7.3.2. Influence of diffusion and dispersion

Foregoing details it may be stated that diffusion and dispersion always lead to a flattening of the front, in a manner very similar to the formation of a 'diffusion' front in a static liquid column (figure 7.4c). In contrast to the 'tilting' effect caused by non-favorable exchange - which produces a front spreading proportional to V and hence to x_p - here the front spreading

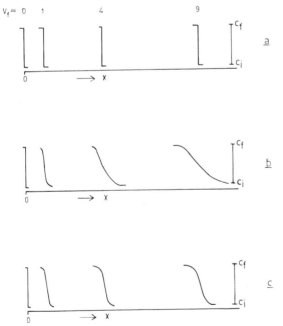

Fig. 7.4. Penetration of a step-feed of an adsorbed cation.
a. Linear or favorable exchange isotherm, diffusion/dispersion negligible.
b. Non-favorable exchange isotherm, diffusion/dispersion negligible.
c. Linear exchange isotherm, with diffusion/dispersion.
$V_f = J^V \cdot t$, i.e. the total volume fed into the column.

increases with $\sqrt{V}$ (at constant J^V).

7.3.3. Order of magnitude of the front spreading effects

Again referring to part B for details, some estimates may be given of the magnitude of the above effects. In accordance with equation (7.3a) and (7.4) one finds for non-favorable-exchange that the ratio x_{c_i}/x_p, i.e. penetration depth of the 'toe' of the front at $c = c_i$ relative to the mean penetration depth of the front, equals:

$$(1 + \bar{R}_D \cdot \Delta N/\Delta f)/(1 + \bar{R}_D \cdot N_i')$$

It may be verified with the help of e.g. the Gapon equation (cf. also section 7.4, below) that this ratio may easily reach values of three or larger in the case of Na-ions replacing Ca-ions. Thus the sodication process (cf. chapter 9) follows a very flat front, with part of the Na-ions penetrating much beyond x_p. It may also be shown that in such a case the front spreading due to diffusion/dispersion is definitely a second order effect.

Conversely the reclamation of sodic soils with Ca-salts (cf. chapter 9) produces rather steep fronts, as the 'steepening' effect of the favorable exchange generally outweighs the diffusion/dispersion effect.

For linear isotherms (which is a safe assumption for the adsorption of trace amounts of e.g. pollutants) the spreading effect - often mainly due to dispersion - may be estimated at about $0.4\sqrt{x_p}$ (cf. part B), i.e. for $x_p = 1m$, the front would reach from 0.8 to 1.2 m.

The above estimate is valid only if J^V falls in the range of 0.01 to 1 m per day. At lower values of J^V the spreading is dominated by molecular diffusion and may be larger. It is also noted that x_p must be specified in m.

7.4. SOME PRACTICAL EXAMPLES

7.4.1. Reclamation of Na-soils

If a solution of gypsum is leached through a soil containing a high percentage of exchangeable Na-ions, the soil will be transformed into Ca-soil according to the pattern described above. Using $\gamma = 20$ meq per 100 grams of soil, a total electrolyte level of the leaching solution, $C = 0.01$ normal and a moisture content at field capacity equal to 20 cm^3 per 100 grams of soil, one finds $\bar{R}_D = 100$.

A Gapon exchange equation may be expressed in terms of N_{2+} (fraction adsorbed of Ca) and f_{2+} (same in solution) as:

$$(1 - N_{2+})/N_{2+} = K_G\sqrt{2C} \cdot (1 - f_{2+})/\sqrt{f_{2+}} \qquad (7.5)$$

The corresponding normalized exchange isotherm for $K_G = \frac{1}{2}(1/mol)^{\frac{1}{2}}$ and $C = 0.01$ normal is plotted in figure 7.5.

If $N_i = 0.1$ (i.e. 10% Ca and 90% Na on the soil), one finds $\Delta N/\Delta f$ over the range from $N_i = 0.1$ to $N_f = 1.0$ at about 0.90 and $(1+\bar{R}_D.\Delta N/\Delta f)$ thus equals 91. The depth of the zone of reclamation (to 100% Ca) is only slightly more than 1 cm per m of penetration of the leaching liquid. Expressed per m of applied water (with gypsum), one finds for a bulk density of 1500 kg/m³ - which yields $\theta = 0.30$ m³/m³ - that the depth of reclamation is about 3½ cm. The necessary application of gypsum per m of applied water and per ha would then correspond to about 8.5 tons. Reclamation of a shallow rooting zone of 35 cm would thus require the prohibitive amount of 85 tons of gypsum per ha (and 10 m of irrigation water).

A worthwhile consideration in such a case (if reclamation is attempted at all) is to admix NaCl with the gypsum (the NaCl may be present in the irrigation water). This implies incomplete exchange of the Na-ions and accordingly the gypsum will spread over a greater depth. As an example it is considered to bring the total normality of the leaching solution to 0.09 normal, consisting of 0.01 normal $CaSO_4$ plus 0.08 normal NaCl. Reconstructing the normalized isotherm for this value of C, one finds the second curve as shown in figure 7.5. Recovery is now to 64% Ca and $\Delta N/\Delta f$ equals about 5.4. Because of the increase of C, $\bar{R}_D$ now equals 100/9, so $\bar{R}_D.\Delta N/\Delta f$ equals 60. Accordingly the same amount of Calcium will reach to about 1½ times the depth calculated for the previous case.

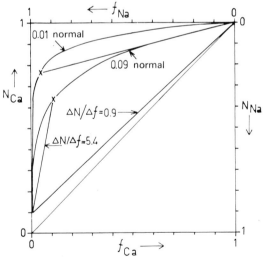

Fig. 7.5. Normalized exchange isotherms following the Gapon equation for mono-divalent exchange with $K_G = \frac{1}{2}$ $(1/mol)^{\frac{1}{2}}$, corresponding to total electrolyte levels of 0.01 normal and 0.09 normal, respectively.

Though the examples given are rather extreme, they demonstrate the principle of this type of calculations. They also point to the fact that the preference for Ca even at 0.1 normal electrolyte is so high, that only a very small fraction remains in solution as compared to the amount adsorbed. The extremely high values of N' at low values of N (cf. figure 7.5) imply that at the 'toe' of the front the 'steepening' effect is very strong. This will effectively offset the front spreading due to diffusion/dispersion. Together these considerations lead to the conclusion that a penetrating Ca-front is a very sharp front with only a minute fraction present in the solution phase.

7.4.2. The sodication process

Inasmuch as de-sodication constitutes an extreme example of favorable exchange (leading to a sharp front during passage), the sodication process must be an extreme example of non-favorable exchange. Though the spreading due to diffusion/dispersion is now not counteracted, its effect remains negligible in comparison to the severe front spreading caused by the very low values of N'_{Na} at the 'toe'-end of the penetrating Na-front. Making use again of figure 7.5, now reading $N_{Na} \equiv 1 - N_{Ca}$, etc. one finds for $C = 0.01$ normal, that f_{Na} equal to 0.95 leads to $N_{Na} = 0.23$. Thus a leaching solution containing 0.0095 normal Na-salts and 0.0005 normal Ca-salts will give 23% Na-ions on the exchange complex.

This composition may also be calculated from the original Gapon equation, yielding:

$N_{Na}/[1-N_{Na}] = \frac{1}{2} \cdot 0.0095/\sqrt{0.00025} = 0.30$ or $N_{Na} = 0.23$.

The value of $\Delta N_{Na}/\Delta f_{Na}$ over the range from 0 to 23% equals $0.23/0.95 = 0.24$. An estimate for the 'toe' region from 0 to 5% gives N'_{Na} at about 0.10. Leaching the soil with a solution of the above composition then produces a Na-front with $x_p = (V/\theta)/(1+24)$ while the 'toe' with about 2½% Na-ions reaches a depth equal to $(V/\theta)/(1+10)$. Thus for 1 m of irrigation water (penetrating to 3.3 m) one finds 2½% Na-ions on the complex at 30 cm depth, while x_p is about 13 cm.

Without going into further details, the result of the calculation of the entire front according to the above procedure is shown in figure 7.6. It is noted in passing that for non-linear isotherms the curves for N and f will not coincide.

7.4.3. The penetration of trace components into soil

Typical examples are here certain cationic pollutants which are fed into the soil at low concentrations. This situation is fairly simple as these small amounts will hardly influence the overall composition of the soil exchange complex. Thus assuming that the exchange complex has roughly a fixed composition with regard to the major occupants i.e. Ca(+Mg) and Na (+ K), one may relate the amount adsorbed of the minor occupant (i.e. the pollutant) to that of a major cation. Obviously one then takes the major ion of the same valence as the minor one. Thus for a divalent cation one finds with the help of the Kerr equation $\overset{+}{\gamma}_{2+}/\overset{+}{\gamma}_{Ca} = K \cdot c_{2+}/c_{Ca}$, and the distribution ratio for this cation then equals:

$$R_{D,2+} \equiv \overset{+}{\gamma}_{2+}/Wc_{o,2+} = K \cdot \overset{+}{\gamma}_{Ca}/Wc_{o,Ca} = K \cdot R_{D,Ca}$$

As $R_{D,Ca}$ is fixed (for this major cation), the distribution ratio for the minor cation is also constant during leaching. The transport of traces of 'foreign' cations thus follows a linear exchange isotherm and the foreign cation moves in with a step front (if it is introduced in that manner). Using as an example the movement of radio-strontium, a much feared pollutant in the case of nuclear explosions, one may thus conclude that:

$$x_{p,Sr} = (V/\theta)/(1+K_{Sr/Ca} \cdot R_{D,Ca}).$$

In the above cited example (C = 0.01 normal, γ = 20 meq/100 g), $R_{D,Ca}$ for a soil nearly saturated with divalent cations may be taken at about 100. Using 0.3 m of leaching per year, at θ = 0.3, this implies (K ≈ 1!) that the added Sr moves about 1 cm per year, 99% being located on the adsorption complex and 1% in the accompanying solution. According to this estimate it

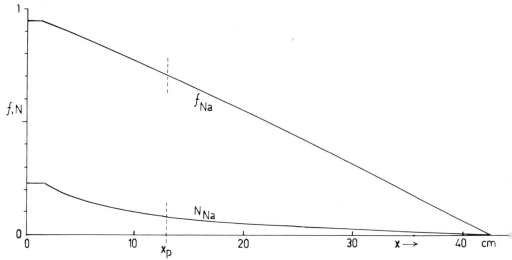

Fig. 7.6. Distribution of Na-ions in the profile following the application of 1m of a solution containing 0.0095 normal Na-salt and 0.0005 normal Ca-salt, assuming the validity of the exchange isotherm of figure 7.5 and neglecting diffusion/dispersion.

would take 100 years to reach groundwater at 1 m depth. In this case, however, diffusion and dispersion will contribute significantly. Using the estimate for the front 'width' as given in section 7.3.3 one would find that the actual front after 100 years would be spread from 0.8 m to 1.2 meter.

As may be shown (cf. part B), the very low percolation rate in this case will cause the diffusion effect to actually exceed the dispersion effect, leading to roughly double the width of the front. In actuality this implies that in the present case one could expect that a concentration of about 10% of the feed value would reach to about 1½ m depth.

7.5. SOME CAUTIONING REMARKS

The above treatment pertains to fairly idealized conditions. In practice many complications may arise, which will be elaborated upon in part B. Obviously a badly cracked soil may lead to instantaneous penetration of water to great depths. In the 'theory' this could be brought out by using larger values for the dispersion coefficient. Another aspect is the rate of equilibration between the adsorption complex and the solution. If the soil is aggregated to large units the rate of exchange will in effect be determined by (slow) diffusion inside the aggregates. This leads again to an increase of the value of the diffusion/dispersion coefficient.

Aside from the above also precipitation reactions between the added component and soil constituents may occur. The ensueing process of 'precipitation chromatography' is difficult to describe in the form of a differential equation. Predictions in such a case will usually involve the numerical solution of a set of conservation equations and flux equations constructed per layer of the profile.

ILLUSTRATIVE PROBLEMS

1. Estimate the penetration of a Zn-front into the soil profile after 10 years immission of ZnO-containing rainfall corresponding to a rate of 125 kg ZnO (Zn = 65, O = 16) per ha per year, using the following data:
   a. Percolation of a sample containing 100 g solid phase and 20 ml solution at a total electrolyte level of 0.02 normal (sample at field capacity) with excess of a normal solution of NH_4NO_3 delivers: T_{Ca} = 17.0 meq, T_{Na} = 2.0 meq, $T_{H,Al}$ = 1.0 meq, T_K = 0.4 meq, T_{Cl} = 0.2 meq, T_{SO_4} = 0.2 meq.
   b. The bulk density of the dry soil equals 1500 kg/m^3.
   c. The net drainage in the area equals 300 mm/year, which excess moves downward on the average at field capacity.

(Answer: Rough guess gives $\bar{R}_D$ = 50, carrier moves down with 1 m per year, so Zn^{2+} has penetrated to about 20 cm in 10 years. Estimating with Gapon's equation the value of $c_{o,Ca}$ at about 0.007 normal at field capacity gives $R_{D,Ca}$ = 130 so, following the divalent species, Zn^{2+} would have penetrated to about 8 cm depth).

2. Calculate the distribution of Zn in the profile for the above situation.
   (Answer: The average concentration of Zn^{2+} in 300 mm drainage water at the above rate of immission of 3keq/ha per year is about 10^{-3} normal. With $R_D \approx 130$ at field capacity this gives $\overset{+}{\gamma}_{Zn} \approx 2.6$ meq/100 gr = 13% exchangeable Zn. Over a depth of about 8 cm this corresponds again to the total immission of 30 keq/ha in 10 years.)

CHAPTER 8

CHEMICAL EQUILIBRIA AND SOIL FORMATION

N. van Breemen and R. Brinkman

8.1. INTRODUCTION

8.1.1. Soil formation and soil forming factors

Soils may be studied from different points of view. A soil, or a soil horizon, is considered here as the result of processes acting on an assemblage of mineral and organic matter at the earth's surface. Generally, the organic matter and part of the minerals in soils are thermodynamically unstable, since they originated under conditions different from those prevailing at present (e.g. at a higher pressure and temperature, or under strongly reduced conditions, cf. also section 1.1).

In order to arrive at a working model for the study of chemical aspects of soil formation a soil will be treated here as a rigid framework of mineral particles in contact with mobile liquid and gas phases. Thus, the effects of biological homogenization or nutrient cycling, or physical movement of soil particles (as in the formation of an argillic horizon) will be left out of consideration. Instead, the discussion will emphasize chemical changes of materials in situ. An example of such changes is the oxidation of organic matter (which consists of highly reduced compounds of C, H, O, N, S and P) in the presence of oxidants such as oxygen or manganese and ferric oxides. Another example is the weathering of 'primary' minerals (e.g. feldspars, augites, hornblendes and micas). These were formed at temperatures and pressures considerably higher than under earth-surface conditions and may subsequently decompose in the soil by hydrolysis, resulting in the formation of secondary phases such as clay minerals.

Soil formation is usually considered as the resultant of the factors climate, relief, gravity and biosphere, acting on a parent material in the course of time. The relation of these different soil forming factors to the present approach on the basis of chemical processes will be touched upon briefly. As to the parent material, attention will be focused almost exclusively on the inorganic fraction, consisting mainly of silica, silicates, ferric oxides and in certain cases aluminum oxides, carbonates, gypsum and more soluble salts. With respect to climate, especially temperature and rainfall are important. Both high temperature and sufficient moisture favor mineral transformations,

whether purely chemical or biologically mediated. Climate and relief determine to a large extent whether the net water movement in the profile is downward (precipitation > evapotranspiration), upward (precipitation < evapotranspiration) or nil (precipitation = evapotranspiration). Water is indeed essential for weathering and soil formation for four main reasons:
a. water participates in most weathering reactions;
b. water may contain a number of dissolved substances (such as carbonic acid, other mineral acids, complexing organic compounds) that are active in weathering processes;
c. saturation with water inhibits diffusion of gases and may lead to soil reduction;
d. flowing water carries away or introduces dissolved or suspended matter and thus leaches or enriches the weathering profile. The latter may be the case under ponded conditions or if the ground water is near to the surface while evapotranspiration exceeds precipitation.

The biosphere is responsible, directly or indirectly, for most other weathering agents:
a. all the oxygen of the atmosphere, which brings about the oxidative conditions prevailing in most weathering environments, was evolved by photosynthesis;
b. carbon dioxide, quantitatively the most important 'agressive' weathering agent, is produced during respiration of plants and animals and combustion of fuels;
c. metabolic uptake of certain elements by plants and micro-organisms induces a strong driving force for reactions leading to the release of these elements into the soil solution;
d. organic matter produced in the biosphere may serve as a reducing agent, and certain water-soluble organic substances, formed by decomposition of organic matter or excreted by plants and micro-organisms, are capable of complexing and mobilizing metal ions;
e. many micro-organisms catalyze oxidation and reduction processes that would proceed at extremely slow rates in their absence.

The effects of hydrosphere and biosphere are of tremendous importance to practically all weathering phenomena. Weathering comes to a virtually complete standstill if water and living matter cannot exert their influences. Deserts and the moon landscape are examples of such conditions.

8.1.2. The use of water analysis in the study of soil formation

Most soil forming processes are so slow that their effects are demonstrable by morphological, chemical or mineralogical studies of soils and soil samples only after hundreds or thousands of years. The relatively low solubility of most soil minerals is often responsible for the sluggish character of soil formation. Only few processes, such as soil reduction, may produce clear morphological effects within a few months. The composition of the interstitial water, however, is modified rapidly and considerably, even by the slow dissolution of silicates and other minerals. This suggests an important and very effective tool for the study of weathering processes: the use of water analysis. If concentrations are recalculated into activities, ground water samples can be used to check for equilibrium, supersaturation or undersaturation with respect to different minerals, and thus indicate what minerals may be formed or may dissolve. Due to the very high mass ratio between solid and dissolved matter in any soil-water or rock-water system, immeasurably small changes in the amount of solid matter may result in easily detectable changes in the composition of the interstitial water. This method has the additional advantage that the observed changes refer to actual (present) processes. In contrast, comparison of weathered material (soil) with parent rock always involves an assumption about the homogeneity of the parent material that is often difficult to check. Additionally, the effects of former (fossil) soil forming conditions may be reflected in the present soil profile. For these reasons, the main emphasis in this chapter will be on interactions between minerals and aqueous solutions.

8.1.3. A landscape model

The chemical aspects of soil formation will be discussed here in relation to a hypothetical toposequence (cf. figure 8.1). This toposequence is characterized by steep slopes at the highest elevations (A), a gently or very gently sloping pediment surface or piedmont plain (B), an essentially level old terrace (C) and a recent floodplain with levees (D) and basins (E) in the lowest parts.

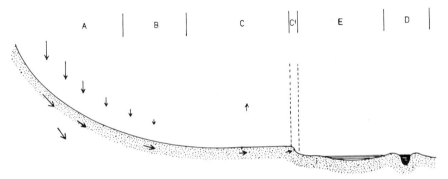

Fig. 8.1. Hypothetical toposequence of soils. Arrows: net flow of water.

As a rule the rainfall increases with increasing elevation. On the other hand, temperature and evapotranspiration generally decrease in the same direction. Therefore it will be assumed that at A the precipitation exceeds evapotranspiration, i.e. the moisture regime is leaching, and that non-leaching and evaporative regimes prevail at B and C, respectively. Seasonal inundation may occur at C (flooding by rain water) and at E (flooding by river water). Depending on the nature of the deeper strata, the subsurface water flow leading from A may move to very great depths, or the water may reappear near to the surface at B, C or E. Dissolved constituents carried away from site A may thus influence the soils at lower elevations.

In tropical and subtropical regions, this landscape configuration may give rise to soil catenas (toposequences of soils) characterized by deeply weathered Oxisols or Ultisols at A. In Oxisols, most alkali and earth alkali ions and a large part of the silica derived from silicates have been leached away, leaving a residue of iron and/or aluminum oxides, the two-layer clay mineral kaolinite and some quartz. If the climate is characterized by a distinct dry season, less extreme weathering ensues. Also, clay migration is likely to occur, often leading to the development of Ultisols.

Under a non-leaching or slightly evaporative regime, soils rich in smectite* (Vertisols) may develop, especially if the ground water contains appreciable amounts of silica and magnesium. If the climate is semi-arid or arid, saline and/or sodic soils (Aridisols) may be formed at B and C (due to evapotranspiration of water draining from A) and at E, along the fringes of the river basins (due to evapotranspiration of water from river floods).

In temperate climates, soils typically found at A may be Spodosols (podzols), Alfisols or Inceptisols, depending on rainfall, temperature and parent material. Under a non-leaching regime, Mollisols (chernozems), characterized by a thick black A_1 horizon relatively rich in organic matter, may be formed at B.

Both in the tropics and in temperate climates, seasonal flooding or water saturation may give rise to gley soils at E and to pseudogley soils at C. Gley soils have a perennially water-saturated and reduced subsoil, while alternating reduction and oxidation take place near the surface. In pseudogley soils the subsoil remains oxidized throughout the year, but alternating reducing and oxidizing conditions occur at shallower depths. Peat soils will develop in basins if these are perennially wet (e.g. at E). In topographic positions as at C and C', iron-rich ground water draining from A may give rise to horizons with an absolute accumulation of iron oxides (generally hard nodules or

---

*smectites are a group of swelling three-layer clay minerals including, e.g., montmorillonite and beidellite.

indurated horizons, both in the tropics and in temperate climates).

Although the situations sketched above are highly simplified, and though complete sequences are relatively rare, the landscape model is a useful conceptual framework for the discussion of the chemical aspects of soil formation.

In section 8.2, mineral weathering reactions will be discussed in order to explain the main soil forming processes at A and the composition of the soil solution. Section 8.3 deals with oxidation and reduction processes helpful in unravelling the formation of gley, pseudogley and peat soils. The so-called 'reverse weathering' processes are dealt with in section 8.4. The latter are useful in understanding the genesis of vertisols, saline and sodic soils, and iron oxide accumulations.

8.2. WEATHERING OF SOIL MINERALS

8.2.1. Congruent and incongruent dissolution

At A (figure 8.1) the interactions between rain water and soil minerals are particularly important. Rain water normally contains only a few ppm of dissolved matter, but the solution quickly becomes more concentrated after entering the soil. This is mainly due to the dissolution of CO_2 (in the soil the CO_2 pressure is normally much higher than in the open atmosphere) and the solvent action of $H_2CO_3^*$ [1] thus formed. If the parent material is calcareous, complete (congruent) dissolution of calcite may occur, according to the reaction:

$$CaCO_3 + H_2CO_3^* \rightarrow Ca^{2+} + 2HCO_3^- \tag{8.1}$$

In basalt, congruent dissolution of the common mineral forsterite (an Mg-olivine) would take place as follows:

$$Mg_2SiO_4 + 4H_2CO_3^* \rightarrow 2Mg^{2+} + 4HCO_3^- + H_4SiO_4 \tag{8.2}$$

Both calcite and forsterite produce equivalent amounts of metal ions and bicarbonate, but the silicate also releases dissolved silica as H_4SiO_4.

Most silicates are aluminous and generally show a less simple type of weathering. Examples are the weathering of K-feldspar to kaolinite or to gibbsite, which may be represented by:

$$2KAlSi_3O_{8\,(K\text{-feldspar})} + 2H_2CO_3^* + 9H_2O \rightarrow$$
$$Al_2Si_2O_5(OH)_{4\,(kaolinite)} + 4H_4SiO_4 + 2K^+ + 2HCO_3^- \tag{8.3}$$

---

[1] $H_2CO_3^*$ refers to the sum of hydrated and unhydrated dissolved CO_2, cf. section 6.1.

and:

$$KAlSi_3O_{8\,(K\text{-feldspar})} + H_2CO_3^* + 7H_2O \rightarrow$$

$$Al(OH)_{3\,(gibbsite)} + 3H_4SiO_4 + K^+ + HCO_3^- \qquad (8.4)$$

In these reactions, equivalent amounts of cations and bicarbonate ions are formed, together with uncharged H_4SiO_4 - as in reaction 8.2 - but in reactions (8.3) and (8.4) solid residues are left behind (incongruent dissolution). The residual minerals contain less K^+ and silica and more aluminum than the feldspar. This illustrates the low mobility of aluminum in most soils: Al is conserved in the solid phase.

Fe is also conserved in the solid phase in most well-drained conditions. Although ferrous iron is rather mobile, it is readily oxidized to ferric iron, which forms oxides with very low solubility. The weathering of fayalite (an Fe(II)-olivine), for example, may proceed as follows:

$$Fe_2SiO_4 + 4H_2CO_3^* \rightarrow 2Fe^{2+} + 4HCO_3^- + H_4SiO_4$$

$$2Fe^{2+} + \tfrac{1}{2}O_2 + 2H_2O \rightarrow Fe_2O_3 + 4H^+$$

$$4H^+ + 4HCO_3^- \rightarrow 4H_2CO_3^*$$

$$\underline{\hspace{10cm}} +$$

$$Fe_2SiO_4 + \tfrac{1}{2}O_2 + 2H_2O \rightarrow Fe_2O_3 + H_4SiO_4 \qquad (8.5)$$

This weathering reaction is 'neutral', i.e. less 'alkaline' than the silicate reactions discussed earlier, because the HCO_3^- released is consumed by the oxidation of Fe^{2+}. An example of a truly 'acidic' reaction is the oxidation of a ferrous sulfide such as pyrite:

$$2FeS_2 + 7\tfrac{1}{2}O_2 + 4H_2O \rightarrow Fe_2O_3 + 8H^+ + 4SO_4^{2-} \qquad (8.6)$$

8.2.2. Solubility and stability relationships

As was discussed in sections 2.4 and 2.6, the solubility of all minerals can be calculated if the standard free enthalpies of formation of all relevant substances or the thermodynamic equilibrium constants for the reactions are known. Table 8.1 lists a number of important reactions and the corresponding thermodynamic equilibrium constants. Figure 8.2 shows the solubility curves for some minerals as a function of pH calculated from tables 8.1 and 6.1. With the exception of gypsum, quartz, gibbsite and kaolinite, the solubility of the minerals shows a monotonic increase with a decrease in pH. The circles denote the solubility and the equilibrium pH at $P_{CO_2} = 10^{-2}$ bar (as found by substituting the respective equation for the solubility line in equation (8.14), at $P_{CO_2} = 10^{-2}$ bar). Under these conditions primary minerals such as albite, anorthite and forsterite appear to be more soluble than gypsum! The secondary minerals gibbsite, kaolinite, Mg-beidellite (a Mg-saturated smectite with a high structural Al content) and, to a lesser extent, calcite have considerably lower solubilities. Especially congruently dissolving minerals

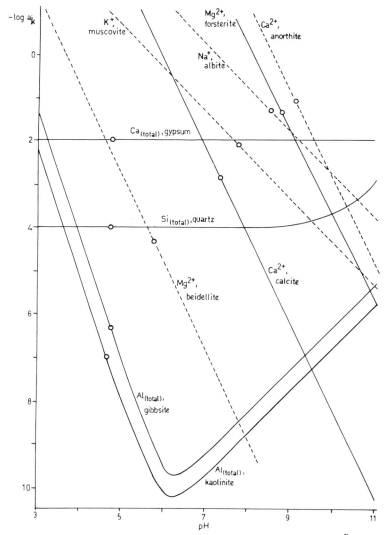

Fig. 8.2. Solubility of various minerals in aqueous solution at 25°C, 1 bar total pressure, $P_{CO_2} = 10^{-2}$ bar and $(H_4SiO_4) = 10^{-3}$.
Solid lines: congruent dissolution; Broken lines: incongruent dissolution; Circles: equilibrium pH without added acid or base; Al(total) = (Al^{3+}) + $(AlOH^{2+})$ + $(Al(OH)_4^-)$; Ca(total) = (Ca^{2+}) + $(CaSO_4^\circ)$; Si(total) = $(H_4SiO_4^\circ)$ + $(H_3SiO_4^-)$.

TABLE 8.1

Equilibrium constants for the hydrolysis of silicates and various oxides and hydroxides at 25 °C and 1 bar total pressure (values after Helgeson 1969, *Amer.J.Sci.* 267 p.729 and Christ et al 1973, *Amer.J.Sci.* 273, p.65).

Nr.	Mineral	Reaction	log K°
1.	amorphous SiO_2	$SiO_2 + 2H_2O \rightleftharpoons H_4SiO_4°$	−2.70
2.	quartz	$SiO_2 + 2H_2O \rightleftharpoons H_4SiO_4°$	−4.00
3.	microcline	$KAlSi_3O_8 + 4H^+ + 4H_2O \rightleftharpoons K^+ + Al^{3+} + 3H_4SiO_4°$	1.29
4.	high albite	$NaAlSi_3O_8 + 4H^+ + 4H_2O \rightleftharpoons Na^+ + Al^{3+} + 3H_4SiO_4°$	5.00
5.	anorthite	$CaAl_2Si_2O_8 + 8H^+ \rightleftharpoons Ca^{2+} + 2Al^{3+} + 2H_4SiO_4°$	24.69
6.	kaolinite	$Al_2Si_2O_5(OH)_4 + 6H^+ \rightleftharpoons 2Al^{3+} + H_2O + 2H_4SiO_4°$	7.63
7.	muscovite	$KAl_3Si_3O_{10}(OH)_2 + 10H^+ \rightleftharpoons K^+ + 3Al^{3+} + 3H_4SiO_4°$	17.05
8.	illite	$K_{0.6}Mg_{0.25}Al_{2.3}Si_{3.5}O_{10}(OH)_2 + 8H^+ + 2H_2O \rightleftharpoons 0.6 K^+ + 0.25Mg^{2+} + 2.3Al^{3+} + 3.5H_4SiO_4°$	10.34
9.	annite	$KFe_3AlSi_3O_{10}(OH)_2 + 10H^+ \rightleftharpoons K^+ + 3Fe^{2+} + Al^{3+} + 3H_4SiO_4°$	23.3
10.	Mg-chlorite	$Mg_5Al_2Si_3O_{10}(OH)_8 + 16H^+ \rightleftharpoons 5Mg^{2+} + 2Al^{3+} + 3H_4SiO_4° + 6H_2O$	73.2
11.	K-beidellite	$3K_{0.33}Al_{2.33}Si_{3.67}O_{10}(OH)_2 + 22H^+ + 8H_2O \rightleftharpoons K^+ + 7Al^{3+} + 11H_4SiO_4°$	18.32
12.	Mg-beidellite	$6Mg_{0.167}Al_{2.33}Si_{3.67}O_{10}(OH)_2 + 44H^+ + 16H_2O \rightleftharpoons Mg^{2+} + 14Al^{3+} + 22H_4SiO_4°$	36.60
13.	clino-enstatite	$MgSiO_3 + 2H^+ + H_2O \rightleftharpoons Mg^{2+} + H_4SiO_4°$	11.36
14.	diopside	$CaMg(SiO_3)_2 + 4H^+ + 2H_2O \rightleftharpoons Ca^{2+} + Mg^{2+} + 2H_4SiO_4°$	19.68
15.	talc	$Mg_3Si_4O_{10}(OH)_2 + 6H^+ + 4H_2O \rightleftharpoons 3Mg^{2+} + 4H_4SiO_4°$	19.01
16.	chrysotile	$Mg_3Si_2O_5(OH)_4 + 6H^+ \rightleftharpoons 3Mg^{2+} + 2H_4SiO_4° + H_2O$	32.20
17.	fayalite	$Fe_2SiO_4 + 4H^+ \rightleftharpoons 2Fe^{2+} + H_4SiO_4°$	17.22
18.	forsterite	$Mg_2SiO_4 + 4H^+ \rightleftharpoons 2Mg^{2+} + H_4SiO_4°$	29.28
19.	brucite	$Mg(OH)_2 + 2H^+ \rightleftharpoons Mg^{2+} + 2H_2O$	16.78
20.	sepiolite	$Mg_4Si_6O_{15}(OH)_2 \cdot 6H_2O + H_2O + 8H^+ \rightleftharpoons 4Mg^{2+} + 6H_4SiO_4°$	31.8
21.	gibbsite	$Al(OH)_3 + 3H^+ \rightleftharpoons Al^{3+} + 3H_2O$	7.69
22.	cryptocryst. gibbsite	$Al(OH)_3 + 3H^+ \rightleftharpoons Al^{3+} + 3H_2O$	9.23

yielding species containing Al and Fe(III) (gibbsite, kaolinite and, for example, hematite), give very low equilibrium concentrations at a near-neutral pH. Figure 8.2 is in agreement with the fact that rocks rich in Ca-feldspar and olivine (basalt, for example) weather more rapidly than felsic rocks (e.g. granite) consisting mainly of quartz, muscovite and Na-feldspars, and that phases such as gibbsite, kaolinite and hematite or goethite are characteristic weathering residues.

Although a calculated solubility is a useful yardstick, it does not necessarily correspond to the actual 'weatherability' of a mineral, because it gives no information about the reaction kinetics. Figure 8.2 shows, for example, that gypsum and muscovite have approximately equal solubilities at $P_{CO_2} = 10^{-2}$ bar and near-neutral pH. In fact, muscovite is very inert and weathers much more slowly than gypsum. Physical aspects related to the exposed surface area (degree of compactness, fracturing, grain size, etc.) are also very important in this respect (cf. section 2.9).

Besides focusing attention on the different solubilities of individual minerals, the solubility data may be used for the construction of diagrams that show the stability relationships between various minerals. This can be done conveniently by studying the minerals and the aqueous solution in the system Al_2O_3-SiO_2-H_2O-XO or Y_2O, where Y stands for Na or K and X for Ca or Mg.

As can be seen from figure 8.2, in the pH range typical for soils (4 to 9) the concentrations of dissolved aluminum are very low compared to those of most other solutes. Therefore, essentially all Al is conserved in the solid phase and the reactions necessary for describing the stability relationships are of the type:

mineral I + H^+ ⇌ mineral II + Y^+ + aH_4SiO_4

Hence the thermodynamic equilibrium constants contain the variables $\log(Y^+)^*$ + pH, or $\log(X^{2+})$ + 2pH, and $\log(H_4SiO_4)$. This permits the use of two-dimensional diagrams.

Although $H_2CO_3^*$ is the most important weathering agent in nature, it is more useful to write reactions such as represented by this equation with H^+. This yields the readily measurable pH as one of the variables, and moreover makes the resulting diagram more general in that other proton donors besides $H_2CO_3^*$ may be considered (e.g. stronger inorganic acids, complexing organic acids). The use of pe in pe-pH diagrams has a similar advantage: the oxidation-reduction intensity represented by pe is independent of the particular electron donors and electron acceptors that may be present.

Figure 8.3 is a stability diagram for the system Al_2O_3-SiO_2-H_2O-K_2O, with the phases microcline (K-feldspar), muscovite (K-mica), kaolinite, cryptocrystalline gibbsite and a hypothetical K-beidellite. The field boundaries

---

*(Y^+) denotes the activity of Y^+.

were calculated from the equilibrium constants and reactions in table 8.1. At the boundary of microcline and kaolinite, for example, both minerals are in equilibrium with the same solution (and thus with each other). Hence, reactions 3 and 6 from table 8.1 must apply simultaneously. Combining both reactions in such a way that no dissolved Al^{3+} appears in the final reaction equation, one finds:

$$
\begin{array}{ll}
2\,\text{Microcline} + 8H^+ + 8H_2O \rightleftharpoons 2K^+ + 2Al^{3+} + 6H_4SiO_4 & 2\times 1.29 \\
2Al^{3} + 2H_4SiO_4 + H_2O \rightleftharpoons \text{Kaolinite} + 6H^+ & -7.63 \\
\hline
2\,\text{Microcline} + 2H^+ + 9H_2O \rightleftharpoons \text{Kaolinite} + 2K^+ + 4H_4SiO_4 & -5.05
\end{array}
\qquad (8.7)
$$

with the thermodynamic equilibrium constant:

$$\log K^\circ = -5.05 = 2\log(K^+) + 2pH + 4\log(H_4SiO_4) \qquad (8.8)$$

Besides the boundaries of the various mineral stability fields (solid lines),

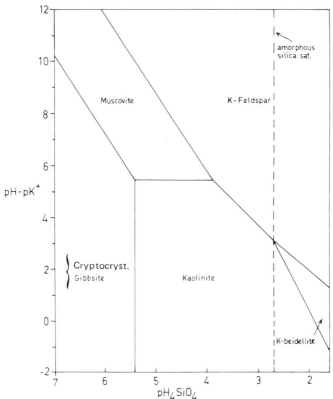

Fig. 8.3. Stability diagram of minerals in the system Al_2O_3-SiO_2-K_2O-H_2O at 25 °C and 1 bar total pressure.

figure 8.3 also indicates the H_4SiO_4 activity in equilibrium with amorphous silica (broken line). In most natural waters the silica activity is less than this value. Amorphous silica would readily precipitate at higher activities (reaction 1, table 8.1).

A similar diagram for the system Al_2O_3-SiO_2-H_2O-MgO with the minerals Mg-chlorite, Mg-beidellite, kaolinite and gibbsite is given as figure 8.4. For example, the boundary between the stability fields of Mg-beidellite and kaolinite is found by combining reactions 6 and 12 from table 8.1:

$$6\text{Mg-beidellite} + 2H^+ + 23H_2O \rightleftharpoons 7\text{Kaolinite} + Mg^{2+} + 8H_4SiO_4 \qquad (8.9)$$

$$\log K° = 36.6 + 7 \times (-7.63) = -16.8 =$$

$$\log(Mg^{2+}) + 2pH + 8\log(H_4SiO_4) \qquad (8.10)$$

The diagrams indicate that under relatively alkaline conditions and high K^+ and Mg^{2+} activities, K-feldspar and Mg-chlorite are stable relative to the

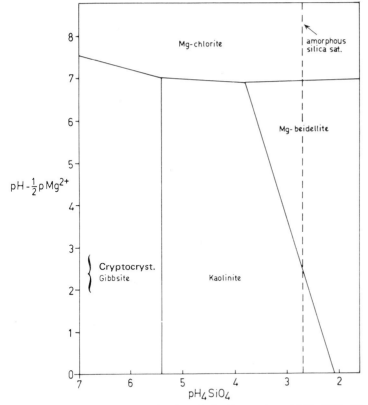

Fig. 8.4. Stability diagram of minerals in the system Al_2O_3-SiO_2-MgO-H_2O at 25 °C and 1 bar total pressure.

other minerals under consideration. At a lower pH and lower K^+- and Mg^{2+}-activities, however, they are unstable with respect to kaolinite or beidellite, depending on the silica concentration. Mg-beidellite becomes unstable with respect to kaolinite with decreasing (H_4SiO_4), (Mg^{2+}) and pH. Only at very low silica concentrations does gibbsite become stable with respect to kaolinite.

Low activities of H_4SiO_4 and K^+ or Mg^{2+} may be expected if strong leaching occurs. Upon removal of K^+ and other cations, HCO_3^- (produced from CO_2 during weathering) is removed simultaneously, so that the pH decreases. Thus in well or excessively drained soils, weathering sequences such as K-feldspar → kaolinite → gibbsite or Mg-chlorite → kaolinite → gibbsite may be expected. The Oxisols (Latosols) of the tropics are typical examples of late stages in such sequences. Where drainage is restricted, dissolved silica, K^+ and Mg^{2+} will not readily reach very low values and weathering may be slowed down or proceed differently. If, however, some special process is responsible for the production of acid (e.g. H_2SO_4 from oxidizing pyrite), intensive weathering may occur even in the absence of leaching. The theoretical mineral sequences discussed above are indeed found in soils.

A 'normative mineral composition' may be calculated from the total chemical analysis according to a simple set of rules (the Norm) which depends on the type of material. The calculation often is a gross simplification of the real system, but may point to soil forming processes that might otherwise have remained obscure. The normative mineral compositions of two well developed Oxisols, one from Angidapuram (India) derived from a hornblende granulite and the other from the Kukusan mountains (S.E. Kalimantan, Indonesia), derived from a serpentinized peridotite rich in chrysotile and clinoenstatite, as shown in tables 8.2 and 8.3 suggest the following.

TABLE 8.2.

Normative mineral composition (weight per cent) of the Angidapuram profile (after Mohr, van Baren and van Schuylenborgh 1972, Tropical soils).

Horizon	Q	Or	Plg	Bi + Ho	Mt	Ill	Kaol	Go	Misc.
I	29	-	-	-	-	10	28	27	5
II	27	-	-	-	-	4	51	14	4
V	16	12	2	10	-	37	-	12	11
VI	3	42	41	10	2	-	-	-	2

I = upper zone; II = laterite horizon; V = 'rotten rock'; VI = unweathered rock (= hornblende granulite). Q = quartz; Or = K-feldspar; Plg = plagioclase; Bi = biotite; Ho = hornblende; Mt = magnetite; Ill = illite; Kaol = kaolinite; Go = goethite; Misc. = miscellaneous.

TABLE 8.3.

Normative mineral composition (weight per cent) of the Kukusan profile (after Mohr, van Baren and van Schuylenborgh 1972, Tropical soils).

Horizon	Serp	Pyr	Sp	Mt	Q	Chl + Mm	Kaol	Gibb	Go	Misc.
A_1	-	-	-	-	-	-	16	8	74	2
B_1	-	-	-	-	-	1	6	18	71	4
B_2	-	-	-	-	-	3	3	17	74	4
B_3	-	-	-	-	-	22	2	4	68	4
C_1	-	-	-	-	6	33	-	-	57	4
C_2	13	-	-	-	22	23	-	-	40	3
R	44	40	4	7	-	-	-	-	4	1

Serp = serpentine (chrysotile); Pyr = pyroxenes (clino-enstatite); Sp = spinell; Mt = magnetite; Q = amorphous SiO_2; Chl = chlorite; Mm = montmorillonite; Kaol = kaolinite; Gibb = gibbsite; Go = goethite.

In both soils the primary minerals have disappeared to a large extent in the horizons just above the parent rock. The secondary minerals formed initially are chlorite and montmorillonite in the Mg-rich parent material, and illite (closely related to K-mica) on the K-rich rock. Higher in the profile, these are replaced in their turn by kaolinite (Angidapuram) and gibbsite (Kukusan). Both the appearance of gibbsite and the strong enrichment in ferric oxide (goethite) in the Kukusan profile point to extreme leaching. The soils discussed above are very old (of the order of several 100.000 of years). Similar, but less extreme, effects are also observed in younger soils in the tropics.

8.2.3. The concept of partial equilibrium

Diagrams such as figures 8.3 and 8.4 refer primarily to overall equilibria. They predict, for example, that the minerals of the following pairs are incompatible in a stable environment: K-beidellite and muscovite, gibbsite and feldspar, gibbsite and beidellite, gibbsite and amorphous SiO_2. They can be used for partial equilibrium systems as well, however. In a cation exchange experiment on a clay mineral, for example, one would not normally be concerned about the long-term effect of the slow transformation of the clay mineral which may be thermodynamically unstable. The clay mineral structure would generally be considered as stable for the duration of the experiment, while equilibrium is attained rapidly between the ions on the exchanger and in solution. Similarly, in a soil containing gibbsite or kaolinite in the clay fraction, the soil solution may be close to equilibrium with these phases, and at the same time far undersaturated with respect to the occasional sand-size grain of potassium feldspar that may be present. The partial equilibrium of the soil solution with gibbsite and kaolinite is a valid approximation in such a case. One normally works with partial equilibria whenever the reaction rates of minerals not in equilibrium within the system are too low to attain overall equilibrium within the time span considered.

Moreover, the diagrams may be used to indicate the pathways of irreversible reactions, by assuming partial equilibrium with one or more of the minerals produced. If a K-feldspar is placed in pure water (pH 7), the initial release of Al(OH)$_4^-$ (along with K$^+$ and H$_4$SiO$_4$) will eventually lead to saturation with, and precipitation of, gibbsite. Then the solution composition falls somewhere in the gibbsite field of figure 8.3. From then onwards, further dissolution of K-feldspar will proceed incongruently leaving a residue of gibbsite. At the same time (H$_4$SiO$_4$) and (K$^+$) continue to increase, so that eventually the gibbsite--kaolinite boundary will be reached. As long as (partial!) equilibrium between gibbsite and kaolinite is maintained, the solution composition remains along the gibbsite-kaolinite boundary: i.e. (H$_4$SiO$_4$) is fixed. The H$_4$SiO$_4$ released from the K-feldspar is used to transform gibbsite into kaolinite. Only after all gibbsite formed initially has been consumed, the soil solution composition will leave the gibbsite-kaolinite boundary and traverse the kaolinite field in the direction of the feldspar field, which will be reached eventually (overall equilibrium). This situation may occur in nature, for example in weathering rock, where drainage is impeded. Normally, however, the solution remains permanently undersaturated with respect to a weathering primary mineral. The degree of undersaturation depends to a large extent on the leaching rate.

8.2.4. Weathering products

Returning to location A in figure 8.1, one may conclude that under conditions of fairly restricted drainage, rock weathering leads to the development of various secondary phases such as illite, smectite or chlorite, depending on the kind of parent material and the leaching rate. In the same location, but under conditions of moderate to excessive leaching, kaolinite or gibbsite will be formed almost irrespective of the type of parent material. In the latter case, parent materials are important mainly since they determine the amounts of minerals that may be residually enriched, such as ferric and aluminum oxides, titanium oxide as well as coarse-grained quartz.

The effects of weathering are best expressed on gentle slopes (where the rate of production of soil material from parent rock exceeds the rate of erosion) under a strongly leaching regime. Such optimal weathering conditions give rise to Oxisols, which show little profile differentiation apart from the development of a weakly expressed A$_1$ horizon on top of the weathered material (the B horizon).

On steeper slopes, soils generally have a 'younger' character due to the continual removal by erosion of solid weathering products. If the climate is characterized by a dry season and if the mechanical stability of the soil is not high (unlike in Oxisols, which are stable due to cementation by ferric oxides), textural B horizons may be formed, giving rise to Ultisols.

8.2.5. Decay of organic matter, humification and chelation

The decay of freshly produced organic matter involves complete or partial oxidation, resulting in the formation of carbon dioxide and water or intermediate products, such as soluble organic acids. These comprise low-molecular aliphatic (e.g. citric, oxalic) and aromatic acids (e.g. vanillic, hydroxy benzoic), as well as 'fulvic acids' with much higher molecular weights and containing both phenolic and carboxylic groups. Biological agents are very important in mediating these processes. Mineralization of organic matter (biological oxidation to carbon dioxide and water with liberation of the mineral nutrients) is favored by high temperatures and presence of oxygen and moisture, for example in well-drained tropical soils.

As was mentioned in section 1.1.2 this breakdown process may be accompanied by polymerization and condensation of the decay products into less soluble compounds. The transformation of fresh organic matter to such highly polymerized compounds is termed humification. This process is favored by relatively high activities of divalent (or trivalent) metal cations. Such conditions are found in Mollisols (chernozems) and Vertisols, for example, where there is little if any leaching (location B in figure 8.1). On the other hand, if conditions are unfavorable for mineralization or humification (because of low temperatures, periodic wetness, extremely low nutrient status), strongly chelating, water-soluble organic acids may be formed. These considerably enhance mineral weathering, and cause the preferential dissolution and migration of ferric iron and aluminum characteristic for the formation of Spodosols (podzols). These are common on poor soils in temperate rainy climates, but they also occur on extremely poor parent materials (quartz sands) or in cool mountainous regions in the tropics, particularly where there is seasonal or perennial water saturation.

Many Spodosols occur on relatively young surfaces. The time span available for their formation would be much too short to explain the observed migration of Fe and Al by simple dissolution of their sesquioxides (or, in case of Al, by dissolution from alumino-silicates) in the form of Fe^{3+}, Al^{3+} and their various hydroxy complexes. In the presence of chelating compounds, the concentrations of soluble Fe and Al chelates may be several orders of magnitude higher than the total concentration of dissolved ionic forms. This raises the total solubility of Fe and Al to levels as high as, or higher than that of Si, resulting in their rapid removal from the eluvial horizons. Within the framework of this chelate hypothesis, there are two variants. In one theory, compounds of the water-soluble fulvic acid (F.A.) fraction are invoked (F.A. is part of the humus that is soluble in both alkali and acid). The compounds in question are capable of forming water-soluble complexes with different metals. Complexes with Fe(III) and Al are among the most stable ones. The molar ratio of metal to F.A. may vary from 1 to as high as 6, and the solubility of the complexes decreases as this ratio increases. A fulvic acid solution percolating downwards through a soil profile would form complexes with Fe and Al until a critical metal-F.A. ratio is reached, whereupon the complex would precipitate. Thus, this theory explains the formation of both the eluvial A_2 (albic) horizon and the illuvial B (spodic) horizon. Microbial oxidation of part of the organic matter could further raise the metal-fulvic acid ratio and, especially in the lower part of the B horizon, produce some free sesquioxides. These, Al as well as Fe, have in fact been identified in many cases. Once started, the accumulation process tends to be self-perpetuating, since the free oxides already present tend to precipitate further additions of soluble metal chelates.

In the other variant, water-soluble organic acids of low molecular weight such as citric, oxalic, vanillic and p-hydroxy benzoic acid are assumed to be the important complexers. The presence of such acids has been demonstrated in actual podzols and some of them are known as strong complex formers with Fe and Al. This theory thus explains the removal of Fe and Al from the A_2 horizon. Precipitation in the B horizon is then explained by a presumed increase in pH with depth. If the pH becomes sufficiently high, iron or aluminum oxides would precipitate due to hydrolysis as shown below (H_2L stands for the complexing organic acid):

$$FeL^+ + 3H_2O \rightarrow Fe(OH)_3 + H^+ + H_2L \tag{8.11}$$

or:

$$FeOHL^\circ + 3H_2O \rightarrow Fe(OH)_3 + H_2L \tag{8.12}$$

The fact that the pH increases with depth in most (but not all) podzols suggests that this mechanism might operate in some cases. Experiments indicate that these simple organic acids, too, are partly mineralized in the B horizon.

External factors besides microbial activity and pH aid accumulation in certain cases. Near a stationary interface between (soil) air and soil solution at a textural discontinuity or a lower limit of wetting, for example, continued microbial oxidation may cause considerable accumulation of sesquioxides.

8.2.6. Composition of the soil solution

Whereas solids and dissolved matter are removed from A (figure 8.1) by erosion and leaching, the reverse is the case at B, due to sedimentation and evaporation of ground water. Weathering at A supplies the ground water mainly with dissolved silica, Na^+, K^+, Ca^{2+} and Mg^{2+}. If dissolved CO_2 is the main weathering agent and salt concentrations are relatively low, HCO_3^- is the main dissolved anion (cf. equations (8.1) through (8.4)). If insufficient oxygen is present to oxidize ferrous iron released from primary minerals, Fe^{2+} may also be an important constituent of the ground water. In areas with volcanic activity, acids such as H_2SO_4 and HCl may be more important than CO_2, leading to high concentrations of dissolved SO_4^{2-} and Cl^-. Oxidation of pyrite, which is present in small quantities in most igneous and metamorphic rocks, also gives rise to H_2SO_4 and sulfate in ground waters.

The composition of the soil solution is of particular interest, because it yields information not only about the weathering processes at A, but also about the possible effects of evaporation at B, as discussed in section 8.4 below. If both the primary and the secondary minerals at A are known, the composition of the soil solution may be used to estimate the contributions of the various minerals to the dissolved load. For example, water draining from a pyritic limestone area which contains 0.9 mmole Ca^{2+}, 1.6 mmole HCO_3^- and 0.1 mmole SO_4^{2-} per liter has presumably evolved as a result of the interaction of 0.1 mmole H_2SO_4 (formed by the oxidation of 0.05 mmole FeS_2) and 0.7 mmole dissolved CO_2 with 0.9 mmole $CaCO_3$ per liter of drain water. This can be readily seen from the weathering reactions:

$$0.05 FeS_2 + 0.1875 O_2 + 0.1 H_2O \rightarrow 0.1 H_2SO_4 + 0.025 Fe_2O_3$$

$$0.1 H_2SO_4 + 0.1 CaCO_3 \rightarrow 0.1 SO_4^{2-} + 0.1 Ca^{2+} + 0.1 H_2CO_3^*$$

$$0.8 H_2CO_3^* + 0.8 CaCO_3 \rightarrow 1.6 HCO_3^- + 0.8 Ca^{2+}$$

$$\begin{aligned}&0.05 FeS_2 + 0.1875 O_2 + 0.1 H_2O + \\ &+ 0.7 H_2CO_3^* + 0.9 CaCO_3 \rightarrow 0.025 Fe_2O_3 + 0.1 SO_4^{2-} + \\ & \qquad\qquad\qquad\qquad\qquad\qquad 1.6 HCO_3^- + 0.9 Ca^{2+}\end{aligned} \quad (8.13)$$

Similar relationships may be worked out for waters draining from silicate weathering zones, provided the compositions of the relevant minerals are known. This approach leads to (crude) estimates of the actual weathering rates of various minerals.

Of all constituents of the soil solution, HCO_3^- is of particular relevance, because together with P_{CO_2} it may determine the pH (cf. section 6.1) according to:

$$pH = \log(HCO_3^-) - \log P_{CO_2} + 7.81 \quad (8.14)$$

In uncontaminated rain water, electroneutrality requires that $[HCO_3^-] =$

$= [H^+]^*$. Because activity corrections are negligible one may calculate directly from equation (8.14) that the pH of rain water is 5.65 at the P_{CO_2} of the atmosphere ($10^{-3.5}$ bar). In soils P_{CO_2} is normally higher, but at the same time (HCO_3^-) may be considerably increased due to the interaction between H_2CO_3 and soil minerals (equations (8.3) and (8.4)). If weatherable minerals are present, the net effect is generally a rise in pH above 5.65. Very high pH values (up to 9 or 10) are frequently found in partly weathered rock, still rich in primary minerals (see also the equilibrium pH values in figure 8.2). These are the results of high concentrations of metal cations (Na^+, K^+, Ca^{2+} and Mg^{2+}) and the associated HCO_3^-, as well as the low CO_2 concentrations caused by the consumption (and slow replenishment) of CO_2 during weathering.

8.3. SOIL REDUCTION AND OXIDATION

In section 2.7.2. oxidation-reduction reactions were discussed in general terms, leading to the conclusion that for a given redox couple the position of the equilibrium depends on the locally prevailing value of pe - i.e. the minus logarithm of the (relative) electron activity, cf. p. 33 - as compared to pe° of the redox couple (cf. equations 2.50 and 2.50a). In soils the prevailing pe value may vary over a wide range, bounded on the 'high' side by the simultaneous presence of water and free O_2 under well aerated conditions, and on the 'low' side by water plus free H_2 in water logged soils rich in organic matter (cf. discussion below).

8.3.1. Environmental requirements for soil reduction

If periodic inundation takes place at C (by rain water) or at D (by river water), reduction and oxidation processes need to be considered. Besides water saturation, organic matter and temperatures suitable for microbial activity are prerequisites for soil reduction. All organic matter is highly 'reduced': photosynthesis, trapping light energy and converting it to chemical energy, produces reduced states of higher free energy and thus non-equilibrium concentrations of C, N and S compounds. The production of organic matter from CO_2 and H_2O can be conceived as a process producing localized centres of highly negative pe. For the conversion of CO_2 to glucose the following equations apply:

$$CO_{2(g)} + 4H^+ + 4e^- \rightleftharpoons \tfrac{1}{6} \text{ glucose} + H_2O \tag{8.15}$$

$$pe = \tfrac{1}{4}\log K° - pH + \tfrac{1}{4}\log P_{CO_2} \tag{8.16}$$

---

*$[H^+]$ represents the concentration of H^+ in mol/liter.

Since $\log K° = -0.8$, equilibrium at pH 7 and unit activities of other reactants and products requires $pe = -7.2$. This is very near to the lower stability limit of water, where H_2O will be reduced to H_2.

A number of redox equations relevant to soil reduction are listed in table 8.4. Inspection of this table shows that in addition to electrons, protons are transferred in all the redox reactions mentioned. Therefore it is convenient to depict equilibria diagrams for redox processes in terms of both pe and pH, see figure 8.5. The two dashed lines in this pe-pH diagram show the stability limits of water at 0.2 bar oxygen and 1 bar hydrogen pressure, respectively. Conditions for equal activities in solution of the pairs nitrate--nitrite and sulfate-sulfide are given by dotted lines. The solid lines represent the boundaries for 0.001 molar Mn^{2+} and 0.001 molar total dissolved iron ($Fe^{3+} + Fe^{2+}$) in equilibrium with different manganese and iron oxide minerals. The bottom (dotted) line gives the pe-pH conditions where CH_2O (organic matter) is in equilibrium with CO_2 at 1 bar pressure. The diagram indicates that 'organic matter' is the most reduced of all substances considered. Hence, oxidation of organic matter by O_2, NO_3^-, various iron and manganese oxides, SO_4^{2-} and even by H_2O would be thermodynamically possible.

Figure 8.5 indicates that 'dead' organic matter is bound to be oxidized, e.g. according to equation (8.15) proceeding from right to left. If oxygen is available this may function as electron acceptor, and the overall reaction can be written as:

$$O_2 + \tfrac{1}{6} \text{ glucose} \rightarrow CO_2 + H_2O \tag{8.17}$$

If oxygen is not available, other reducible compounds may function as electron acceptors. Under soil conditions nitrate, Mn(IV), Mn(III), Fe(III) and S(VI) are important in this respect. Reduction of these compounds by organic matter is thermodynamically possible under soil conditions, but most of the reactions are extremely slow. However, many non-photosynthetic micro-organisms catalytically decompose the unstable products of photosynthesis through these energy yielding redox reactions, and can use this energy both to synthesize new cells and to maintain existing ones. Therefore, reduction may take place if the following conditions are met simultaneously: presence of organic matter, absence of an oxygen supply, and presence of anaerobic micro-organisms in an environment suitable for their growth. The requirement that an oxygen supply is absent may be fulfilled if the soil becomes saturated with water: because the diffusion of gases in water is approximately 10^4 times slower than in the gas phase, the exchange of gases is virtually stopped by flooding or water saturation. The soil more than a few mm away from the soil-air interface becomes depleted of oxygen, because this is rapidly consumed by the aerobic organisms and cannot be replenished quickly enough to prevent its virtual disappearance. Generally within one or two days after flooding, essentially all free oxygen has been consumed; the activity of anaerobic micro-organisms increases, and soil reduction starts.

8.3.2. *The sequential appearance of reduction products upon flooding*

Besides stability relationships, figure 8.5 also suggests the sequence of events expected upon flooding an originally aerobic soil. Reduction of remaining oxygen will take place first, followed by nitrate, then manganese in neutral soils; or manganese, then nitrate in acid soils. Later, ferrous iron may appear and still later one may expect formation of sulfide and even hydrogen (and methane - not indicated in figure 8.5). This theoretical sequence is indeed found in nature.

TABLE 8.4.

Equilibrium constants for half reactions important in submerged soils (25 °C).*

Reaction no.	Half reaction	log K°	Source
1.	$\frac{1}{4}O_{2(g)} + H^+ + e^- \rightleftharpoons \frac{1}{2}H_2O_{(l)}$	20.77	1
2.	$H^+ + e^- \rightleftharpoons \frac{1}{2}H_{2(g)}$	0	4
3.	$\frac{1}{2}NO_3^- + H^+ + e^- \rightleftharpoons \frac{1}{2}NO_2^- + \frac{1}{2}H_2O_{(l)}$	14.15	1
4. pyrolusite	$\beta MnO_2 + 4H^+ + 2e^- \rightleftharpoons Mn^{2+} + 2H_2O$	42.2	2
5. manganite	$\gamma MnOOH + 3H^+ + e^- \rightleftharpoons Mn^{2+} + 2H_2O$	24.6	2
6. hausmannite	$Mn_3O_4 + 8H^+ + 2e^- \rightleftharpoons 3Mn^{2+} + 4H_2O$	61.1	2
7. pyrochroite	$Mn(OH)_2 + 2H^+ \rightleftharpoons Mn^{2+} + H_2O$	15.2	2
8. hematite	$\frac{1}{2}Fe_2O_3 + 3H^+ + e^- \rightleftharpoons Fe^{2+} + 1\frac{1}{2}H_2O$	12.57	3
9. magnetite	$Fe_3O_4 + 8H^+ + 2e^- \rightleftharpoons 3Fe^{2+} + 4H_2O$	33.1	3
10.	$SO_4^{2-} + 10H^+ + 8e^- \rightleftharpoons H_2S + 4H_2O$	40.6	3
11.	$SO_4^{2-} + 9H^+ + 8e^- \rightleftharpoons HS^- + 4H_2O$	33.8	3
12. 'organic matter'	$\frac{1}{4}CO_{2(g)} + H^+ + e^- \rightleftharpoons \frac{1}{4}CH_2O + \frac{1}{4}H_2O$	-1.2	1

1: Stumm, W. and Morgan, J.J., 1970. *Aquatic chemistry*. Wiley-Interscience, New York.
2: Mohr, E.C.J., Baren, F.A.. van, Schuylenborgh, J. van, 1972. *Tropical soils*. 3rd. ed. Mouton-Ichtiar Baru-Van Hoeve, The Hague-Paris-Djakarta.
3: Garrels, R.M. and Christ, C.L., 1965. *Solutions, minerals and equilibria*. Harper--Weatherhill, New York-Tokyo.
4: by convention

*Cf. see also Table 6.1.

The reduction of NO_3^- (mainly to N_2) starts immediately after flooding, also because many micro-organisms are able to use nitrate as an electron acceptor. After the disappearance of NO_3^-, the concentrations of Mn^{2+} and, somewhat later, of Fe^{2+} generally show a distinct increase to peak values during the first weeks after the onset of reduction, followed by a gradual decrease to a fairly constant level. The peaks are steeper and higher as the pH is lower and the organic matter content is higher. If sufficient easily oxidizable organic matter is present, the very low pe levels suitable for sulfate reduction or production of marsh gas (a mixture containing methane and hydrogen) may be reached. Under very acid conditions, however (pH below 4 to 4.5), reduction may be slowed down or inhibited, probably because the conditions are unfavorable for the growth of microorganisms.

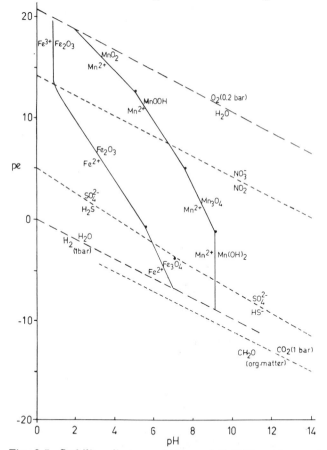

Fig. 8.5. Stability diagram of Mn and Fe(III) oxide species at 10^{-3} molar activities of Mn^{2+} and total dissolved ionic iron ($Fe^{3+} + Fe^{2+}$); lines of equal activities for the couples $NO_3^- - NO_2^-$ and SO_4^{2-} - total sulfide ($H_2S + HS^-$); and stability limit of organic matter at 1 bar CO_2 pressure (25 °C).

8.3.3. Soil reaction and production of alkalinity during reduction

If the reduction process is studied by measuring the redox potential, it will be found that for most soils the Eh^{*1} drops from values around 0.5 to 0.7 Volts under aerobic conditions to values of 0.2 to −0.2 Volts after a period varying between one week to several months. If the content of easily reducible Mn oxides is high or if conditions for microbial growth are unfavorable (low organic matter content, extremely low pH), Eh may remain positive for six months or longer.

Following inundation, the pH normally increases in acid soils and decreases in alkaline soils. After 4 to 12 weeks the pH reaches fairly stable values between 6.5 and 7.0 in most soils. The pH increase is mainly the result of the increase in (carbonate) alkalinity^{*2} accompanying most reduction processes in soils. The following equations, in which CH_2O stands for organic matter, illustrate the formation of alkalinity upon reduction:

$$5CH_2O + 4NO_3^- \rightarrow 4HCO_3^- + CO_2 + 2N_2 + 3H_2O \tag{8.18}$$

$$CH_2O + 2Fe_2O_3 + 7CO_2 + 3H_2O \rightarrow 4Fe^{2+} + 8HCO_3^- \tag{8.19}$$

$$2CH_2O + SO_4^{2-} \rightarrow 2HCO_3^- + H_2S \tag{8.20}$$

The stabilization of the pH after the rise to a near-neutral value should be ascribed to the precipitation of Fe(II) and Mn(II) hydroxides and carbonates. For instance, $Fe(OH)_2$ may be precipitated according to the equation:

$$Fe^{2+} + 2HCO_3^- \rightarrow Fe(OH)_2 + 2CO_2 \tag{8.21}$$

By such reactions, a certain amount of alkalinity is transferred from the solution (HCO_3^-) to the solid phase (OH in $Fe(OH)_2$), thus counteracting any further pH increase. Such processes may also account for the observed decrease in (Mn^{2+}) and (Fe^{2+}) in the later stages of reduction.

In calcareous soils, the pH under aerobic conditions is rather high (generally between 7 and 8.5, cf. chapter 6). The reduction of iron and manganese oxides is inhibited at high pH by the low mobility of Fe^{2+} and Mn^{2+} under those conditions. Therefore, the production of alkalinity, e.g. according to equation (8.19) is insignificant. On the other hand, the production of CO_2 by the anaerobic decomposition (fermentation) of organic matter has an acidifying effect, as can be seen from equation (8.14). As a result, the pH of most calcareous soils drops 0.5 to 1 pH unit upon submersion.

[*1] Eh (Volts) $= 0.059$ pe (at 25 °C)

[*2] Alkalinity, often indicated as (Alk), here refers to the sum:
  $[HCO_3^-] + 2[CO_3^-] + [OH^-] - [H^+]$ (cf. chapters 6 and 9).

8.3.4. Water regimes in hydromorphic soils

Figure 8.6 shows four examples of water regimes typical for hydromorphic and flooded soils. Type I is rarely or never flooded, but the ground-water table occurs at a shallow depth throughout the year. Type II is more poorly drained and periodically submerged. These situations are common in recent riverine landscapes (Type I at D and type II at E in figure 8.1) and in recent marine sediments. Types III and IV have perched water tables during a certain period of the year, due to temporary water saturation or flooding by rain water. An argillic horizon, a fragipan, an iron pan, an abrupt change in texture or even an air cushion formed during rapid sheet flooding may be responsible for the predominantly aerobic conditions below the water-saturated topsoil. Perched water tables are common in soils of older alluvial terraces (location C in figure 8.1) that have already undergone distinct soil formation, e.g. in soils with argillic horizons. They are also found in artificially terraced paddy soils, e.g. in Oxisols on volcano slopes in Indonesia.

The permanently water-saturated horizons shown in figure 8.6 are normally homogeneously gray, because ferric oxides (which cause the yellowish, brownish or reddish tinge in most mineral soils) are either absent or masked

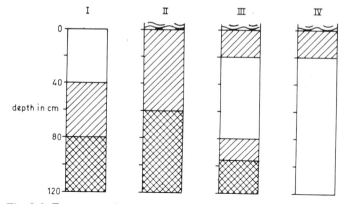

Fig. 8.6. Four examples of water regimes in hydromorphic soils.
I : moderately deep, fluctuating ground-water table, no flooding (gley, ground-water gley)
II : shallow ground-water table, periodic water saturation or flooding (gley, ground-water gley)
III : moderately deep ground-water table, periodic water saturation or flooding by a perched water table (pseudogley, surface-water gley)
IV : ground water at great depth, periodic water saturation or flooding by a perched water table (pseudogley, surface-water gley)
blank : rarely or never water-saturated horizons
hatched : periodically water-saturated horizons
cross-hatched : permanently water-saturated horizons

by greenish to black ferrous compounds. Most periodically water-saturated horizons show strong mottling ('hydromorphic characteristics') of brownish to reddish ferric compounds in a grayish matrix, low in iron. Mottling probably forms as the long-term result of alternate reduction and oxidation because precipitation of newly oxidized ferric compounds takes place preferentially on already existing ferric oxide surfaces.

In types I and II, the brown mottles accumulate especially on ped faces and along cracks or biogenic channels, while the gray colours predominate in the interiors of the structural elements. This is because oxidation of Fe^{2+} and precipitation of ferric oxide take place at the ped surfaces as soon as these are exposed to the air, while the ped interiors remain water-saturated for a longer time, e.g. by capillary recharge from ground water. Moreover, many sediments contain considerable amounts of finely divided organic matter, which may lead to reduction throughout the soil mass in water-saturated conditions. In contrast, in the surface horizons of types III and IV, water penetration and reduction are predominant along the voids (partly also because roots tend to grow preferentially along the ped faces, which locally raises the organic matter content), whereas oxidized conditions prevail for relatively long periods inside the structural elements. This results in strongly mottled and mainly brownish ped interiors and paler, grayish vertical cracks. The term 'gley' refers to the morphology and genesis of types I and II, and the term 'pseudogley' to types III and IV.

In typical flooded soils (paddy soils) which have a reduced surface horizon, the top few mm or cm of the soil are generally oxidized because some O_2 can penetrate continuously. Oxidizing conditions may also persist in the subsoil, where reduction proceeds much more slowly than in the topsoil as a result of a lower organic matter content. At the interface of predominantly reduced and oxidized environments, Fe^{2+} and Mn^{2+} may be oxidized and precipitated as Fe(III) and Mn(III,IV) oxides. The concentration gradient that develops stimulates the diffusion of Fe^{2+} and Mn^{2+}, and thus a distinct accumulation may take place. Each of the four types of water regimes leads to a specific pattern of iron-manganese accumulation horizons at the boundary between predominantly reduced and predominantly oxidized horizons. Irrespective of the type of oxide formed, iron precipitates more readily (at a lower pe) than manganese upon oxidation (see figure 8.5). This accounts for the Fe-Mn separation commonly observed in accumulation zones.

8.3.5. Weathering under seasonally reduced conditions

During the waterlogged stage, dissolved ferrous iron replaces part of the exchangeable cations, as can be inferred from the increase in dissolved Ca^{2+},

Mg^{2+}, K^+ and Na^+ normally following soil reduction. Upon the return of aerobic conditions, the adsorbed Fe^{2+} is oxidized again to ferric oxide. The protons released during this oxidation may be neutralized by the HCO_3^- formed earlier during solution-reduction of ferric oxide, and in this case the displaced basic cations would again become exchangeable. As an example, the reduction and oxidation of ferric oxide in the presence of clay high in exchangeable Na^+ is considered.

reduction stage:

$$Fe_2O_3 + 0.5CH_2O + 3.5CO_2 + 1.5H_2O \rightarrow 2Fe^{2+} + 4HCO_3^- \quad (8.22)$$

$$2Fe^{2+} + 4\text{ads-Na} \rightarrow 4Na^+ + 2\text{ads-Fe} \quad (8.23)$$

$$Fe_2O_3 + 0.5CH_2O + 3.5CO_2 + 1.5H_2O + 4\text{ads-Na} \rightarrow 4Na^+ + 4HCO_3^- + 2\text{ads-Fe} \quad (8.24)$$

oxidation stage:

$$2\text{ads-Fe} + 0.5O_2 + 2H_2O \rightarrow Fe_2O_3 + 4\text{ads-H} \quad (8.25)$$

$$4\text{ads-H} + 4Na^+ + 4HCO_3^- \rightarrow 4\text{ads-Na} + 4CO_2 + 4H_2O \quad (8.26)$$

$$2\text{ads-Fe} + 0.5O_2 + 4Na^+ + 4HCO_3^- \rightarrow Fe_2O_3 + 4\text{ads-Na} + 4CO_2 + 2H_2O \quad (8.27)$$

The net effect of the whole reduction-oxidation cycle (8.24 + 8.27) is the oxidation of half a mole of CH_2O (organic matter) to CO_2 and H_2O per mole of ferric oxide involved, but otherwise the initial and final situations are similar. However, if part of the displaced cations are removed from the profile together with HCO_3^-, either by leaching or by diffusion into the surface water, the net effect of the cycle also includes the formation of a hydrogen clay (8.24 + 8.25). The acid H-clay is chemically unstable and partly decomposes (releasing some H_4SiO_4) to form e.g. an Al-Mg-clay (cf. section 4.6). If repeated over and over again this cycle leads to serious acidification and pronounced weathering of clay minerals. This process has been termed 'ferrolysis'.

The effects of ferrolysis are mainly confined to the upper horizons of the rain-fed hydromorphic soils of types III and IV (figure 8.6), and the process probably explains the formation of many acid hydromorphic soils, e.g. part of the Aqualfs, Aquults, Albolls (pseudogley soils, Solodic and other Planosols). These soils generally have acid, silty upper horizons with a low structural stability and a low cation exchange capacity of the clay fraction, abruptly overlying a strongly mottled B horizon higher in clay. (The B horizon may be strongly alkaline Solodic Planosols). They are typical for physiographic positions indicated by C in figure 8.1. In recent floodplains (locations D and E in figure 8.1), a continual supply of dissolved alkalinity (present in most river waters) may prevent ferrolysis, while a regular supply of fresh sediment may mask its effects. The presence of $CaCO_3$ inhibits ferrolysis in two ways: its dissolution supplies alkalinity to the soil solution, while the high equilibrium pH keeps the Fe^{2+}-activity at very low levels.

A process closely resembling ferrolysis is prevalent in recent marine soils used for paddy cultivation. These soils contain much water-soluble sulphate, and during submersion sulphate reduction in the topsoil (rich in organic matter) leads to the formation of FeS. Part of the alkalinity (HCO_3^-) formed during sulphate reduction is removed by diffusion into the surface water, while potential acidity is retained in the soil: solid FeS, which after drainage will be oxidized to H_2SO_4. On a much larger scale, a similar process takes place in most tidal flats with a luxuriant vegetation (reed marshes in temperate climates, mangroves in the tropics). Reduction of dissolved sulphate from sea water in the surface horizons produces FeS and later FeS_2, while dissolved alkalinity is removed and carried to the sea by tidal action. If the sediments in question are low in $CaCO_3$ (which is frequently the case, especially in the tropics), they will seriously acidify after drainage. The so--called acid sulphate soils ('cat clays') formed in this way are characterized by yellow mottles of jarosite, $KFe_3(SO_4)_2(OH)_6$, pH values below 4, and the presence of soluble Al.

8.4. REVERSE WEATHERING

8.4.1. Vertisols, calcium carbonate, salinity and high pH

Whereas removal of materials dominates the soil forming processes at A (figure 8.1), supply is characteristic for the soils at B if local rainfall does not exceed evapotranspiration. Evaporation leads to concentration of the soil solution and eventually to precipitation of various minerals, including readily soluble salts if the climate is arid enough. Because HCO_3^- is also concentrated, the pH too has a tendency to rise under those conditions. Evaporation not only leads to the precipitation of soluble salts: silicates may also be affected. If the pH and the concentrations of H_4SiO_4 and Mg^{2+} are high enough, resilication of kaolinite to beidellite may take place according to:

$$7\text{Kaolinite} + 8H_4SiO_4 + Mg^{2+} \rightarrow 6\text{Mg-beid.} + 2H^+ + 23H_2O \qquad (8.28)$$

The protons released by this reaction are used immediately to protonate

HCO_3^- to the weak acid $H_2CO_3^*$, with a resulting release of CO_2:

$$H^+ + HCO_3^- \rightarrow H_2CO_3^* \rightarrow H_2O + CO_2 \qquad (8.29)$$

The overall process can be represented by:

$$7\text{Kaol.} + 8H_4SiO_4 + Mg^{2+} + 2HCO_3^- \rightarrow 6\text{Mg-beid.} + CO_2 + 24H_2O \qquad (8.30)$$

Likewise, evaporation may lead to the precipitation of a carbonate, such as calcite:

$$Ca^{2+} + 2HCO_3^- \rightarrow CaCO_3 + CO_2 + H_2O \qquad (8.31)$$

Note that equations (8.31) and (8.28) are exactly the same as the weathering reactions (8.1) and (8.9) discussed earlier, but written in the opposite direction: hence the term 'reverse weathering'. Weathering (dissolution) and reverse weathering (precipitation) may also involve readily soluble minerals such as gypsum, $CaSO_4 \cdot 2H_2O$, and halite, $NaCl$. In contrast to silicates and carbonates, which may be considered as 'salts' of weak acids and strong bases, most of these soluble minerals are salts of strong acids which do not consume HCO_3^- upon precipitation.

The kind of reverse weathering processes taking place at B depends mainly on the supply and concentration of the ground water, the mineralogical composition of the sediment and the rate of water loss due to evapotranspiration. If mafic ('basic') igneous rocks are present at A, supplying much Mg^{2+}, Ca^{2+} and H_4SiO_4 to the drainage water, both smectite and calcite may be formed during evapotranspiration of ground water at B. This process would result in the formation of calcareous Vertisols. Other minerals that could be formed under similar conditions are dolomite, $CaMg(CO_3)_2$, magnesite, $MgCO_3$, and palygorskite (an Mg-Al-silicate). The formation of calcite, dolomite, palygorskite and smectite takes place at pH values between 6.5 and 8.5 Felsic ('acid') rocks such as granite produce waters relatively rich in Na^+ and HCO_3^-. Nahcolite, $NaHCO_3$, and trona, $Na_3H(CO_3)_2 \cdot 2H_2O$ are minerals that may form during reverse weathering involving Na^+. The high solubility of the Na carbonates requires high concentrations of Na^+ and HCO_3^-, and hence very high pH values, before they may be precipitated (cf. section 9.5). Of the silicate minerals, the Na-zeolite analcime, $NaAlSi_2O_6 \cdot H_2O$, is also formed in certain strongly alkaline soils (pH > 9).

The presence of analcime explains the anomalously high apparent exchangeable Na percentage of several soils which may have relatively favorable physical (and reclamation) characteristics. The Na in analcime is partly exchangeable by the conventionally used K and NH_4 ions, but virtually nonexchangeable by Ca or Mg ions. Since zeolites are easily soluble in acid, the conventional acid pretreatment of samples for mineralogical analysis probably delayed the detection of analcime (and possibly other zeolites) in sodic soils until a decade ago. The excess Na exchangeable with Na isotope, K or NH_4 over the amount exchangeable with Ca or Mg in a soil sample may be used to estimate the Na 'trapped' in zeolites.

During continued evapotranspiration of water containing bicarbonate, equivalent amounts of divalent cations and HCO_3^- are removed by reverse weathering. Therefore, the very high pH values associated with high alkalinities may only be attained by evaporation of waters containing alkalinity in excess of divalent cations. The concept of 'residual alkalinity':

residual alkalinity = Alk - $2[Ca^{2+}]$ - $2[Mg^{2+}]$

was developed as an indicator for the potential occurrence of a very high pH under strongly evaporative conditions (cf. chapter 9). The residual alkalinity of potentially alkaline water will be lowered if it flows through gypsiferous material: because $CaCO_3$ is less soluble than $CaSO_4.2H_2O$, calcite will be formed at the expense of gypsum and the ground water changes from a sodium(bi)carbonate to a sodium sulfate type. The residual alkalinity is also lowered when such waters traverse Ca-saturated soil material, by exchange reactions of the type:

$$\text{ads-Ca} + 2Na^+ \rightarrow 2\text{ads-Na} + Ca^{2+} \tag{8.32}$$

Reactions as shown in equation (8.32) increase the exchangeable sodium percentage of the soil materials traversed by waters containing any kind of sodium salts. Bicarbonate and carbonate are particularly effective in this regard, however. High concentrations of dissolved (bi)carbonate limit the calcium concentration in solution to very low levels due to reverse weathering ($CaCO_3$ precipitation). The resulting very high Na/Ca ratio in solution leads to a high Na/Ca ratio on the exchange complex. Soils and horizons with more than 15 per cent exchangeable Na are termed 'sodic' (formerly 'alkali'). Strongly alkaline soils are almost invariably sodic, but sodic soils need not be strongly alkaline. Sodic soils are dealt with extensively in chapter 9. Potassium carbonates are very soluble just like sodium carbonates, but the mobility of K^+ is very low compared to that of Na^+ due to the relatively low solubility of K-bearing silicates as well as the ready incorporation of K^+ in secondary minerals (illite). Therefore, alkalinity due to potassium carbonate is very rare, while alkalinity due to sodium carbonate is common.

Evapotranspiration of waters draining from rocks containing sodium silicates, combined with the dominance of CO_2 as a weathering agent, gives rise to the formation of strongly alkaline soils. If A (figure 8.1) is situated in an area with volcanic activity, however, acids such as H_2SO_4 and HCl may be the main weathering agents. Salinization by chlorides and sulphates would then occur at B without the development of strongly alkaline conditions. Mainly sodium salts would accumulate if the waters drain from felsic rocks, and mainly Ca and Mg salts in the case of waters from mafic rocks. Salts are not necessarily derived directly from silicate weathering alone. Possible sources of salts in soils and sediments also include salt beds of marine and continental origin, the sea and salt lakes, as well as (in low concentrations) liquid brine inclusions in mineral grains and interstitial brines in rocks. Besides the obvious mode of transport by water, salts may also be transported through the air in the form of aerosols and brought down as solids or in rain water.

The reverse weathering processes discussed above normally do not occur together in one location. The differential solubility of various minerals formed by reverse weathering results in a spatial distribution of the different processes. Precipitation of $CaCO_3$ takes place at relatively low concentrations, and hence closer to A (and at greater depths below the evaporating surface) than the precipitation of the more soluble gypsum. Salinization with or without the development of high pH may be dominant still further from A. In this respect the soil toposequence (in a lateral direction) as well as the soil profile itself (in a vertical sense) closely resemble chromatograms. The widths of the various accumulation zones depend largely on the dissolved load: drainage water from mafic rocks tends to produce a wide belt of calcareous soils with pH 8.5 or less and at most a narrow strip of strongly alkaline soils further downstream. The same sequence may be found below felsic rocks, but there the zone with calcareous soils is narrow and the strongly alkaline zone very wide.

In many cases, weathering products released under a leaching moisture regime do not become involved in reverse weathering in soils, but are carried into rivers and further to the sea. Notably in the humid tropics large areas exist where reverse weathering processes are absent (basins of the Amazon and Congo rivers). Reverse weathering associated with the material thus removed takes place in the sea and is not discussed here.

8.4.2. Absolute accumulation of iron oxide

Under certain conditions, dissolved ferrous iron may be transported over considerable distances by subsurface flow. This is the case if organic matter is present in a water-saturated horizon with net downward water flow, and if oxygen is virtually absent along the further subsurface flow path. As was discussed under 'Water regimes in hydromorphic soils' in section 8.3.4, iron may migrate over short (vertical) distances, for example in a profile, by diffusion with or without flow. Where the solution comes into contact with oxygen again, generally near the soil surface or in a root zone, Fe^{2+} is oxidized and ferric oxide is formed according to:

$$2Fe^{2+} + 4HCO_3^- + 0.5O_2 \rightarrow Fe_2O_3 + 4CO_2 + 2H_2O \tag{8.33}$$

Accumulation of ferric oxide in a certain horizon or zone is common in terrace soils and terrace escarpments (locations C and C' in figure 8.1). The source of the ferrous iron may be ferrous silicate minerals, or ferric oxides reduced by organic matter in water-saturated soils at higher elevations in the landscape, as discussed previously.

Not all accumulations of ferric oxides may be ascribed to reverse weathering: in Oxisols, for example, iron oxides have accumulated in a relative, not absolute, sense. In this case, incongruent dissolution of aluminosilicates has left a solid weathering residue, as shown by example in equation (8.5) and described in section 8.2.

ILLUSTRATIVE PROBLEMS

Garrels, R.M. and Mackenzie, F.T., 1967. Origin of the chemical compositions of some springs and lakes. In: *Equilibrium concepts in natural water systems.* W. Stumm (ed.). Advances in Chemistry series no. 67, Amer.Chem.Soc., p. 222-242

Garrels, R.M. and Mackenzie, F.T., 1971. *Evolution of sedimentary rocks.* Norton, New York. Chapters 4,6 and appendix B.

Mohr, E.C.J., Baren, F.A. van and Schuylenborgh, J. van, 1972. *Tropical soils.* 3rd ed. Mouton - Ichtiar Baru - van Hoeve, The Hague - Paris - Djakarta.

EXEMPLARY PROBLEMS

1. In interstitial solutions of soils affected by sulfuric acid (acid sulfate soils, colliery spoils, soils polluted by SO_2), the term pAl + pSO_4 - pH appears to be constant. Equilibrium with which compound could explain this constancy (give the stoichiometric formula)? (Answer: $AlOHSO_4$).

2. Why is it unlikely that gypsum occurs in soils with pH > 8.5 ? (Answer: $CaSO_4 \cdot 2H_2O$ + $+ CO_3^{2-} \rightarrow CaCO_3 \downarrow + SO_4^{2-} + 2H_2O$).

3. Water of composition A (see below) is purified by infiltration in dune sand containing pyrite and $CaCO_3$ (shells) near the infiltration site and abundant organic matter and ferric oxide a few meters from the site. Water from a well still further from the infiltration point has composition B.

	O_2	SO_4^{2-}	HCO_3^-	Ca^{2+}	Fe^{2+}	NO_3^{2-}	pH
			mmol/liter				
A	0.245	0.791	3.231	2.350	0.00	0.149	7.6
B	0.00	0.922	3.640	2.612	0.036	0.056	7.7

(a) What processes have affected the water composition (i) in the calcareous, pyritic zone, and (ii) in the zone with organic matter and ferric oxide? (Answer: in (i):oxidation of pyrite to Fe_2O_3 and H_2SO_4, and solution of $CaCO_3$; in (ii): reduction of NO_3^- and of Fe_2O_3).

(b) Write reaction equations of these processes with the proper reaction coefficients ('how many moles of P react with how many moles of Q to produce so many moles of R per liter, etc.) to explain the changes in the composition of the water. (Answer: e.g. $0.065\ FeS_2 + 0.093\ NO_3 \rightarrow 0.093\ HCO_3^- + 0.023\ CO_2 + 0.047\ N_2 + 0.070\ H_2O$).

(c) Calculate the difference in dissolved $H_2CO_3^*$ between A and B using $H_2CO_3^* =$ $= HCO_3^- + H^+$, $pK^\circ = +6.35$ and assuming that activity corrections can be neglected. (Answer: $[H_2CO_3^*]$ decreases by 0.018 mmol/liter.).

(d) Compare the change in dissolved CO_2 under (c) with that expected from the reaction coefficients found under (b). (Answer:$[H_2CO_3^*]$ decreases by 0.040 mmol/liter.

CHAPTER 9

SALINE AND SODIC SOILS

A. Kamphorst and G.H. Bolt

In many regions of the earth, notably in the subtropics, the annual rainfall is not enough to meet the evaporative need of a complete vegetative cover on the soil. Any addition of water from other sources to these soils, either by capillary ascent from the groundwater (if present at shallow depth) or by irrigation, necessarily implies the addition of salts to the soil profile, as all natural waters contain varying amounts of salts. These salts are primarily the most soluble ones, viz., the chlorides of the common cations Na, Ca, Mg and K and also the sulphates of these cations. In as far as the water added locally to a soil is used entirely for consumptive use of the vegetation, salinization of the soil profile is unavoidable. Thus all arid and semi-arid region soils should be considered to be subject to potential salinization.

As plant growth is severely inhibited once the concentration of soluble salts in the soil solution exceeds certain limits, the maintenance of agricultural production in these regions necessitates a set of measures which could be termed 'adequate salt and water management'. While leaving a detailed description of these measures to existing texts on Irrigation and Drainage Engineering, it is pointed out here that basically these amount to removal by leaching of the salts added to the profile. At the same time this indicates the need for an appropriate 'sink' for the salts thus leached, as otherwise the salt removed from one location will move with the groundwater to somewhere else, which simply implies shifting the problem from one spot to another. Keeping in mind that the overall solution of the problem of salinization in a certain region goes beyond the study of the processes occurring locally in a certain profile, only the latter processes will be the subject of consideration in this chapter.

If salinization of a soil profile is defined as an increase in the concentration of salts in, and eventually precipitation of salts from, the soil solution, one may distinguish in addition to the above 'zonal' type (i.e. associated with arid zones) local azonal forms of salinization. These are often due to salts derived from the sea, e.g. salinization following flooding by seawater or wind-blown salts near the ocean coast. A more detailed review of the occurrence of saline soils belongs to the field of Soil Survey and will not be considered in the present text.

Aside from their influence on the concentration of the soil solution, the addition of salts to the soil profile may also lead to an alteration of the composition of the exchange complex. Such an alteration is typically in the direction of an increase of the percentage of exchangeable Na-ions, because Na-salts are the most soluble salts occurring in nature. The gradual increase in the sodium saturation of the exchange complex is termed the process of sodication (formerly alkalization). The ensueing high-sodium soils are referred to as sodic (formerly alkali) soils. Obviously the rate of the sodication process depends on the composition and concentration of the water supplied to the profile, the amount of water added per year and the CEC of the soil.

While deferring a more detailed discussion of the sodication process to section 9.3 it is pointed out here that sodium saturation of the adsorption complex takes place when the soil comes in contact with a solution having a high value of the 'reduced ratio' of the concentrations of Na-ions to divalent cations, $c_{o,+}/\sqrt{c_{o,2+}/2}$ (cf. equation 4.10). Such a condition is likely to occur upon concentrating the soil solution (as was typical for the salinization process). If a solution containing Na- and Ca-ions is 'concentrated' with a factor x, the reduced ratio $c_{o,Na}/\sqrt{c_{o,Ca}/2}$ increases with a factor $\sqrt{x}$! Thus it may be concluded that salinization in arid zone soils is likely to lead also to sodication.

The above effect may be enhanced if the anions in the system are sulphates, as the solubility of $CaSO_4$ is rather limited, while Na_2SO_4 has a high solubility. This 'secondary' effect of the anions on the alkalization process shows up most strongly in the case of irrigation water containing substantial amounts of carbonates. These substantial amounts must be alkali- (i.e. Na-) carbonates and/or bicarbonates, and such waters are thus recognized by a high pH-value (> 8). The solubility of Ca-carbonates at such pH-values being virtually negligible it is concluded that water high in carbonates will have high values of the reduced ratio $c_{o,Na}/\sqrt{c_{o,Ca}/2}$ even if the total salt content is not very high. If a rise in pH of the soil due to the introduction of water containing carbonates would be termed 'alkalinization' (i.e. the rise of the pH to alkaline values) it could thus be stated that alkalinization of the soil is invariably accompanied by alkalization or sodication (cf. also section 8.4.1 and 9.3.2). Inverting the above reasoning it is concluded that sodication of soils is as a rule caused by either salinization or by alkalinization, or by a combination of the two.

The agricultural problems associated with saline and sodic soils may be divided in disturbances of the water- and nutrient uptake by plants and effects resulting from undesirable physical characteristics of the soil. As regards water uptake the osmotic effect of the total salt concentration is mentioned. Due to the presence of semi-permeable barriers in the route of water from the soil to the plant tissue large negative values of the 'osmotic potential' (cf. Soil Physics texts) of the soil water tend to inhibit water uptake. Disturbances in plant nutrition are due primarily to unfavorable ionic ratios. E.g. in saline soils the sodium ion usually has a dominant position, which may adversely affect the uptake of other cations. Apart from this, mention should be made of the toxic effect of certain anions on plants, e.g. boron and high concentrations of chloride. In the special case of the 'alkalinized' soil the high pH-values caused by the presence of dissolved carbonates will strongly affect the solubility of many other ions, notably certain micro-nutrients but also the phosphates.

As was discussed already in section 3.3 sodic soils tend to exhibit very poor physical properties, at least at low electrolyte levels and if the clay content is fairly high. Both aeration and drainage properties of non-saline sodic soils are usually very unsatisfactory. As the reclamation of these soils always implies leaching procedures, several difficulties may be encountered in improving such soils.

9.1. CHEMICAL CHARACTERIZATION OF SALINE AND SODIC SOILS

The concentration of the soluble salts in the soil solution is influenced by the moisture content. With the purpose in mind to characterize the salt content of the soil it is logical to measure this concentration at a standardized moisture content of the soil. To this purpose the salinity of the soil is usually characterized by means of the electrical conductivity of the saturation extract (EC_e, in mmhos cm^{-1}). This extract is obtained by suction-filtration of a water-saturated paste of the soil.

In fact, the electrical conductivity or specific electrical conductance of a solution is the reciprocal value of its specific electrical resistance. The latter is the resistance of a 'column' of the solution with a cross-section of 1 cm^2 and a length of 1 cm. Hence the specific electrical resistance is expressed in Ohm cm^{-1} cm^2 or Ohm cm and its reciprocal value in Ohm^{-1} cm^{-1}, which is generally written as Mho cm^{-1}. In practice the EC-value is given in millimhos cm^{-1} (mmhos cm^{-1}) or sometimes micromhos cm^{-1}. Obviously the salt concentration is only approximately defined by the EC-value, mainly because the mobilities of different ions are only approximately the same. However, for practical purposes a rough estimate on the basis of EC-values often suffices. For the present it is recalled that the ionic mobilities are around 50-70 mho cm^{-1} per eq. cm^{-3}. Taking into account that the activity coefficients are usually less than unity one may accept 50 as a guide number. This implies that a solution containing 10^{-6} eq. of monovalent salt per cm^3 has a conductivity of:

$(50 + 50)/10^6$ mho cm^{-1}.

Thus a 10^{-3} normal solution has an EC-value of about 0.1 mmho cm^{-1}. Conversely a conductivity of 1 mmho cm^{-1} corresponds to about 0.01 normal salt. For more accurate values reference is made to relevant graphs in USDA Handbook No. 60.

The U.S. Salinity Laboratory proposed a classification of saline soils with respect to expected salt damage to crops. In table 9.1 this classification is presented in terms of EC_e-values and the corresponding values of the salt concentration and the osmotic pressure. Obviously such classification criteria can serve as guide numbers only, as there exists only a fair correlation between the 'average' moisture contents under field conditions and the moisture contents of the saturated pastes. These 'average' moisture contents in the field will correspond to about ½ (fine textured soils) to ¼ (coarse textured soils) of the saturation percentages (SP). The division of soils into saline and non-saline soils is now taken at EC_e equal to 4 mmho/cm. Expressed in the osmotic pressure at field capacity this corresponds to approximately 5 bars.

It would be better if the electrical conductivity could be measured in an extract of a soil sample with a moisture content near the 'average' value encountered in the field, i.e. somewhere between the moisture contents at field capacity and at the permanent wilting point. However, such an extract is very difficult to obtain. The saturation percentage is the nearest moisture content that allows for an easy extraction of an amount of solution sufficient for analysis.

TABLE 9.1

Expected salt effect on crop growth, as a function of the concentration in the saturation extract (according to USDA Handbook No. 60).

EC_e (mmho/cm)	0		2		4		8		16
Effect on crop		Salinity effects mostly negligible		Yields of very sensitive crops may be restricted		Yields of many crops restricted		Only tolerant crops yield satisfactorily		Only a few very tolerant crops yield satisfactorily
Concentration of saturation extract (normal)	0.0		0.02		0.04		0.08		0.16
Osmotic pressure (bar)	0		1		2		4		8
Proposed classification		non-saline		slightly saline		moderately saline		very saline		extremely saline

With regard to the division of soils into sodic and non-sodic soils one generally accepts an *Exchangeable Sodium Percentage* (ESP-value = = $100 \cdot \bar{\gamma}_{Na}/\gamma$) of 15 as the critical value. Also this value is rather arbitrary, because the physical reaction of a soil to a certain ESP-value depends very much on its salt content, types and amounts of clay minerals and humus, etc., while different crops react quite differently to the same ESP-value.

Referring to the fact mentioned above that the process of sodication is sometimes accompanied by alkalinization, it appears logical to make also a distinction between alkaline and non-alkaline (sodic) soils. As will be commented on in section 9.4 one often uses a pH-value of 8.5 as a division line in this respect. Alternatively, a division between alkaline and non-alkaline may be based on the concentration of carbonates in the soil solution. The total concentration of CO_3^{2-}, HCO_3^- and the (usually negligible) excess of OH^- over H^+ is termed the alkalinity (concentration) of the soil solution (meq/l). In a well-aerated soil the pH will rise above 8.5 if the alkalinity is more than approximately 1.5 meq/l.

9.2. SALINIZATION OF SOILS UPON IRRIGATION

The origin of saline and sodic soils through 'natural' causes belongs to the field of soil genesis and will be left out of consideration here. However, also irrigation practices may give rise to salinization and/or sodication. The degree to which this will occur depends on the composition of the irrigation water and also on the balance between supply of irrigation water to the soil surface and removal of drainage water from the lower boundary of the profile. As to the latter aspect it can be stated that salinization of a given profile depth must occur if the product of the average concentration and the amount of the drainage water emerging from the bottom of the profile is less than the corresponding product for the irrigation water supplied from above; i.e.:

$$EC_{dw} \times D_{dw} < EC_{iw} \times D_{iw} \tag{9.1}$$

In this equation D_{dw} and D_{iw} represent the amounts of drainage water and irrigation water during the period under consideration, expressed in cm waterlayer. The corresponding EC-values are then average values for these amounts. In practice it is often convenient to consider periods of one year, as will be done in the following text. In that case, the annual amounts of irrigation water supplied and drainage water removed are related to each other by the annual 'water balance':

$$D_{iw} + D_{rw} = D_{dw} + D_{cw} \tag{9.2}$$

in which D_{rw} is the annual rainfall and D_{cw} the annual consumptive use (evapotranspiration) of the crop rotation.

The rate of salinization in the situation represented in equation (9.1) can be expressed in the annual increase of the average EC_e-value of the profile, ΔEC_e, which equals:

$$\Delta EC_e = \frac{(EC_{iw} \times D_{iw} - EC_{dw} \times D_{dw})}{D_{soil} \times SP \times \rho_b/100} \qquad (9.3)$$

in which D_{soil} is the depth of the profile under consideration (cm), SP is the average moisture content of the saturated paste (Saturation Percentage, expressed as a percentage by weight of the dry soil) and ρ_b is the average bulk density of the profile (g cm^{-3}).

The above salt balance equation (9.3) indicates that, given a more or less constant annual average concentration and annual quantity of the irrigation water supplied, the rate of salinization is highly dependent on the amount of drainage water removed from the bottom of the profile. This becomes at once clear if we consider situations where $D_{dw} = 0$ (stagnant high groundwater or impermeable subsoil layer) or where $D_{dw} < 0$ (groundwater seepage). In these cases the rates of salinization (ΔEC_e) are obviously higher.

Equation (9.3) also shows that the average concentration of the drainage water removed is also important. Under natural conditions this concentration cannot be measured or controlled easily. Moreover, it increases while salinization progresses, because the drainage water may be considered (as a first approximation) to be the soil solution leaking from the soil at moisture contents just above field capacity. Hence:

$$EC_{dw} \approx EC_{FC} \qquad (9.4)$$

Assuming equation (9.4) to be sufficiently valid, salinization will gradually come to a standstill as soon as:

$$EC_{FC} \approx EC_{dw} = \frac{D_{iw}}{D_{dw}} \times EC_{iw} \qquad (9.5)$$

This follows from substitution of equation (9.4) in equation (9.3) with $\Delta EC_e = 0$. Then the soil salinity will have reached a constant (i.e. steady state) level according to:

$$EC_e = \frac{FC}{SP} \times EC_{FC} \approx \frac{FC}{SP} \times \frac{D_{iw}}{D_{dw}} \times EC_{iw}, \qquad (9.6)$$

where FC is the average moisture content at field capacity and SP is the average moisture content of the saturated paste of the soil.

It is mentioned here that the above model, in which EC_{FC} is equated to EC_{dw}, is a gross oversimplification of the process occurring in most field situations. This is due to two effects, viz.

a. Even if the drainage water leaking from the bottom of the profile were in fact the soil moisture held in the profile at a suction near field capacity, EC_{dw} would have to be equated to the EC_{FC} at the bottom of the soil profile instead of to the EC_{FC} of the whole profile. As the salt concentration in the profile tends to vary strongly with depth and time in these salinization situations (cf. figure 9.1), this simplification introduces a severe error.

b. Particularly under irrigation systems where the irrigation water is ponded on the soil surface (basin irrigation, border irrigation), water does not really move only through the smaller pores that hold moisture near field capacity. Part of the irrigation water applied is transported downward through larger pores, without removing much salt from the profile. A simplified correction for this effect of hydrodynamic dispersion (cf. chapter 7) can be made by assuming that only a fraction of the drainage water ($0 < f < 1$) is water actually held at field capacity and that the remaining part, 1-f, is irrigation water moving through the bigger pores straight from the soil surface to the bottom of the profile. Hence:

$$EC_{dw} = f.EC_{FC} + (1-f)EC_{iw}$$

The fraction f is called the leaching efficiency factor. This factor can be determined experimentally by measuring the quantities and salt concentrations of irrigation and drainage water, along with soil salinities, in leaching experiments. Naturally these experimentally determined leaching efficiency factors will also incorporate the effect mentioned under a.

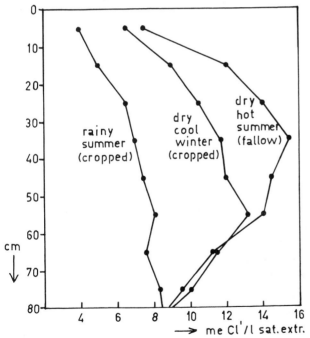

Fig. 9.1. Salt profiles in a double-cropped irrigated soil during different seasons of the same year (after Arjan Singh, India)

Furthermore, the leaching efficiency varies under different irrigation systems. Under drip or sprinkler irrigation, where the downward percolation does indeed take place at suction values very near to field capacity, the leaching efficiency factor does approach unity. Under border- or basin-irrigation, on the other hand, its value may vary between 0.6 in a sandy soil with a quite uniform pore-size distribution and 0.3 in a cracking clay soil.

Equation (9.6) can be used to calculate the excess amount of irrigation water (D_{dw}) to be supplied above the amount required for consumptive use ($D_{cw} - D_{rw}$) in order to keep the salinity of a soil below a level corresponding to a specific EC_e-value. This level depends on the salt tolerance of the crops to be grown. In practice this excess amount is often expressed as a fraction of the total amount of irrigation water required and is then called the leaching requirement, which can be calculated by rewriting equation (9.6) as follows:

$$LR = \frac{D_{dw}}{D_{iw}} \approx \frac{FC \times EC_{iw}}{SP \times EC_e} \qquad (9.7)$$

Together with the water balance equation (9.2) the leaching requirement equation may serve to calculate the total amount of irrigation water to be supplied in a given situation (with known D_{cw}, D_{rw}, EC_e and EC_{iw}). Drainage engineers also use it for the calculation of the capacities of subsoil drainage systems, to be installed in irrigated land for the discharge of the leaching requirement.

In case the leaching efficiency factor is smaller than 1, the above equation will be:

$$LR \approx \frac{EC_{iw}}{f \cdot \frac{SP}{FC} \cdot EC_e + (1-f)EC_{iw}} \qquad (9.7b)$$

If the EC_{iw} is small as compared to the EC_e, this equation may be further simplified to:

$$LR \approx \frac{FC \times EC_{iw}}{f \times SP \times EC_e} \qquad (9.7c)$$

The excess irrigation to be given, naturally, may not exceed the drainage capacity (cm/year) of the soil. Aside from limitations in the supply of irrigation water, this drainage capacity often constitutes the limiting factor in meeting the leaching requirement. In case the impeded drainage is caused by an impermeable material in the profile or subsoil, chemical or mechanical measures (application of gypsum, subsoiling) will be required to improve the structure of this material. If it is due to the presence of a high groundwater table in a stagnant condition or seepage situation, artificial subsoil drainage (by tiles, ditches or tubewells) will be necessary.

In practice the annual excess irrigation can often be given only during a limited period of the year. This is the case, e.g., when water is in short supply during part of the cropping season. Then some salinization will occur temporarily, even if the annual salt budget remains in balance. Also, in such a situation the drainage capacity of the soil will have to be higher, because the annual leaching requirement will have to be met in a short period. Figure 9.1 illustrates a situation where leaching takes place almost exclusively during a rainy season of less than three months duration.

9.3. SODICATION OF SOILS UPON IRRIGATION

Aside from an increase of the salt concentration in case of insufficient leaching, which was discussed in section 9.2, the actual composition of the solution following irrigation practices must be considered. If the total amounts of ions added by irrigation over periods of many years are considerably larger than the amounts adsorbed in the soil, one may infer that on the long run this composition of the soil solution will become identical with that of the irrigation water. This process is accompanied with a gradual adjustment of the composition of the adsorption complex. The final ESP-value or 'yield value' of the complex composition may then be estimated from the composition of the irrigation water *and* the total concentration of the soil solution under average field conditions.

As detailed information about the composition of the soil solution during the growing season is usually not available, one may attempt to make a rough description of the situation along lines of reasoning similar to those employed before (cf. also the small print sections on pages 177 and 183). This rough description will not take into account the possible formation of an exchange 'front' as discussed in chapter 7, for reasons which will be elaborated upon in the small print section on page 189.

Assuming that the 'average' moisture content of the soil under irrigated conditions is about 2/3 of the moisture content at field capacity, one may estimate $c_{o,Na}$ and $c_{o,Ca}$ by multiplying the concentrations of these ions at field capacity by a factor 1.5. The average concentrations in the profile at field capacity may in turn be estimated by multiplying the average concentrations in the irrigation water by a factor D_{iw}/D_{dw} (cf. equation 9.7). Hence a reasonable estimate is:

$$c_o = 1.5 \times \frac{D_{iw}}{D_{dw}} \times c_i {}^* \tag{9.8}$$

in which c_o is the average concentration of an ion species in the soil solution under average field conditions and c_{iw} is its concentration in the irrigation water. To evaluate the hazard of sodication one then uses the reduced concentration ratio of Na and Ca, i.e., $c_{o,Na}/\sqrt{c_{o,Ca}/2}$. Substituting this ratio in a suitable exchange equation one finds the corresponding 'yield value' of the complex. The U.S. Salinity Laboratory uses for this purpose a similar reduced

---
* if necessary divided by the leaching efficiency factor, cf. p. 177.

ratio, in which the sum of the Ca- and Mg-concentrations is entered in the denominator as the total concentration of divalent cations. Moreover, the concentrations are expressed in mmol/liter. This reduced ratio is given the name *Sodium Adsorption Ratio* (SAR-value). Thus, with c_o in meq/l:

$$\text{SAR} = \frac{c_{o,Na}}{\sqrt{(c_{o,Ca} + c_{o,Mg})/2}} (\text{mmol/l})^{1/2} \tag{9.9}$$

The name indicates that the SAR determines the composition of the adsorption complex.

The relation between the reduced ratio of the molar concentrations, $c_{o,+}/\sqrt{c_{o,2+}/2} \equiv r$, introduced in equation (4.10), and the above SAR-value is: $\text{SAR} = r\sqrt{1000}$. When deriving an exchange equation based on the SAR-value one should thus divide the 'exchange coefficient' (Gapon constant) by the same factor $\sqrt{1000}$. Using the suggested value for $K_{Na/Ca} = 1/2(\text{mol/liter})^{-1/2}$ one finds:

$$\frac{\text{ESP}}{100\text{-ESP}} = \frac{\overset{+}{\gamma}_{Na}}{\overset{+}{\gamma}_{Ca}} = \frac{1}{2\sqrt{1000}} \cdot \text{SAR, or:}$$

$$\frac{\text{ESP}}{100\text{-ESP}} = 0.015\,\text{SAR} \tag{9.10}$$

The above equation appears to be satisfactory also for a soil containing Mg. As was indicated before, Ca and Mg are adsorbed with roughly the same preference, whereas (100-ESP) accounts also for the adsorbed Mg.

The U.S. Salinity Laboratory determined the correlation between both sides of equation (9.10). A satisfactory linear relationship was found between the left hand side of the equation and the SAR, for a number of different soils (mostly from the U.S.). The exchange constant calculated from this relationship was reported as 0.01475 (mmol/liter)$^{-1/2}$, which agrees very well with the value suggested above.

The amount of irrigation water and thus the time period needed to attain the 'yield value' of the complex depends on, amongst other factors, the salt concentration of the irrigation water. At high concentrations this value is reached faster than at low concentrations. On the other hand, the adverse physical effects of a high ESP-value will appear earlier at low salt concentrations (cf. figure 9.2). This implies that swelling and peptization will occur particularly if water with a high concentration and a high value of the reduced concentration ratio is followed by water with a low electrolyte concentration. In practice this may occur if 'poor quality' water is used for supplementary irrigation in an area which also possesses a rainy season. The

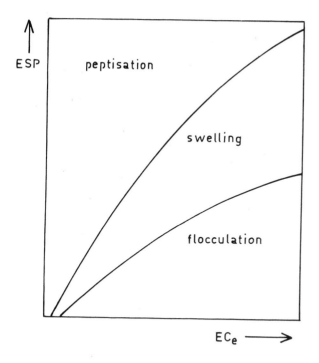

Fig. 9.2. Schematic diagram illustrating clay behaviour as a function of ESP and EC_e

adverse effects are then noticed in the rainy season. A similar situation arises if soils which were inundated with seawater (EC > 100, SAR = 150) are leached with rainwater, as may occur in coastal regions of the temperate and humid tropical zones.

Generalizing, the quality of irrigation water may be judged from its EC--value and its SAR-value. Based on these a simplified classification of irrigation water is given in table 9.2. Obviously this is for orientation purposes only, as the dangers of salinization and sodication depend also on the irrigation excess used (cf. equations (9.6) and (9.8)).

TABLE 9.2

Classification of irrigation waters (simplified scheme, after USDA Handbook No. 60)

		0.25		0.75		2.25	
EC, mmho/cm: salinization hazard:	low		medium		high		very high
SAR, mmol $^{1/2}$/liter$^{1/2}$:		7		13		20	
sodication hazard:	low		medium		high		very high
RSC, meq./liter:		1.25		2.5			
alkalinization hazard:	low		medium		high		
classification:	safe		marginal		unsuitable		

When judging the danger of sodication also the anionic composition of the water must be considered. If a considerable amount of HCO_3^- or CO_3^{2-}-ions are present, these may give rise to removal of Ca-ions through precipitation of Ca-carbonate upon concentration of the solution in the soil. In this case the SAR-value gives too favorable a picture. As a first approximation, a correction can be made for this by subtracting from the concentration of $(Ca^{2+} + Mg^{2+})$ in the irrigation water an amount equivalent to the concentration of $(CO_3^{2-} + HCO_3^-)$. This is a safe correction, because it assumes that all CO_3^{2-} and HCO_3^- present will precipitate in the soil with Ca^{2+} and Mg^{2+}. Occasionally part of it will remain in solution, as the solubility of $MgCO_3$ is considerably higher than of $CaCO_3$.

Also high quantities of gypsum in the irrigation water may lead to a loss of Ca^{2+}-ions by precipitation in the soil, although the solubility of gypsum in salt solutions is 30 meq/l or more.

Figure 9.3 shows that the ionic product $[Ca^{2+}][SO_4^{2-}]$ varies between 600 and 1600 meq^2/l^2 in salt solutions with ionic strengths in the range between 0.05 and 0.5 mol/l (Van Beek, 1973). The EC-values corresponding to these ionic strengths (see Griffin and Jurinak, 1973) are plotted on the ordinate at the bottom of the graph. The graph can be used to determine how much Ca^{2+} and SO_4^{2-} will precipitate and form the ion pair $CaSO_4^°$ and

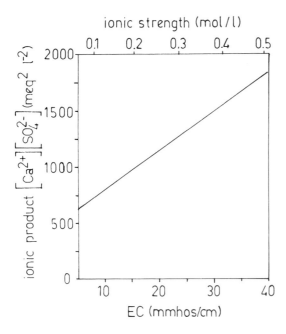

Fig. 9.3. Ionic product of $CaSO_4$ as a function of the ionic strength and electrical conductivity.

thus estimate the amounts of Ca^{2+} and SO_4^{2-} that will remain in solution. This is done by subtracting from the $c_{o,Ca}$ and c_{o,SO_4} calculated with equation (9.8) equal concentrations of Ca^{2+} and SO_4^{2-}, such that the product of the remaining Ca^{2+}-and SO_4^{2-}-concentrations equals the ionic product corresponding to the prevailing ionic strength or EC-value.

All the estimates described in sections 9.2 and 9.3 have been based on calculated average salt concentrations of the whole profile depth under consideration. As was already pointed out in section 9.2, in reality the salt concentrations may vary strongly with depth. This is an additional reason why the estimates discussed must be viewed to be very rough approximations. A refinement of the estimates can be obtained only if the salt- and water-balance model described in section 9.2 is applied to separate depth intervals within the profile or rooting zone. This leads to complicated and time-consuming calculations, which can be carried out only with the aid of a computer. However, the calculations for the complete profile depth will at least yield rough predictions to be used for the evaluation of practical field situations.

9.4. ALKALINIZATION UNDER IRRIGATION

Whereas the alkalization (or sodication) described in section 9.3 is the progressive saturation of the soil complex with 'alkali' ions (mainly Na), the term alkalinization is used here to indicate a rise of the pH to 'alkaline' values. In this respect it must be mentioned that high sodium-saturation is not necessarily accompanied by high pH-values as is the case, e.g., in most solonetzes. Many sodic soils have a near neutral pH, i.c. when the Na^+-ions in solution are present as neutral salts such as $NaCl$ or Na_2SO_4. Some are even acid in reaction (solods and solodized solonetzes).

The strongly alkaline reactions of most solonetzes and many other sodic and saline sodic soils is due to the presence of appreciable concentrations of carbonate and bicarbonate ions in the soil solution. Hence the sum of (HCO_3^-) and (CO_3^{2-}) plus the (usually negligible) excess of (OH^-) over (H^+) is referred to as the alkalinity concentration (Alk, cf. chapter 6 and 8).

As explained in section 6.1, the pH of the solution may now be expressed as a function of the carbon-dioxide pressure in the gas phase (P_{CO_2}) and the alkalinity concentration (Alk). Figure 9.4 presents this relationship for a well-aerated and a poorly aerated soil. It can be seen in this figure that the two alternative criteria for the distinction of alkaline and non-alkaline soils (section 9.1) coincide only if P_{CO_2} = 0.3 mbar, i.e., in well-aerated soils. Under poorly aerated conditions, which occur temporarily shortly after an irrigation gift and more permanently under wet rice cultivation, an alkalinity concentration of 1.5 meq/l causes a pH-value considerably lower than 8.5. This phenomenon is also responsible for the fact that pH-values measured in saturated pastes (a standard practice in the analysis of saline and sodic soils) may vary with the aeration status of the paste. Thus the pH-values measured

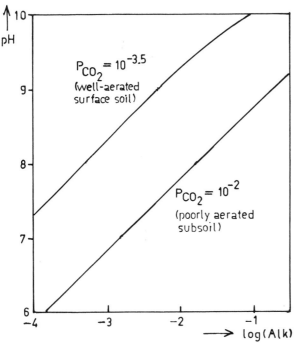

Fig. 9.4. pH as a function of the alkalinity concentration (mainly CO_3^{2-} and HCO_3^-, in eq/l) at two CO_2-pressures (bar) (after Van Beek and Van Breemen, 1973)

directly after the preparation of the paste are generally higher than those measured a day later. Therefore, the alkalinity concentration appears to be a safer and more reliable criterion for judging the alkalinity status of a soil than the pH-paste.

It was discussed already in chapter 6 that the concentrations of Ca^{2+} and Mg^{2+} in the soil solution are kept at a very low level in the presence of dissolved carbonates. Figure 9.5 shows that in a well-aerated soil the concentration of $CaCO_3$ is approx. 1 meq/l. It can be shown that in the presence of 1.5 meq/l dissolved carbonate the Ca-concentration decreases to about 0.7 meq/l. As the total salt concentration in (semi-)arid zone soils invariably surpasses 10 meq/l, one may conclude that this low Ca^{2+}-concentration is accompanied by at least 10 meq/l of other cations. Usually these cations are almost exclusively sodium, which yields an SAR-value of 17 and an ESP-value of approximately 20 %. Thus it can be seen that alkalinity in a well-aerated soil is almost invariably accompanied by sodicity.

A notable exception to this general rule is found where the Mg^{2+}-concentration in the irrigation water or ascending groundwater is higher than the Ca^{2+}-concentration. Then an appreciable part of the Mg^{2+} will remain in solution, as the solubility product of $MgCO_3$ under well-aerated conditions may be estimated at approximately 10 meq^2/l^2. This can lead to the formation of so-called 'Mg-Na-solonetzes', which have a high Mg^{2+}-saturation combined with an ESP-value below 15%. Some of these are known for their poor physical behavior. In most natural waters, however, the Ca^{2+}-concentration surpasses the Mg^{2+}-concentration. In that case practically all Mg^{2+} may be precipitated in the form of the highly insoluble dolomite: CaMg(CO$_3$)$_2$.

In conclusion it can be stated that alkaline sodic soils are formed if the irrigation water contains an excess of (CO$_3^{2-}$ + HCO$_3^-$)-ions over the (Ca^{2+} + Mg^{2+})-ions present. This excess is called the *R*esidual *S*odium *C*arbonate (RSC) or residual alkalinity (meq/l). It is an important additional value used for classifying irrigation waters, because it determines the alkalinization hazard (cf. figure 9.4 and table 9.2). When all the Ca^{2+} and Mg^{2+} added by an irrigation have precipitated in the soil as Ca- and Mg-carbonates, the excess (CO$_3^{2-}$ + HCO$_3^-$) will be present as dissolved Na$^+$ (and K$^+$)-carbonates and -bicarbonates. On the long run, the excess carbonates and bicarbonates added will also precipitate Ca^{2+} and Mg^{2+} exchanged from the soil, until the adsorption complex is almost completely saturated with Na$^+$ (+ K$^+$). Thereafter, continued addition of irrigation water will cause accumulation of dissolved Na (+ K)-carbonates, leading to pH-values as high as 10 in well-aerated soils (cf. figure 9.4).

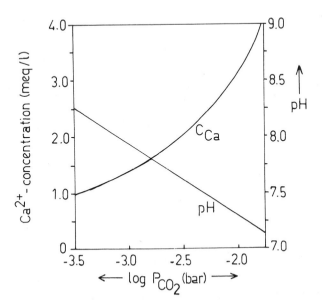

Fig. 9.5. Solubility of calcite and resulting pH-value in water at different CO$_2$-pressures. (after Van Beek, 1973)

A final conclusion that can be drawn from the above description of the alkalinization process is that truly alkaline soils (pH > 8.5 or Alk > 1.5 meq/l) are almost invariably sodic, because the carbonate and bicarbonate ions preclude the simultaneous presence of considerable Mg- and Ca- concentrations. Hence the alkalinity should be viewed as the cause rather than the result of sodication.

There are indications that the carbonate and bicarbonate-ions present are not always added to the soil by irrigation water (or ascending groundwater). Under certain sets of conditions they may be formed in situ in the soil profile. An example is the formation of carbonate accompanying sulphate reduction in water-logged soils (cf. chapter 8).

9.5. CHEMICAL ASPECTS OF THE RECLAMATION OF SALINE AND SODIC SOILS

For saline soils the method of repair is leaching (combined with measures to improve the drainage characteristics, if necessary), which has hardly any soil chemical aspects. For sodic soils the adsorbed Na-ions must be exchanged against Ca-ions, which is effectuated by leaching with water of a low SAR-value. To this purpose one may add soluble Ca-salts to the water used for leaching. Lime is practically useless for this purpose, due to its low solubility.

Only under certain conditions lime can be instrumental in lowering the ESP of strongly sodic soils to some extent. The first of these conditions is that the lime must be present in the soil in a finely distributed form. Secondly, the leaching must take place under a crop grown in a submerged environment (e.g. rice). The explanation for this is probably that the root activity will cause a strong increase of the CO_2-pressure. As a result $CaCO_3$ will become more soluble (cf. figure 9.5 and also figure 6.3). Part of the Ca^{2+} thus dissolved will exchange with adsorbed sodium, which causes some further dissolution of $CaCO_3$, provided that the carbonate ions thus accumulating are continuously removed by leaching. Nevertheless this process has a low efficiency per unit of leaching water applied, because the Ca^{2+}-concentration can hardly exceed 10^{-3} molar even at a CO_2-pressure of 0.01 bar. This has been confirmed by field experiments, where the decrease of the ESP in strongly sodic soils became noticeable after some years, but then only in the surface soil. Addition of farmyard manure and other organic amendments appears to enhance the process, which may be due to the CO_2-production caused by their decomposition.

Generally gypsum ($CaSO_4$) is used, because as a raw material it is often rather cheap. A situation may arise, however, where the solubility of gypsum is still too low for the set purpose. This depends on the relation between the factors which favor the development of extended double layers and thus cause swelling, i.c. the percentage Na on the exchange complex and the total electrolyte concentration. For a given ESP-value there exist critical concentrations below which swelling and peptization will occur. The higher the ESP-value, the higher are these critical concentrations (cf. figure 9.2). For a soil with a very high ESP-value (e.g. above 60) the electrolyte concentration of a saturated gypsum solution may be too low to prevent swelling. Application of gypsum with 'high quality' irrigation water (i.e. low salt concentration) may then result in swelling and structure decay. This will occur specifically in the deeper layers, which upon the application of a gypsum solution are initially percolated by the leachate from the upper layers. Following the exchange in the upper layers this leachate still has the high SAR-value corresponding to the original ESP-value of the unreclaimed soil, combined with a salt concentration equal to that of the leaching water used.

Such swelling in the subsoil may be catastrophal, because the impeded drainage prevents further leaching with the gypsum solution. It may thus be necessary to run test-percolations with gypsum. In those cases where gypsum proves to be unsatisfactory a salt with a higher solubility must be applied (e.g. $CaCl_2$). Although this is often a rather expensive cure it may, nevertheless, be necessary.

If local circumstances do not permit the application of Mg- or Ca-salts other than gypsum, it may be possible to start the leaching treatment with water in which the concentration of sodium salts has been increased (temporarily). The initial decrease of the very high ESP-value is then executed at a rather high salt content, which suppresses swelling. Following a drop of the ESP-value the concentration of the added salts may be (gradually) decreased. To this purpose one might adjust the concentration of the leaching water by changing the ratio of high quality water and low quality water (high EC-value) in a mixture used for leaching (while applying gypsum). The disadvantage of this method is the decreased efficiency of the gypsum applied, which is due to the competition of the other (usually Na-)salts present. Because of this reduced efficiency excessive quantities of water may be needed for the reclamation. These quantities may not be available or exceed the drainage capacity of the soil or the area.

The final value of the ESP-value to be attained by a reclamation operation depends on the sodicity tolerance of the crops to be grown and on the physical reaction of the soil to different ESP-levels. In turn this final ESP-value prescribes the SAR-value of the water to be used for leaching (cf. equation 9.9). Using this SAR value it is possible to estimate the amount of gypsum to be added per unit irrigation water with the equation:

$$\text{SAR} = \frac{c_{i,+}}{\sqrt{(c_{i,2+} + x)/2}}, \qquad (9.11)$$

where $c_{i,+}$ and $c_{i,2+}$ are the original concentrations of the monovalent and divalent cations in the irrigation water and x is the amount of gypsum to be added (meq/l). As the reclamation process described here is a case of favorable exchange chromatography (cf. chapter 7), there is a sharp exchange front between a reclaimed top soil (with final ESP) and the not yet reclaimed lower part of the profile (with initial ESP). The depth of soil to be reclaimed (compare penetration depth x_p in chapter 7) is determined by the optimal rooting depth of the crops to be grown. Once this reclamation depth is decided upon, the extra amount of divalent cations to be fed into the soil with the leaching water (in keq/ha) may be estimated with the equation:

$$(Ca + Mg) = (ESP_i - ESP_f) \times CEC \times \rho_b \times D_{soil}/100, \qquad (9.12)$$

where ESP_i is the initial ESP-value of the unreclaimed soil, ESP_f is the final ESP-value (yield value) to be obtained, CEC is the cation exchange capacity (meq/100 g) and ρ_b is the bulk density (g/cm^3) of the soil. The figures to be filled out in this equation are averages over the soil depth to be reclaimed (D_{soil} in cm).

The total amount of leaching water required in mm waterlayer is found if $10(Ca + Mg)$ in keq/ha is divided by $(c_{i,2+} + x)$ in meq/ml.

If the reclamation is performed by intermittent irrigation (e.g. under a crop) the required gypsum concentration in the irrigation water (x in meq/l) will have to be calculated, taking into account that the ionic concentrations in the irrigation water will have to be increased according to equation 9.8.

The gypsum requirement thus estimated must be corrected for any amount of solid gypsum already present in the depth of soil to be reclaimed. This correction may be quite large for soils that have become sodic mainly due to the precipitation of gypsum from irrigation water or ascending groundwater (cf. section 9.3). In some cases such soils may, upon leaching, change from saline sodic soils to non-saline non-sodic soils without any addition of gypsum to the leaching water.

Once the amount of water required for reclamation is calculated, it is also possible to make a fair estimate of the time needed to complete the reclamation operation. This period is determined by the amount of water added per annum, which in turn depends on the leaching method applied (cf. Boumans at al. 1963). Percolation resulting from ponding of water on the soil surface (e.g. under a rice crop) is more rapid than under intermittant irrigation. Furthermore, the amounts of water and gypsum required and thus the time period in which the reclamation is completed have to be divided by the leaching efficiency factor if the latter is not equal to unity. Apart from this, the infiltration and drainage capacities of the soil have to be considered. They are often the limiting factors determining the downward percolation of water, particularly in sodic soils.

As a final remark it is mentioned that, given a total amount of gypsum applied, the reclaimed depth and the ESP-reduction effectuated are inversely proportional. In practice this means that the reclaimed depth per amount of gypsum applied can be manipulated by adjusting the SAR-value of the water used. This can be done, e.g., by controlling the amount of gypsum dissolved per liter of irrigation water (application to the irrigation water instead of to the soil surface) or even by adding Na^+-salts. In this way one can choose from a whole range of posibilities between 'high intensity' reclamation of a shallow surface layer or 'low intensity' reclamation over a greater soil depth. Which of these alternatives is to be preferred depends on the specific requirements (sodicity tolerance and rooting depth) of the crops to be grown.

Thus it is seen that the possibility to calculate and manipulate the penetration depth of the desodication front is a valuable tool in the control of reclamation operations. In the text on sodication on page 179 it was already mentioned that no such possibility exists in case of the sodication process. This is only partly due to the fact that sodication is a case of unfavorable exchange, with a diffuse exchange front. It should also be recognised that desodication as described here is a one-time operation, whereas sodication is a long-term process lasting several years. During these years the soil is subjected to many cycles of wetting and drying, accompanied by alternate upward and downward transport of water. In such a complicated situation a first approximation can hardly go beyond assuming a fairly homogeneous distribution of the different ions over the rooting zone.

ILLUSTRATIVE PROBLEMS

1. Calculate the SAR-value of a saturation extract containing 25 meq/l Na^+, 10 meq/l $(Ca^{2+} + Mg^{2+})$, 20 meq/l Cl^-, 14 meq/l SO_4^{2-} and 1 meq/l $(HCO_3^- + CO_3^{2-})$. Estimate the EC_e-value and the ESP of the sample and classify the soil in terms of salinity, sodicity and alkalinity. (Answer: SAR = 11.2 $meq^{1/2}l^{-1/2}$, ESP $\approx$ 14%, $EC_e \approx$ $\approx$ 3.5 mmho cm^{-1}; classification: slightly saline, non-sodic, non-alkaline).

2. A non-saline soil (ρ_b = 1.6 g cm^{-3}, SP = 75%, FC = 25%) is brought under an irrigation scheme providing 100 cm irrigation water per year, which is just sufficient to meet the consumptive use of the crops, (rainfall is negligible). The irrigation water contains 4 meq/l salts. Estimate how long it will take before the yields of salt-tolerant crops with a rooting depth of 50 cm go down. Also estimate how much extra irrigation water has to be supplied annually in order to keep the salt concentration in the rooting zone below the level harming very salt-sensitive crops, assuming a leaching efficiency factor equal to 0.5. (Answer: 24 years, 14 cm/year).

3. An illitic soil ($K_{Na,Ca}$ = ½ $mol^{-1/2}l^{1/2}$) is irrigated with water containing 6 meq/l NaCl, 3 meq/l $CaCl_2$ and 1 meq/l $MgCl_2$. The amount of irrigation water supplied is 10% more than the amount required for the consumptive use (rainfall is negligible). Classify the irrigation water in terms of salinization and sodication hazard and calculate the yield value of the adsorption complex, assuming a leaching efficiency factor equal to unity. (Answer: high salinization and low sodication hazard; ESP $\approx$ 21%).

3a Also estimate the yield values in the situation described above for the following compositions of the irrigation water (meq/l). Assume the soil to be well-aerated.

$C_{i,Na}$	$C_{i,Ca}$	$C_{i,Mg}$	$C_{i,Cl}$	C_{i,SO_4}	C_{i,HCO_3^-}
4.0	1.5	0.5	5.0	1.0	-
4.0	1.5	0.5	1.0	5.0	-
4.0	1.0	1.0	4.0	1.0	1.0
4.0	0.5	1.5	4.0		2.0

(Answer: 19%, 24%, 25% and 44%, respectively).

4. An illitic soil (CEC = 8 meq/100 g, $K_{Na,Ca}$ = 0.5 mol$^{-1/2}$l$^{1/2}$, SP = 60%, FC = 20%) is irrigated with a sprinkling system (leaching efficiency factor ≈ 1). The irrigation water contains 3 meq/l Ca^{2+}, 2 meq/l Mg^{2+}, 5 meq/l Na$^+$, 5 meq/l Cl$^-$ and 5 meq/l SO$_4^{2-}$. Rainfall is negligible and the consumptive use of the crop equals 800 mm/year. How much irrigation water must be supplied annually in order to keep the salinity level of the soil just below EC$_e$ = 2 mmho/cm ? Estimate the yield value of the soil under this irrigation system. Investigate the possibilities to reduce this yield value to ESP = 10% (e.g. by the application of gypsum or by increasing the annual irrigation supply). (Answer: 960 mm/year gives ESP ≈ 13%. Reduction of yield value by application of gypsum only is not possible, because the soil solution is already saturated with Ca^{2+}- and SO$_4^{2-}$-ions. The yield value can be reduced to ESP ≈ 10% by supplying 1100 mm irrigation water).

5. A gypsum-free sodic soil (ρ_b = 1.8 g/cm^3, CEC = 20 meq/100 g, $K_{Na,Ca}$ = = 0.5 mol$^{-1/2}$l$^{1/2}$) is reclaimed by continuous leaching under ponding of water (leaching efficiency factor ≈ 0.5). Calculate how much (Ca^{2+} + Mg^{2+}) is required (keq/ha) to lower the ESP of a 50 cm thick topsoil from an initial value of 20% to a required value of 5%. If given that the available leaching water contains 10 meq/l Na$^+$ and 3 meq/l (Ca^{2+} + Mg^{2+}), mainly as chloride, then calculate how much gypsum has to be added to this water (meq/l). Also calculate the total amount of water (cm) and gypsum (tons/ha) required to effectuate this reclamation. Finally estimate the time period needed to complete the operation, if given that the drainage capacity of the soil equals 1 cm/day. (Answer: 270 keq/ha; 13 meq/l; 334 cm; 39 tons/ha; 300 days).

CONSULTED LITERATURE

Arjan Singh, 1969. *PhD-Thesis*, Haryana Agric. University, Hissar (Har), India.
Beek, C.G.E.M. van, 1973. Chemische processen in zoute gronden. Voordracht 73e wetenschappelijke bijeenkomst van de Ned. Bodemkundige Vereniging, Amsterdam.
Beek, C.G.E.M. van, and van Breemen, N., 1973. The alkalinity of Alkali Soils. *J. Soil Sci*, Vol. 24, No. 1.
Boumans et al., 1963. *Reclamation of salt-affected soils in Iraq*. Publ. No. 11. Int. Inst. for Land Reclamation and Improvement, Wageningen, The Netherlands.
Bower, C.A., 1961. Prediction of the effects of irrigation waters on soils. *Proc. Unesco Arid Zone Symp.*, Teheran, p. 215-222.
Griffin, R.A. and Jurinak, J.J., 1973. Estimation of activity coefficients from the electrical conductivity of natural aquatic systems and soil extracts. *Soil Sci.* Vol 116, No.1 (Utah State Univ. Logan).
Molen, W.H. van der, 1973. Salt Balance and Leaching requirement. In: *Drainage Principles and Applications*, Publ. 16, Vol II, Int. Inst. for Land Reclamation and Improvement, Wageningen, The Netherlands.

Reeve, R.C. and Doering, E.J., 1966. The high salt-water dilution method for reclaiming sodic soils. *Soil Sci. Soc. Amer. Proc.* 30: 498-504.
U.S. Salinity Laboratory Staff, 1954. *Diagnosis and improvement of saline and alkali soils* USDA Agric. Handbk. No. 60.

RECOMMENDED LITERATURE

Reeve, R.C. and Fireman, M., 1967. Salt problems in relation to irrigation. In: *Irrigation of Agricultural Lands, Agronomy Series No 11*, Amer. Soc. of Agron., pp. 988-1008.
U.S. Salinity Laboratory Staff, 1954. *Diagnosis and Improvement of saline and alkali soils* USDA Agric. Handbook No. 60.

CHAPTER 10

POLLUTION OF SOIL

F.A.M. de Haan and P.J. Zwerman

Very briefly soil pollution could be typified as the malfunctioning of soil as an environmental component following its contamination with certain compounds particularly as a result of human activities. Unfortunately this statement throws up more questions than it answers. This is particularly the case when it is attempted - in view of the presumed undesirability of soil pollution - to specify limits as to permissible and non-permissible human interference with soil. Such limits are the necessary prerequisites of any legislative action undertaken for protective purposes.

The reason for the above lack of clarity is all too obvious. In order to establish present - or predict future - malfunctioning of soil, one would have to know precisely how soil functions as an environmental component, both for 'natural' and 'man-made' conditions. In addition it would be required to extrapolate this knowledge to all those situations involving the presence of 'contaminants' in order to see whether these could possibly interfere, and if so at what levels. Further dissection of the problem then shows that the functioning of soil as an environmental component is manyfold - granted that its role as a 'support' for the growth of plants is a major one - while the term 'contaminant' is often ill-defined as many compounds which are present regularly in particular soils and are even necessary in small amounts, may become inhibitive beyond certain limits. Finally the phrase 'resulting from human activities', though inferring a possibility of terminating such activities if adversely affecting the functioning of soil, does not necessarily point to the desirability of stopping these, as many human activities were designed to enhance the functioning of soil in certain aspects, though admittedly they could lead to undesirable effects in others.

A typical example is here the introduction of irrigation practices. Designed to counter the malfunctioning of soil because of a deficit of water for sustaining satisfactory crop growth, these often tend to lead to the accumulation of salts, locally or regionally, which are in fact contaminants leading to malfunctioning. As such, soil salinization as a result of irrigation practice could be regarded as one of the oldest examples of soil pollution, in contrast to forms of salinization occurring at natural conditions without human influence.

Clearly, any effort to discuss the pollution of soil within one chapter must remain limited in scope. To this purpose it appears reasonable to discuss first - rather superficially - the functioning of soil as an environmental component. Next some general aspects of malfunctioning due to the presence of contaminants will be mentioned. The bulk of this chapter will then be devoted to a compound-wise discussion of possible contaminants, mentioning source and fate (in the soil system) and elaborating somewhat on their specific role in the functioning of soil.

10.1 SOIL AS AN ENVIRONMENTAL COMPONENT

Next to air and water - both seen in a global sense and not as gaseous and liquid phases, respectively, of the soil system- soil is generally considered as the third main environmental component. Aside from providing a 'platform' for the activities of land-based animals, including human society, soil has an unique function as a medium supporting many forms of life, with plant growth as a prime example. In practice soil provides the basis, also in a literal sense, of a substantial part of the collected life on earth via the capture of sun-energy by green plants. While the above stresses the functioning of soil with particular reference to the global carbon cycle, most other elements involved in the sustenance of life are subject to more 'down-to-earth' cycles. In this respect soil provides the proper environment for a, usually fairly rapid, breakdown of dead material particularly via microbial action to elementary compounds that may subsequently re-enter the cycle, again particularly via the vegetation.

At this point the intimate contact between the soil solid and liquid phases must be considered. Besides being a prime prerequisite for all forms of life, water also serves as the main carrier of many compounds through soil. In this way water may exert a moderating effect, e.g. by carrying away excess salts to the oceans. Conversely soil often acts in practice as a purifying 'filter' on water containing dissolved and colloidal constituents, particularly also organic materials which may mineralize during the passage through the aerated top soil. This filtering capacity of soil may intentionally be made use of in waste disposal systems.

The above picture of soil as an environmental component, although grossly oversimplified, nevertheless summarizes its functioning in terms of an intricate semi-steady state process, fluctuating in response to external factors. The dual functions of soil in this process, viz. stressing plant growth and other forms of life while acting in the mean time as a sink for nature's refuse - including materials locally entering the soil with flood water - are intimately connected. Such a system in its natural condition tends to have a fairly large buffering capacity with respect to these external factors. And this then leads immediately to the heart of the problem of soil pollution: lately situations have arisen where the buffering capacity of the system became exceeded.

In this context it must be realized that the primary functions of soil mentioned above may become conflicting to some degree if either one of these is pushed to an extreme. Thus the excessive addition of fertilizers in an attempt to increase crop yield lead to an impaired filtering function of soil, the drainage water containing excess fertilizer compounds, much alike excessive irrigation with poor quality water may lead to salinization elsewhere. Conversely the local addition of waste water could lead to impaired crop growth due to nutrient imbalance. The answer in such cases of 'conflicting interests' must obviously be sought in the direction of optimalization schemes. The natural functions of soil may and should all be used, provided care is taken not to upset grossly the balance. Fortunately the buffering capacity of the soil as a steady state system appears to be fairly large.

10.2. RECOGNITION AND PREDICTION OF SOIL POLLUTION

Much in contrast to the other two main environmental components, viz. air and water, pollution of soil is not easily measured in terms of a chemical composition. As will have become overly clear in the preceding chapters, a 'pure' soil is undefinable. Granted that in a number of cases the presence of a particular compound may be traced back to local human activities (by comparison with similar non-exposed soil samples), this in itself does not prove that the functioning of soil is impaired.

Turning to the other extreme, impaired functioning is sometimes easily observed, that is, unsatisfactory crop yields both in a quantitative and a qualitative sense, or undesirable constituents in drainage water may relatively easily be detected. Unfortunately, at the time of such observations it may be too late for easy repair. This is particularly so because the reaction time of an involved steady state system like soil tends to be long.

It is realized that a number of biological indicators can be used as early warning signals of changes in the composition of the natural soil ecosystem. In this regard not the reliability of the coherence between chemical and biological composition of soil but much more the absolute evaluation of the biotic soil fraction constitutes a problem. As a rather artificial but elucidating example in this respect it may be said that the absence of earth worms in soil as a consequence of excessive Cu-levels can obviously be evaluated as disadvantageous. The adaptation of specific genera of earth worms to high Cu-levels, however, constitutes a very specific problem which is hard to evaluate.

Accordingly it seems imperative that the handling of potential pollution problems in soil must be based on the *prediction* of likely or possible impairment of the functioning of soil. In practice this implies in the first place knowledge of the composition of the influx. Next the fate of the compounds

present in the influx when passing through the soil system must be predicted. Thus the central theme of soil pollution phenomena is knowledge of transport and accumulation processes in soil, particularly of hazardous compounds. In this respect it is evident that *accumulation* and *mobility* are greatly governed by interactions of the compound of interest with the soil solid phase and in specific cases by its degradability. Such interactions can in a preliminary way be enumerated as:
1. Positive adsorption as induced by electrostatic attraction between charged compounds and oppositely charged soil constituents.
2. Electrostatic repulsion when the electric charge of compound and soil consituent is of the same sign. This is usually the case with certain anions and the predominantly negative charge on e.g. clay minerals.

   These phenomena have been covered sufficiently in the chapters 4 and 5, respectively. As has been pointed out there positive adsorption and electrostatic repulsion of ions usually occur simultaneously in soil. Especially with respect to anion interactions they use to be of the same order of magnitude. Whether the final combined effect results in a net positive adsorption or exclusion entirely depends on the relative magnitude of both phenomena at the prevailing conditions.
3. Chemisorption. This interaction mechanism can often hardly be distinguished from electrostatic positive adsorption except for the value of the adsorption energy. This is usually considerably higher in case of chemisorption. Moreover, and actually as one of the consequences of this high adsorption energy, chemisorption is usually characterized by a very limited exchangeability with other compounds.

   Although the above three mechanisms constitute the major interactions between ions and soil constituents, different bonding mechanisms may prevail for specific compounds. So, for instance, in the case of organic chemicals also London-van der Waals attraction, H-bridge formation, salt bridges and metal ion bridges may contribute to the bonding in soil.
4. Precipitation and dissolution reactions. These have been described to considerable extent in chapters 2 and 6, dealing with chemical equilibria in soil systems. They may play a predominant role in governing the mobility of certain compounds like heavy metals and phosphorus.

   The considerations as presented in the above mentioned chapters are extremely useful in the calculation of prospective equilibrium concentrations as a function of the main factors which govern the solubility of a certain compound of interest. However, they do not allow any conclusions as to the rates of reactions involved. In many soil pollution problems especially these reaction rates, and thus reaction kinetics, are of prime consideration.

5. Decomposition and turnover reactions. Many compounds, when present in soil, are subjected to reactions as (photo)chemical degradation, microbial degradation, or a combination of these. An example as such is provided by the fate of pesticides, or more generally organic chemicals including e.g. certain soil conditioners. Then the persistence in soil, air or water, as governed by the degradability of the compound, is one of the main factors with respect to possibly hazardous effects on the environment.

The above interaction mechanisms may either occur separately or in combination with each other, either successively, simultaneously or even alternating. Rather than attempting to present a complete description of all these phenoma, the discussion is limited in the following pages to a general description of the behavior in soil of a number of compounds which may be of interest with respect to soil or groundwater pollution. As follows from earlier considerations, it fully depends on the concentration of the compound whether or not it should be looked upon as a pollutant.

Certain criteria in soil pollution evaluation are to be derived from the required standards of drainage water reaching the groundwater stream; these in turn may depend on the geographic location with respect to water 'harvesting' for consumptive use and/or the vicinity of open water. As to the aspect of malfunctioning of soil in sustaining plant growth, the expected effect will have to be based on existing (or newly collected) knowledge of plant reaction to 'abnormal' composition of soil, in particular of the soil solution. These considerations may comprise poor growth but also the accumulation of undesirable compounds in the plant. Depending particularly on the modes of interaction of a compound under consideration with plant and soil, the malfunctioning of soil may or may not coincide with a malfunctioning in sustaining plant growth.

Granted that the control of soil pollution is obviously the task of a multidisciplinarily oriented team of scientists comprising e.g. environmentalists, soil scientists, microbiologists, plant physiologists, toxicologists, ecologists, etc., the present text is primarily centered on soil chemical aspects, thus excluding the very important biotic part of soil. The latter part is only referred to in specific cases, e.g. where microbial activity plays a dominant role in the fate of a compound under consideration.

Because of the above, the compound-wise discussion of possible contaminants has been chosen. The delineation of compounds to be discussed was also based on the consideration that soil may play a significant role in the solution of a number of environmental problems, especially in relation to waste disposal. Thus emphasis is given to nitrogen and phosphorus which are of specific interest with respect to eutrophication, and to a number of compounds which may be of direct importance with respect to human and animal health, like heavy metals and pesticides. Soil pollution as induced by somewhat specific events like oil spills, natural gas leakages and sanitary landfills are described under a miscellaneous heading.

Effort has been aimed more toward a general coverage of the entire field of soil pollution aspects than toward an exhaustive description of the fate and pathway of a number of compounds. As a consequence, this chapter has more or less the character of a review.

10.3 NITROGEN AND PHOSPHORUS IN SOIL

In contrast to the other compounds to be discussed, N and P are necessary prerequisites for the growth of higher plants in comparatively large quantities, thus constituting major plant nutrients. In the existing soil science literature considerable attention has always been paid to the occurrence and behavior of these major plant nutrients. Notwithstanding the fact that this extensive literature was particularly directed toward the finding of solutions for situations where the nutrients were deficient, much of the information collected in the past decennia is of profound interest also in the context of the present discussion. While referring to existing textbooks on soil fertility and also soil microbiology for more detailed information, some of the relevant facts will be reviewed here briefly. Before going into this matter for each of the two elements separately it may be useful to make a few off-hand statements outlining some common features of both as against their differences, from the viewpoint of soil pollution hazards.

Starting with the use of soil for regular agricultural purposes, i.e. growing a crop, malfunctioning of the soil due to excess of phosphates has hardly ever been a problem, while excess of nitrogen occurs only under conditions of imprudent use of chemical fertilizers or excessive use of manure, and is then simply repaired by omitting such excessive gifts.

In contrast malfunctioning of the soil as a filter - or actually its acting as a source - is not uncommon in case of nitrogen compounds, because of the relatively high mobility of the latter in soil (cf. the following section). While this aspect may already be of concern under conditions of regular agricultural use of soil, it constitutes a central problem if soil is used for the disposal of waste products like sewage water and sewage sludges and e.g. waste water from potato starch industry.

It may be of interest to note here that the use of N-containing organic waste products to replace N-gifts in the form of fertilizer, although offering at first sight the possiblity of serving two purposes at once, is not necessarily an ideal approach. This is so because the mineralization of such wastes which necessarily precedes the uptake of their nitrogen by crops (cf. below) constitutes a slow and continuous process, while the uptake is typically tied to a relatively short period of crop growth. Accordingly a sufficient supply of nitrogen derived from organic sources in the period of maximum need will often imply excess leaching, e.g. of nitrates, to the groundwater during the remaining time of the year.

As will be discussed in the sections on phosphates, the latter compounds exhibit a comparatively low mobility in soil. Accordingly chances that excess phosphates (in comparison to the amount used for crop production) reaches the groundwater in deeper layers are usually small.

Contrasting here the present two elements, it may be stated that soil often acts as an excellent filter for phosphates with a huge retention capacity, while its use for the disposal of nitrogen compounds is rather limited and hinges on its capacity to provide a medium for denitrification processes and is as such much more vulnerable.

10.3.1. Pollution effects involving nitrogen

The adverse influence of nitrogen overfertilization on crop quality and yield is a well-known phenomenon in practical agriculture and thus in normal crop production the use of excessive quantities of fertilizer is usually avoided. Leaving a description of the effect of excess nitrogen on plant growth to texts on soil fertility, it is mentioned only that the ensuing depression of quality and/or quantity of crop yield could be weighed against the needs for disposal of certain wastes to reach a decision as to what limits should be imposed. Far more important are the effects of an increase in the concentration of nitrogen compounds in the drainage water leaving the top soil.

The current concern about undesirable nitrogen concentrations in water has a dual base. There are direct health aspects as well as ecological aspects. Nitrate-nitrogen may present a health hazard. Winton et al. (1971) described the circumstances which may induce methemoglobinemia or cyanosis in infants. The main controlling factor in this disease is the daily nitrate intake, and hence the nitrate concentration of drinking water plays an important role. Drinking water standards in the USA and in the Netherlands mention nitrate concentrations of 45 and 100 mg NO_3/l, respectively, as maximally permissible levels. Also livestock may suffer from a number of morbid symptoms caused by too high nitrate-nitrogen levels in the drinking water like methemoglobinemia, vitamin A deficiency, reproductive difficulties and abortions, and loss in milk production.

The second aspect of the concern about increased concentrations of N in water is the fear of eutrophication of surface water. Eutrophication, which means 'enrichment with nutrient elements', may cause rapid growth of aquatic plants. The most commonly known features of eutrophication are the nuisance blooms of phytoplankton (Sawyer, 1947). As was pointed out by Owens and Wood (1968) eutrophication should be looked upon as a natural process which is accelerated by human activities. Examples of such activities are discharges of domestic and industrial waste waters and of treatment-plant effluents, runoff, and leaching from heavily fertilized agricultural land. Vollenweider (1971) presented a classification of water bodies with respect to P- and N-content as given in table 10.1.

TABLE 10.1

Eutrophication classification of water according to Vollenweider (1971).

category	concentration, in mg/m^3 (ppb)	
	total P	inorganic N
1. ultra oligotrophic	< 5	< 20
2. oligo-mesotrophic	5 - 10	200 - 400
3. meso-eutrophic	10 - 30	300 - 600
4. eu-polytrophic	30 - 100	500 - 1,500
5. polytrophic	> 100	> 1,500

Since the exact nature of eutrophication is still insufficiently understood, definite prescriptions to stop or prevent its occurrence are hard to present. Woldendorp (1972) described situations in which each of the nutrient elements phosphorus, nitrogen and carbon should be considered as the limiting factor in the development and growth of algal cultures. It is, however, commonly accepted that P-concentrations below 0.01 mg/l and N-concentrations below 0.2 - 0.3 mg/l are prohibitive for algal bloom.

10.3.2. Sources of (excess) nitrogen in soil

Leaving the 'natural' entry of N into soil via rain water containing some nitrate (as the result of N_2-oxidation during lightning) and as N_2-gas becoming assimilated by certain species of bacteria out of consideration, the application of N-fertilizers and additions via manure form a regular nitrogen source in all areas under agriculture. During recent years considerable discussion has developed about the question whether or not nitrogen fertilization at rates which are considered as required and advisable in agriculture, will contribute to nitrogen enrichment of ground- and surface water (Commoner, 1968; Henkens, 1972; Kolenbrander, 1972). As indicated by Welch (1972) on the basis of balance-sheet calculations for the period 1940 to 1970, for Illinois USA conditions the application of major fertilizer elements, including nitrogen during the third decade exceeded the removal by crops. This situation was found to be opposite to that in a preceding period, when more nutrients were withdrawn in crops than were applied by fertilization. Such a comparison of removal- and application data, however, does not allow conclusions to be drawn about the direct impact of nitrogen on the environment, since over-fertilization does not necessarily lead to an instantaneous leaching of the excess amount. In the long run, however, excessive nitrogen supply will undoubtedly induce translocation of nitrogen through the soil profile and thus increase the probality of an adverse environmental impact.

In agricultural areas nitrogen, no matter from which source it originated, may be present in or pass through the ecosystem in a number of forms (Aldrich et al., 1970). These may, in a first approach, be listed as follows:
a. as constituents of leaves, stems and roots of the crop;
b. as soil constituent in either organic or inorganic nitrogen form;
c. immobilized in microbial tissue resulting from decay of plant residues;
d. returned to the atmosphere as N_2 or as dinitrogen oxide following denitrification.

The last mentioned pathway is favored by anaerobic conditions and thus by the presence of water. Excessive amounts of water will displace nitrogen, especially in the form of nitrate, either by leaching or by surface runoff. An undesirable enrichment of the environment with nitrogen may occur if the above displacement exceeds the rate of denitrification. Zwerman and de Haan (1973) discussed the common misunderstanding that the environmental impact of fertilizer nitrogen would be different from that of other nitrogen sources. As was also pointed out by Aldrich et al. (1970), nitrate ions derived from fertilizer are neither more nor less subject to leaching than nitrate ions from other sources like plant residues, animal or human wastes, soil humus or precipitation.

Which one of the two mentioned displacement mechanisms, leaching or runoff acts as the main contributor to nitrogen pollution of the environment will depend predominantly upon climatic and geographic conditions, as well as on soil characteristics. It is realized that on a world-wide basis the contribution of erosional losses constitutes an important part of the overall soil-originating environmental impact. The present text, however, is mainly confined to leaching as a mechanism of displacement of compounds in soil.

Local and inadvertent addition of N to soil may arise in the neighborhood of industrial activities, e.g. by emissions of N-oxides into the air, which subsequently reach soil with rainfall. Hoeft et al. (1972) showed that for Wisconsin circumstances this form of nitrogen application may amount to an average value of 20 kg total N per ha per year, ranging from 13 to 30 kg/ha per year. Their measurements indicated that NH_4-nitrogen and organic-nitrogen contents of precipitation were higher in areas adjacent to barnyards than in areas removed from barnyards.

Finally waste disposal systems may obviously lead to local excessive additions of N-compounds. In those disposal systems involving crop production, it may appear profitable to apply nitrogen (as constituent of the waste) in certain overdoses as compared to crop needs (cf. also the relevant remarks in the introductory paragraphs of section 10.3.1). Such overfertilization would reduce the land surface area required for disposal, and consequently lower the investment. This decrease in expense may then economically balance the decrease in crop yield or quality that generally results from overfertilization (de Haan et al, 1973).

Having listed the sources of nitrogen it will now be attempted to outline the fate of nitrogen in soil. It will be clear from the above that insight in this fate constitutes also a prerequisite for proper management of waste disposal systems. In this context it is imperative to discuss the different forms in which nitrogen compounds may prevail in soil.

10.3.3. Forms of organic nitrogen in soil

Of the total amount of nitrogen present in most surface soils usually a very large proportion is constituted of organically bound nitrogen. This fraction may well be above 90 percent of total N for topsoils. As indicated by Bremner (1965), certain subsoils may have a considerable part of their total nitrogen, up to 30 and 40 percent, in the form of fixed ammonium. Almost all nitrogen withdrawn by plants from soil is in inorganic forms, although certain plants may be capable of utilizing specific organic nitrogen compounds like e.g. amino acids.

The fact that nitrogen is predominantly stored in soils in organic forms and is available to plants in inorganic forms, is the basis for the profound interest that soil fertility specialists have shown since long for the interrelationships between these forms. Such interrelationships are of renewed relevance in view of the environmental importance of the behavior of nitrogen in soils.

These interrelationships can be summarized by the terms: mineralization and immobilization. Both processes are predominantly microbiological in nature. Mineralization comprises all processes in which organic nitrogen is converted to inorganic forms, whereas immobilization refers to the reverse processes. In the latter case, inorganic nitrogen is being consumed by microorganisms. Conversion of nitrogen, from inorganic to organic forms and vice versa, of non-microbiological nature may occur following nitrogen uptake by plants. It may be reasoned that also fixation in clay crystal lattices and adsorption on e.g. humus complexes could be considered as forms of immobilization. The term is, however, reserved for microbiological conversions only.

A rough and qualitative description of microbial immobilization may read ad follows: when a sufficiently high supply of carbon-containing materials is prevailing, N, P and S may be controlling the expansion of the microbial population, as these elements are - in the above order - main constituents of microbial cells besides carboxylates and water. Following the uptake of these elements by microorganisms they are incorporated in the microbial cell tissue. Upon the addition of fresh organic matter to the soil, the growth of saprophytic bacteria, fungi and actinomycetes is being stimulated, and these microorganisms will attack the organic constituents of the added material (Stevenson, 1974). A part of the carbon disappears as CO_2 and a part is incorporated into biomass. The microbial expansion thus causes an increased loss of carbon and a tendency in the system to develop toward a C:N:P:S ratio which roughly is similar to that in microbial cell tissue. This ratio is commonly found as approximating 100:10:1:1 for the soil organic fraction that can be indicated as 'stable' (which term only means that a certain steady state situation has been obtained). The above reasoning then has an important implication on the fate of organic material when added to the soil: if this material has a C/N ratio considerably smaller than 10, mineralization of organic N will take place, whereas at considerably larger values nitrogen will be immobilized in microbial cell tissue.

From a quantitative standpoint, current knowledge on immobilization is still rather poor. This must be attributed mainly to analytical difficulties in fractionation and determination. Stevenson and Wagner (1970) present a twofold division of organic N compounds in soils. These main groups are: nitrogenous biochemicals which may directly result from enzymic synthesis by microorganisms living on plant and animal residues and a second group indicated as unknown forms of organic N, e.g. residing in the humic substances in soil. The nitrogenous biochemicals are mainly constituted of amino acids and amino sugars; amino acids may account for 20 to 50% of total organic N, whereas 4 to 10% is found as amino sugars. Only a very minor proportion of organic N, namely 1 to 2%, is found as a variety of well-defined organic N compounds like e.g. amines and amides.

From the above percentages it is evident that a substantial proportion of the soil organic nitrogen is present in unknown or ill-defined forms. As postulated by Stevenson and Wagner (1970), nitrogen-containing compounds found in microbial tissue usually can all be listed in the previously mentioned category of well-defined organics. This means that during the process of humification a number of conversion reactions takes place, producing nitrogen compounds of various nature. These are usually indicated as more stable humus forms and in a first subdivision are listed as humic acids and fulvic acids (cf. chapter 8). The chemical characteristics of these acids can be of prime importance in relation to the behavior of compounds which exhibit strong bonding to organic matter, like e.g. the majority of pesticides. Their relevance in this respect will be discussed in a later paragraph.

A number of steps and intermediates can be distinguished in the mechanisms and sequences of the formation of humic substances in soil. So far, plant residues were usually taken as a point of departure in this respect. The increasing problems associated with waste disposal force one to take also other organic sources into account.

There is a gradual enlargement of knowledge about organic matter in soils (e.g. Kononova, 1966; Flaig, 1966). Problems of analysis and identification already mentioned before, combined with the almost infinite number of reactions that may take place, work against rapid progress in this field. It will be obvious that a study of the kinetics of immobilization of nitrogen is even more complicated. Consequently this kind of information is completely lacking (Stevenson, 1970; Stanford et al., 1973).

10.3.4. Forms of inorganic nitrogen in soil

As described before, organic nitrogen compounds are not directly available for uptake by plants, but must be converted first into inorganic nitrogen forms. Since inorganic nitrogen usually amounts to a very minor proportion of total soil nitrogen, the reservoir of nitrogen potentially available for uptake by plants is mainly of organic nature.

The process of mineralization, being the turnover of nitrogen from organic into inorganic forms, is somewhat better understood than is that of immobilization, at least from a qualitative point of view. Inorganic nitrogen compounds are fairly well known and may be listed in the following main groups:

a. Ammonium-nitrogen, NH_4^+, and nitrate-nitrogen, NO_3^-; these are the main forms of nitrogen available for plant uptake.

b. Nitrite, NO_2^-, nitrous oxide, N_2O, nitric oxide, NO, and nitrogen dioxide, NO_2; these are found to occur in soils during short-time periods and under very specific conditions only, as intermediates in processes of microbiological nitrogen transformation like nitrification, denitrification and nitrate reduction.

c. Another, quantitatively less important, group of intermediates in microbial processes, like e.g. hydroxylamine, NH_2OH, and nitramide, $H_2N_2O_2$; these compounds are chemically unstable and occur in a transitory state.

Mineralization has also been studied in a more quantitative way in which special attention was given to the factors which influence the rate of the process. The quantity of nitrogen which can be mineralized within a certain time period depends on a large number of external conditions like temperature, pH, oxygen and water availability, and the presence of other nutrients.

Harmsen and Kolenbrander (1965) found that the temperature influence may be different for different reactions involved. For instance ammonification may continue at temperature levels considerably higher than 35 °C; nitrification was found to be completely stopped above 45 °C. In work by Stanford and Smith (1972) and Stanford et al. (1973) attention was given to the kinetics of mineralization reactions. For a large number of soils they measured the values of nitrogen-mineralization potentials and of the rate constants at different temperatures. Using these data they found Q-10 values of roughly 2. Apparent first order rate constants (weeks^{-1}) were found in the order of magnitude of 0.005 - 0.015 at 5 °C, 0.010 - 0.022 at 15 °C, 0.019 - 0.047 at 25 °C and 0.044 - 0.071 at 35 °C. Completion of the mineralization picture may be obtained when other factors influencing the process are likewise studied in a systematic fashion.

10.3.5. The pathway of nitrogen through soil

Assuming that usually the physical filtering action of a soil layer of sufficient depth allows for an initial complete retention of N-containing organic materials, the description of the displacement of N-compounds through soil centers on the fate of the inorganic N-compounds formed locally by mineralization or added directly to the soil surface.

Starting with NH_4^+-nitrogen, a good overall impression of the different reactions which may take place in a soil system is presented in figure 10.1,

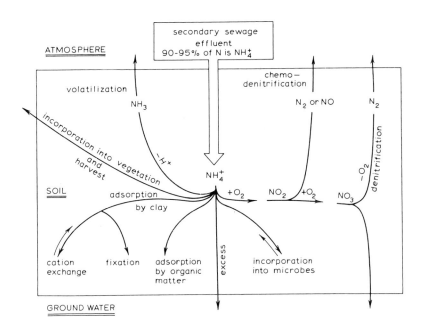

Fig. 10.1. Transformations of nitrogen in soil, departing from a NH_4^+-nitrogen source (from J.C. Lance, 1972.

taken from Lance (1972). He departed from the effluent of a secondary sewage water treatment, but comparable reaction mechanisms will apply when use is made of completely different (waste) materials like e.g. potato starch processing waste water (de Haan et al., 1973), manure (Klausner et al. 1971), septic tank effluents (Walker et al., 1973), or generally any nitrogen source which does contain the nitrogen predominantly in the ammonium form. Taking Lance's picture as a guideline, the different reactions can best be discussed by tracing the various inorganic nitrogen compounds involved.

Under alkaline conditions part of the NH_4^+ entering the soil may revert to NH_3, and losses of N due to volatilization of NH_3 may then become substantial. As a number of plant species may use ammonium as a primary source of nitrogen, part of the ammonium may leave the system in this manner. This may be used to advantage by seeding a disposal area with a suitable plant species which is then harvested and removed periodically (e.g. grass, rice, or in general *Gramineae*). The remaining ammonium will now be displaced through soil subject to several types of adsorption and fixation processes, viz. regular cation exchange, intra-lattice fixation much alike K--fixation (cf. section 4.5.1), and presumably direct chemical incorporation into some humus intermediates.

This phenomenon was already reported on by Mattson and Koutler-Andersson in 1942, and was later confirmed by Broadbent (1960), and Burge and Broadbent (1961). The mechanism of the fixation process, however, is still not well understood. It is usually assumed that this type of fixation consists of the ammonia being built in during the formation of complex polymers with derivatives of lignin and humic substances (Broadbent and Stevenson, 1966).

Referring to chapter 7 it is pointed out that the cation exchange process, though presumably reversible, leads to a retardation of the displacement of NH_4-ions relative to the carrier velocity depending on the relevant R_D-value of the system for this ion.

Finally ammonium may become incorporated into microbial tissue, thus re-entering living organisms, and again be subjected to mineralization upon a decrease of the size of the microbial population.

As indicated in figure 10.1, ammonium not intercepted through one of the above mentioned mechanisms may leach to deeper layers and eventually reach the groundwater. It may, however, also undergo nitrification if the specific requirements for this reaction are met. The main conditions for nitrification are:
a. sufficient amounts of nitrifying organisms
b. sufficient supply of free oxygen
c. presence of suitable C-sources
d. satisfactory conditions with respect to moisture and temperature levels.

Nitrate ions, formed by the nitrification process or added as such to the soil, are usually highly mobile in soil (no significant adsorption of anions !) and would thus become leached down with the carrier stream, unless intercepted by plant roots and taken up as nutrient ions. Typical nitrate leaching curves are presented in figure 10.2, taken from Wetselaar (1962). The different lines represent the nitrate-nitrogen distributions as a function of depth in a clay loam soil, at an initial application rate of 2240 kg $NaNO_3$ per hectare, and following rainfall of 188 mm (a), 391 mm (b) and 602 mm (c), respectively.

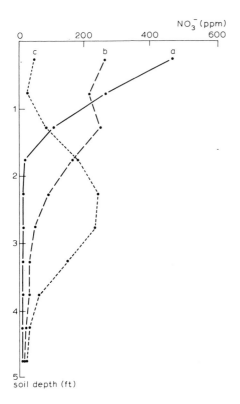

Figure 10.2 Course of nitrate-nitrogen as a function of depth in a clay loam soil following precipitations of 188 mm (a), 391 mm (b) and 602 mm (c), respectively (from Wetselaar, 1962).

The form of the curve and the location of the concentration peak depend mostly upon the amount of nitrate present and the amount of water that is leached through the soil. These factors can be of special importance in liquid waste disposal systems as will be discussed below.

The nitrification process itself consists of different steps. The first is the oxidation of ammonium to nitrite, NO_2, which is performed by *Nitrosomonas*. Usually, the intermediate, nitrite, is rapidly oxidized further to nitrate by *Nitrobacter*. This, however, is not the case under alkaline conditions, where considerable accumulation of nitrite may be found. Presumably such an accumulation must be attributed to the fact that *Nitrobacter* microorganisms are more strongly inhibited by large concentrations of NH_3 than is *Nitrosomonas*. As was described before, alkaline conditions may induce a build-up of NH_3 by shifting the ammonium-ammonia equilibrium in favor of NH_3-formation. If nitrite accumulation takes place, considerable nitrogen loss from the soil may occur by volatilization of gaseous nitrogen compounds. These processes are listed in figure 10.1 under the collective term of chemodenitrification to indicate that no microbial activity is involved in this pathway. The chemical reactions involved here may be described as
a. decomposition of ammonium nitrite according to:

$$NH_4NO_2 \rightarrow 2H_2O + N_2 \text{ (Allison, 1965)}$$

b. the van Slyke reaction, involving nitrous acid and α-amino-acids according to:

$$RNH_2 + HNO_2 \rightarrow ROH + H_2O + N_2$$

c. decomposition of nitrous acid according to:

$$3HNO_2 \rightarrow 2NO + HNO_3 + H_2O$$

These three reactions all require acidic conditions, i.e. pH levels below 5.5 - 5.0. As mentioned by Stevenson (1970) such acidic conditions may e.g. be reached in fertilizer bands after partial oxidation of NH_4^+ to NO_2^- and NO_3^-. Nitric oxide (NO) as arrived at according to the last equation will, in aerated soils, be oxidized and than hydrated to HNO_3.

Mostly, however, nitrification proceeds, via NO_2, to NO_3 by microbial processes. A somewhat closer look at the factors influencing this process seems warranted. Like most biological procedures, nitrification has a temperature optimum, lying somewhere in the range of 25 - 30 $^\circ$C. Low temperatures, as below freezing point, may cause considerable damage to the microbial population. This is one of the reasons why at tropical and subtropical temperatures, nitrification in disposal systems may proceed for years at high rates whereas in temperate regions a lag period has to be taken into account after each cold period. Microbial populations, and especially nitrifiers, are much more sensitive to extreme temperatures than to extreme moisture contents. There is also an optimum moisture content, but the optimum extends over a rather wide range of moisture levels. At a high moisture content, nitrification activity stops, but this is more due to an indirect influence of lack of oxygen supply than to a high water content itself. The nitrifiers are not washed out of the soil when leaching takes place. As soon as the moisture content diminishes far enough to allow sufficient oxygen diffusion to take place, nitrification activity again increases very sharply, provided NH_4^+ ions are available. Recent studies (Gilbert et al., 1974) have indicated that the formerly used concepts about aerobic and anaerobic conditions in soils are ambiguous. It may be shown that in certain microsites in a soil system conditions are sufficiently anaerobic to allow denitrification, whereas in the same soil and at the same time nitrification does occur in other microsites.

A very important deviation from the pathway of nitrogen outlined sofar is the occurrence of microbial denitrification. In contrast to nitrification this process requires reducing conditions, plus an energy source. Accordingly these processes, if occurring both in one system, will occur in a sequence in time and or in space. As furthermore denitrification - leading to the formation of gaseous N_2 - acts only on nitrate ions, nitrification must preceed denitrification if nitrogen enters the soil as ammonium.

> The microorganisms involved must be able to utilize the oxygen from nitrate and nitrite in their metabolism. From the heterotrophic organisms the following genera are known to include denitrifiers: *Achromobacter*, *Bacillus*, *Micrococcus* and *Pseudomonas*. In vitro denitrification may also be induced by the chemoautotrophic species *Micrococcus denitrificans* and *Thiobacillus denitrificans*. It is generally assumed that these are not important for denitrification in soils.

The importance of the denitrification process must not be underestimated, even if locally perhaps of minor magnitude. In a global sense, denitrification constitutes the closing link in the nitrogen cycle, since only in this way nitrogen returns to its ultimate source, viz. N_2 of the earth's atmosphere.

Besides the above mentioned major conditions for denitrification, i.e. the prevalence of nitrate, energy source and reducing conditions, several other factors may influence the rate at which denitrification proceeds, like temperature and pH. At temperatures below 25 °C, denitrification rates decrease and approach zero at 2 °C. Because of the sensitivity of denitrifiers for acidic conditions, denitrification almost ceases at pH values below 5.

From a fertilization point of view, denitrification has always been looked upon as an undesirable process, because it constitutes a loss of nitrogen. It was also realized, however, that denitrification may be considered as desirable when occurring in that part of the soil profile where plant roots cannot reach the nitrate ions, namely below the rooting zone. This is especially true for waste disposal systems in which recycling of nitrogen is promoted. If nitrates present in or originating from waste products are not fully utilized by crops, denitrification may be looked upon as a very useful process. Denitrification may be the only acceptable way for removal of nitrogen from a disposal system. Such a consideration applies e.g. in connection with the disposal of septic tank effluents on or in soil (Walker et al., 1973) or the renovation of secondary effluent by land treatment (Bouwer et al., 1974 a and b).

Considering the denitrification conditions as enumerated above, it is obvious that waterlogging in soils may induce very favorable conditions for denitrification (Tusneem and Patrick, 1971). In a comparable way, surface flooding of soil may also cause an appreciable loss of nitrogen as was shown by Patric and Tusneem (1972). Especially the proper alternation of non-flooding and flooding conditions may meet the requirements for nitrification and denitrification, respectively. In liquid disposal systems such an alternation may be artificially imposed to reach maximum nitrogen losses by volatilization (Lance and Whisler, 1972).

In natural groundwater bodies, however, conditions are usually rather poor for denitrification following nitrate leaching. If this were not so, the contribution of nitrogen to environmental problems would be mainly associated with runoff. Lack of denitrifying microorganisms, which may be found at great depth in soils, is not the main reason for an absence of denitrification, but such an absence is related mainly to a lack of energy material. For optimal denitrification conditions, the adjustment of the energy supply is one of the main problems in the practical management of disposal systems.

Recent laboratory studies (Bollag, 1973) focused attention on the simulation of practical circumstances and on possibilities for system management to obtain maximum denitrification. In soil columns, Lance and Whisler (1974) simulated the situation that is met in a secondary sewage effluent renovation project. They describe different methods that may be used to maintain a C/N ratio in water flowing through the soil at a value about optimum for denitrification. As pointed out in the previous section, the nitrate concentration in the first leachate following a dry period may be quite different from the nitrate content of the water following that first break-through. Hence, one cannot speak of a mean carbon content to fulfill the C/N ratio requirements, but the energy source in the leaching water must be adjusted to the nitrate content of this water. The C/N ratio required depends on the type of C-source that is available. If the denitrification process would be described by the reaction:

$$C_6H_{12}O_6 + 4NO_3 \rightarrow 6H_2O + 6CO_2 + 2N_2$$

this would mean that 1.3 mg of carbon is required for the reduction of 1 mg of nitrate-nitrogen to gaseous nitrogen. This then would refer to a C/N ratio of 1.3. This is a value that has been reported in the literature for denitrification of nitrate-nitrogen in agricultural waste waters (St. Amant and McCarty, 1969). C/N adjustment in the flooding period of liquid disposal systems can be obtained by:
-adding a readily available C-source e.g. in the form of glucose or methanol; from an energetic point of view this seems an approach that should be avoided if possible;
-collecting the first, high-nitrate leachate and bringing it back into the system when the raw waste material is applied; in this case one uses the original carbon of the waste to adjust the C/N ratio;
-diminishing the infiltration rate in such a manner that the break-through curve will not show a sharp nitrate-peak, but will have a much more flattened appearance, which makes it easier to take measures aimed at adjustment of the C/N ratio; moreover the retention time of the water in that part of the soil profile where still some C is available will be longer and chances of denitrification will be improved. Experiments by Lance and Whisler (1974) have indicated that this may possibly be one of the most promising ways to obtain maximum denitrification.

If these and comparable measures are taken to prevent environmental impact, land disposal systems of liquid wastes and of secondary effluents may sometimes be preferable to treatment in waste water treatment plants. The former approach allows the controlled utilization of all intrinsic properties and characteristics of a natural system like a soil.

Summarizing the situation with respect to the management of excess of nitrogen compounds entering locally the soil, it may be stated that undesirable effects are to be expected primarily from excesses of nitrate ions leaving the topsoil and reaching groundwater or open water. Although the soil is capable of retaining temporarily increased levels of nitrogen this is of no avail on the long run as such retention implies a labile storage in organic nitrogen forms. In practice such storage will eventually lead to a new steady state at an increased intensity level where the larger input is accompanied by an increased output of the mineralization and leaching processes, the output via the leaching water - either in the form of nitrate or ammonium - exerting an undesirable impact on the environment. Only if the increase of the input is balanced by a removal via harvested crops or by increased denitrification will it be possible to maintain the output via leaching water at its original level.

10.3.6. Pollution effects involving phosphates

Referring to the introductory section 10.3, cases of malfunctioning of soil in its role as a medium 'supporting' plant growth - and other forms of life - due to excess P are hardly known. Admittedly it has occasionally been inferred that excess P in soil impaires plant growth via indirect action. So e.g. Zn-deficiency symptoms could possibly be traced to high levels of P in a number of cases, but this will only apply if rather soluble phosphates are present. As a rule the major part of phosphate in soil is present in the form of solid phase compounds with low solubility and the activity of phosphate is then independent of the total amount present (cf. chapters 2 and 6).

Thus the situation as regards phosphate is comparatively simple: malfunctioning of soil as a result of excess P is limited in practice to the possible occurrence of undesirable concentrations of P in the drainage water from the soil profile. Even in this aspect the situation is at first sight much less threatening than for nitrate. The low solubility of phosphates in case of the ever present excess of Ca, Al and Fe ions in soil in fact precludes the occurrence of P concentration levels which inhibit the growth of living organisms. This observation leads directly to the sole hazardous aspect with regard to pollution effects resulting from phosphates in soil, viz. the occurrence of phosphate concentrations in surface waters fed with drainage water which favor the excessive growth of phytoplankton. Thus briefly, the polluting action of phosphates, if present, is limited to eutrophication.

Some controversy still exists with respect to the evaluation of the role of phosphorus in eutrophication. It has been suggested that the eutrophication in a number of freshwater lakes might be diminished by limiting the discharges of phosphorus into such lakes. Considering the lower phosphorus concentration limit for the occurrence of algae blooms (10 ppb) and the naturally occurring phosphorus concentration levels in streams and waters, however, the realism of the above suggestion seems at least questionable. Moreover it must be kept in mind that the supply of phosphorus in the underwater deposits of organic matter, silt and clay will provide enough available P for decades to grow most species of aquatic vegetation (Garman, 1973). Particularly because the prevention of eutrophication on short term would depend on the phosphorus level in the water, a prime requirement in the limitation to P input would be to maintain the P level in the discharging streams below 10 ppb. This is usually impossible, not only in agricultural, but also in natural areas (Smith, 1971).

In spite of these uncertainties, or possibly induced by their very existence, considerable attention has been given in recent studies to the P levels in natural water bodies and to the relative contributions of different phosphorus sources. On world-wide scale the contribution by erosion, both in the form of dissolved P in runoff-water as well as in the form of adsorbed P on sediments, is larger than the leaching through soil profiles. This is for P the more so because of its strong bonding with most soil constituents. It is beyond doubt that at comparable conditions as to soil genesis and parent material, the P level in agricultural soils is usually higher than in woodland soils. Data collected by Taylor et al. (1971) in a continuous monitoring system during a three year period indicated for Ohio, USA conditions an average P concentration of 22 ppb in farmland runoff and 15 ppb in woodland runoff. The concentration of P, of course, depends on the distance in space and time from the erosion event. Kunishi et al. (1972) measured concentrations as high as 200 ppb P in the runoff water from a fertile topsoil, whereas this value reduced to 15 ppb as the water moved downstream. The concentration levels in moving streams depend not only on runoff quantities but especially also on the characteristics of the suspended solids, since these determine the possible interactions between dissolved P and sediments, following the runoff event.

10.3.7. Sources of phosphates in soil

The mentioned immobility of phosphate is reflected in its rather localized cycle in 'natural' soils. The phosphates present are then predominantly derived from primary minerals. At any one moment the amount of P available in forms that can be taken up by plants is low and is thus readily removed.

A steady state situation eventually arises where P moves from inorganic forms into organic forms, and back via mineralization. The intensity of such a local cycle builds up to a level depending on the magnitude of the original source of P-containing minerals.

With the introduction of high intensity agriculture P-deficiency in soil became a wide-spread phenomenon. Accordingly the addition of P in the form of manure and later as chemical fertilizer became a standard practice all over the world. The phosphate content of old arable land thus tends to exceed that of natural reserves. Expressed as percentage P_2O_5, soils usually contain 0.1 to 0.25%; under very exceptional circumstances this value may be as high as 0.5%, but this should be considered as an extreme (Dahnke et al. 1964; Black, 1968).

Lately the situation has been shifted to the point where locally P-containing wastes are disposed of on soil. Such disposal systems may be the intentionally designed and well described disposal systems of e.g. raw sewage water or processing waste water (de Haan, 1972) or sewage water treatment plant effluents (Bouwer et al., 1974a). These systems may also be the diffuse and not well registered phosphate sources like e.g. the seepage from numerous septic tanks or manure applications exceeding plant nutrient requirements.

The relative contribution of different sources to the total P burden cannot be described in terms of general validity, because it depends on the specific characteristics of the area under consideration. Kolenbrander (1972) showed for the Netherlands that the annual contribution of urban areas to phosphorus discharges is about sixteen times that of rural areas. This must mainly be attributed to the low degree of phosphate removal from waste waters, which never exceeds 30% of the P input with sewage water except when advanced waste treatment systems are used. Tertiary treatment, however, remains a rare phenomenon so far predominantly due to the doubling of the cost of such waste water treatment. The above relative contribution of urban and rural areas is also in accordance with the low P-concentration in soil drainage water under natural conditions. As may be derived from the information presented in sections 6.4.1 and 6.4.2, the P-concentration level at normal soil pH-values when Ca, Al and Fe serve as coprecipitating cations is somewhere around 10^{-5} molar PO_4. Of course, considerable deviations from this general situation may occur.

As in the case of nitrogen, a prediction whether a soil under a given regime, i.e. under natural vegetation, in use for agricultural production or for waste disposal, may give rise to eutrophication of adjacent surface water must be based on an insight in the pathway of different P-compounds in soil. In this respect it is noted that also here considerable know-how is stored in the existing soil fertility and soil chemistry literature. Thus a vast amount of research has been performed in order to identify, characterize and quantify the reaction mechanisms between soil constituents and phosphorus compounds. These studies all are directed toward plant availability and its counterpart: phosphorus fixation in soils. In present time there is, again in a comparable manner as for nitrogen behavior in soils, a complete dichotomy of purpose in the application of the knowledge gained in this soil phosphorus research (Bailey, 1968). On the one hand it should be realized that for adequate crop production in order to meet food needs of mankind, phosphorus fertilization and thus the application of the above research with the purpose of maximum crop yields still remains (and in the near future will increasingly be) one of the main responsibilities of agriculture. On the other hand the fixation and severe bonding of phosphorus in soils may offer an opportunity for safe phosphorus storage in waste disposal systems in which soils are involved.

10.3.8. The interaction between phosphates and soil

As was pointed out before, the P-cycle in soil involves both organic and inorganic forms of phosphate. The relative proportion of total soil phosphorus that is present in organic form may vary widely. It is evident, however, that the organic form usually increases with increasing organic matter content and hence the organic proportion is higher in topsoils than in subsoils. The percentage organic of total phosphorus has been reported for topsoil samples as ranging from 0.3% to 95% (Black, 1968).

A range of organic phosphorus forms is possible, all of which have in common that one or more of the hydrogen ions of phosphoric acid have been eliminated in ester bonds. Of the organic phosphorus compounds identified so far inositol phosphates form the major part, although to a lesser degree also nucleic acids and phospholipids can be present.

Inositol phosphates are esters of phosphoric acid and inositol, a cyclic hexa-carbon with an alcohol group at each C atom. Because several isomers of inositol exist, the number of possible inositols is at least a sixfold of these since for each isomer from one to six alcohol groups may form an ester linkage with phosphoric acid. Moreover, from the remaining hydrogen ions one or more may be replaced by other cations.

Although a number of difficulties are still met in the determination of soil organic phosphorus (Williams et al.,1970), increasing progress is made in the identification of the organic P forms. The work of Cosgrove (1963, 1964) indicates that soil microorganisms probably are able to form different inositol phosphates because several of these were found in soil whereas only one is known to occur in plant material (Ca-phytate). Omotoso and Wild (1970) identified the whole range of mono-, di-, tri-, tetra-, penta- and hexa-phosphates of inositol in soils, although the lower esters (mono-, di- and tri-) accounted for a few percents only of total organic phosphorus.

Present concern about large amounts of animal manure excesses and their disposal on land has caused an increased interest in organic phosphorus from an environmental point of view. At first sight it would seem obvious that manure application would considerably increase the organic P fraction in soils. Long-term studies described by Oniani et al. (1973), however, indicated that, although the increase in organic P was slightly higher from farmyard manure than from superphosphates, no significant change was found in either case of inositol penta- and hexaphosphates in the soil. This again indicates that the balance of organic phosphorus presumably is much more microbially regulated than source determined. Comparable experiments as the above may lead to different results under different climatic conditions. Thus the inositol phosphate contents under tropical circumstances as regards temperature and rainfall may be considerably less than at temperate climatic conditions, presumably due to the hydrolysis of ferric salts of inositol hexaphosphate, as described by Furukawa and Kawaguchi (1969).

Opposite to the above, Cosgrove (1967) mentioned the evidence that a large fraction of the organic phosphorus of animal manure could be resistant to mineralization. Peperzak et al. (1959) described phosphate conversions for a large number of different manures. They found that storage of the manure caused a substantial decrease in specific organic phosphorus forms although all organic fractions were found to persist in part for a number of years.

Interactions known to occur in soils with inositol phosphate are adsorption and precipitation. As such considerable similarity with the behavior of inorganic orthophosphates seems to exist. Since the pH-solubility dependency of inositol phosphates are very similar to those of the orthophosphates, it may be expected that they can be fixed in soil as the lowly soluble salts of aluminum, iron and calcium. Such insolubility in acidic solutions above pH 2 to 3 has been reported by Jackman and Black (1951) for both iron and aluminum salts of inositol phosphates.

Anderson and Arlidge (1962) measured the adsorption of inositol phosphates on clay minerals as well as on other soil minerals like boehmite, ferric oxide gel and on other sesquioxides. As must be expected the pH was found to have a strong influence on the extent of adsorption. The study by Breeuwsma (1973) indicates comparable effects for the adsorption of inorganic phosphates on hematite.

The adsorption of nucleic acids by clay minerals was studied by Goring and Bartholomew (1952). They determined a number of factors influencing the adsorption, like pH, temperature and cationic composition of the clay. The nucleic acids are, at least partially, adsorbed in interlamellar positions comparable to the fixation of potassium and ammonium ions. Moreover, the nucleic acid adsorption is caused by exchange reactions and by adsorption through the orthophosphate groups.

The concurrency in the behavior of the major organic phosphates in soil (inositol phosphates) and the inorganic phosphorus forms is of importance with respect to the interpretation and prediction of phosphorus transport and accumulation, following disposal of organic wastes. Although specific parameters must be measured separately, the vast knowledge about inorganic phosphorus in soils may thus hopefully also be applied in organic phosphorus studies.

The mineralization of organic P-compounds to inorganic forms is a necessary prerequisite for their reintroduction into the plant and thus for their removal from the system via a harvested crop.

The actual contribution of mineralized organic P-compounds to plant nutrition in practice is hard to evaluate, although it was shown (van Diest and Black, 1959) that easily mineralizable organic P can make a significant independent contribution to the P nutrition of plants. As described by Black (1968), some work supports the view that the plants benefit from the microbial mineralization that occurs throughout the soil; other work, however, supports the idea that the mineralization occurs more rapidly in that part of the soil that is directly adjacent to plant roots, e.g. due to the secretion by plant roots of certain enzymes.

With regard to the prediction of the pathway of P added to soil in organic forms it may be hoped that the overall similarity in behavior of organic and inorganic P-compounds mentioned above makes a precise knowledge of the rate of mineralization unnecessary.

As has been discussed in chapters 5 and 6, inorganic (ortho-) phosphate is subject to strong adsorption on the surface of oxides/hydroxides of Al and Fe (Breeuwsma, 1973) and on the edges of clay minerals (de Haan, 1965). This adsorption has the character of chemisorption and may possibly serve as a precursor of precipitation in the form of Al- and/or Fe-phosphates.

The mathematical descriptions for adsorption isotherms most commonly used are the Freundlich equation and the Langmuir equation. Considerable confusion exists in literature as to which of these equations the experimentally determined phosphate adsorption data do best fit. Sometimes, reported results even indicate that both isotherms are followed (Olsen and Watanable, 1957; Hsu and Rennie, 1962) although in such cases the fit with the Langmuir equation was slightly closer than with the Freundlich isotherm. Kardos et al. (1974) found for sewage effluent disposal systems a more realistic and logical fit with the Freundlich isotherm than with Langmuir; in the latter case the data indicated the probable occurrence of at least two adsorption mechanisms, as derived from the differences in slope of the lines constituting the isotherm. Comparable observations have been described by Chen et al. (1973) in studying the reaction of phosphate with aluminum oxide and with kaolinite. They suggested the adsorption process to consist of a rapid adsorption on reactive surfaces and of the formation of a new solid phase. Van Riemsdijk et al. (1975) identified the new solid phase formed in the reaction of phosphate ions with $Al(OH)_3$ as Sterrettite.

As ferrous phosphate exhibits a much higher solubility than the ferric forms, the bonding of phosphate is sensitive toward the redox potential of the system (Patrick, 1964 ; Patrick and Khalid, 1973).

10.3.9. A characteristic phosphate distribution profile as found on a sewage farm

Phosphate in soil is largely immobile, being either adsorbed or precipitated, while a small fraction may move with leaching water at concentrations which are generally not higher than about 10^{-5} molar.

As indicated by figure 6.11, summarizing the solubility equilibria governing phosphate in soil, Ca-phosphates are likely to be the controlling phases at pH values around 6.5 and above. At low pH-values Al- and Fe-phosphates will play the same role. At near neutral pH-values both controlling mechanisms lead to a concentration of around 10^{-5} molar P (cf. above), which then constitutes a maximum solubility of P for soils containing both Al (Fe) and Ca ions in large quantities. It is interesting to compare this value of 10^{-5} molar $\approx$ 1000 ppb with the earlier given limit of 10 ppb with regard to eutrophication of surface waters (cf. table 10.1).

It should be pointed out that the above considerations pertain to systems of relatively simple composition. So the presence of e.g. fluorapatite may lead to considerably lower P-concentrations (cf. figure 6.11).

The phosphate immobility is demonstrated by observed distribution patterns, e.g. by the fact that the occurrence of high phosphate levels in topsoils can be applied in archaeological tracking of the history of human habitation in certain areas (Cook and Heizer, 1965).

Reports on phosphate distributions with depth thus invariably indicate an accumulation in the topsoil layer. This is even the case in systems of very high phosphate applications as described by Beek and de Haan (1974). Figure 10.3 shows the course of P content with depth, expressed as mg P_2O_5 per 100 g of dry soil, on a sewage farm.

This distribution profile refers to the situation on a farm which is in normal agricultural use as grassland for permanent pasture, while its fields are flooded on rotation basis with raw domestic sewage water at a rate of about 350 - 400 mm each month.

The farm, which has already been used during 50 years for the disposal of large amounts of raw sewage water by surface flooding, is situated on sandy soil containing iron and aluminum oxides and hydroxides. The dotted line in figure 10.3 represents the situation on untreated soils; to this purpose soil profiles were sampled on the forest land soils that surround the farm and from which the sewage farm has been reclaimed. The difference between both situations refers to a total amount of 15,700 kg P_2O_5 per hectare; 90% of this total phosphate storage is obtained in the upper 40 cm and even 65% in the 20 cm top layer.

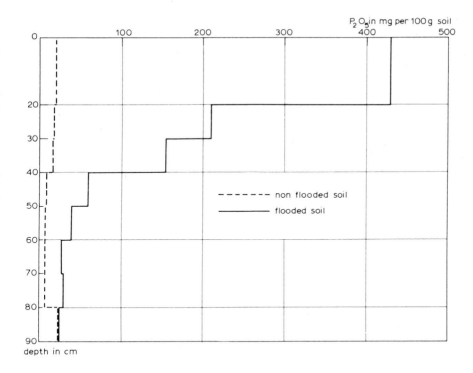

Fig. 10.3. Phosphate distribution with depth on a sewage farm as compared to the non-treated soil (from Beek and de Haan, 1974).

In view of the very limited residence time of the water in the soil (0.5 day) and of the high degree of phosphate removal (96% of total amount applied to the soil), the phosphate bonding mechanism must be an extremely efficient one. This would indicate adsorption of the phosphate ions on soil constituents, rather than precipitation as insoluble phosphate phases. However, the total combined phosphate adsorption capacity of the different soil components can account for a minor fraction only of the total amount of phosphate stored. (It should be realized that in a system as described here the phosphate adsorption capacity is not necessarily constant but may increase as a function of time).

Beek and de Haan (1974) suggested that the bonding mechanism is constituted of an alternation of adsorption and precipitation reactions. In between successive sewage water applications the phosphate ions are removed from the adsorbed state by precipitation, thus delivering new adsorption sites for the phosphate removal at the time of the next flooding. Such a recovery of the phosphate adsorption capacity of soils after a certain time period following the saturation of the adsorption sites has also been measured and described by Ellis and Erickson (1969). This would mean that the short-term removal of phosphate ions from the applied liquid waste in a certain soil layer, and consequently a prevention of break-through of phosphorus for that layer, depends on the phosphate adsorption capacity. The total amount that may be stored on long-term in this layer, however, depends on the total iron, aluminum or calcium content available for precipitation of phosphorus as insoluble phases. The phosphate concentration in the soil water and drainage water of the above mentioned sewage farm was found to equal the equilibrium concentration of aluminum and iron phosphates at the relevant pH value (Beek and de Haan, 1974). This again confirms the contribution of precipitation reactions in the final phosphorus storage in the soil.

A combination of adsorption and precipitation is possibly the explanation for the observation - as sometimes reported in literature - that saturation of phosphate 'adsorption' is found after a very long equilibration time only, if ever. Accordingly, knowledge about the occurrence *and* the kinetics of different phosphate bonding mechanisms is indispensable as a tool in the prediction of phosphate transport and accumulation in soils (Beek et al., 1976; van Riemsdijk et al., 1976).

In conclusion it can be said that the environmental impact of nitrogen and phosphorus is mainly limited to their influence upon eutrophication of surface waters, while nitrogen may also induce direct health hazards under certain conditions. Many soils possess an extremely large storage capacity for phosphate. This may be made use of in land disposal systems of waste waters and primary or secondary effluents of waste water treatment plants. The situation with nitrogen is different in this way that here the increased nitrogen addition must be balanced by a removal via crop harvesting and/or by denitrification in order to forestall future leaching of nitrogen into deeper layers at undesirable levels.

10.4. HEAVY METALS AND TRACE ELEMENTS

This title is used to cover a number of metals (and their ions) which are mostly of high density and belong largely to the group 'transition elements' of the periodic table. Some of these have also been termed trace elements or micronutrients in agriculture, stressing their relatively low abundance in regular soil and the fact that they tend to be essential for plant growth. Other trace elements of non-metallic nature but presumably sometimes involved in pollution problems have also been included in this section.

Disclaiming any authority in toxicology, nevertheless some commonly accepted toxic effects on human beings, animals and plants will be mentioned in passing, making reference to the relevant research information obtained by medical or veterinary workers and plant physiologists. More complete information on these toxic effects should be obtained elsewhere. Instead the discussion will be confined largely to the ranges of abundance found in soil, indicating probable sources and mentioning relevant facts which have a bearing on their mobility in soil. The elements considered here are, alphabetically listed: As, Cd, Co, Cr, Cu, Hg, Mo, Ni, Pb, Se, V and Zn.

Prior to a more detailed discussion of each of these elements it seems appropriate to mention a number of sources which are usually generating a mixture of two or more of them. A large variety of industrial exhausts may be responsible for air pollution with heavy metals and consequently soil pollution when the elements reach the soil surface. An example is the ore smelting industry: severe damage imposed on vegetation and on animals caused by zinc smelters has been reported several times (e.g. Buchauer, 1971; Lagerwerff et al. 1973). Lagerwerff and Brower (1974) recently reported that persons living in an area of high exposure ingested Pb and Cd by their dietary uptake at rates more than 50% above that associated with low exposure areas. The Pb content was found significantly higher in human and animal blood in high than in low exposure areas. Also the combustion of fossil fuel, either in industry or in traffic, constitutes a significant contribution in the trace element impact on the environment.

An extremely difficult source to quantify is brought about by the disposal of solid wastes in sanitary land-fills. This is so because usually no control is and can be held on the materials that are disposed of in this way. Thus these refuse constituents may vary from spilled mercury resulting from broken clinical thermometers to all types of food and beverage cans. A comparable source, although somewhat more defined, is given by the effluents of sewage water treatment plants and by the sewage sludges generated by such plants. More and more it is being realized nowadays that the disposal of these sludges on land should not as much be based on the plant nutritional elements as on the possible increment of heavy metals in the soil (de Haan, 1975). This feature should be based as well on short as on long term considerations (Chaney, 1973). In this respect considerable attention has been given during recent years on the heavy metal composition of sewage sludges. Variations in such compositions, both in place and in time, may be so large because the sludge composition completely depends on the industrial discharges that contributed to a specific sludge batch. A good impression on average concentrations and on the variability of concentrations may be obtained from the data as collected by Page (1974) for different countries (cf. table 10.2). In view of the above, general rules for advisable application rates of sewage sludges are artificial and unwarrented.

TABLE 10.2

Comparison of median and range of trace element concentrations (ppm) of sewage sludge from Michigan, USA; England and Wales; and Sweden[*1] (from Page, 1974).

Element	Median Conc.			Most Common Range[*2]		
	USA (Michigan)	England and Wales	Sweden	USA (Michigan)	England and Wales	Sweden
Cd	12	--	6.7	1-10 (49)	--	1-10 (72)
Cr	380	250	86	10-100 (33)	100-1000 (45)	10-100 (52)
Cu	700	800	560	100-1000 (74)	100-1000 (79)	100-1000 (74)
Ni	52	80	51	10-100 (58)	10-100 (50)	10-100 (70)
Pb	480	700	180	100-1000 (75)	100-1000 (85)	100-1000 (88)
Zn	2200	3000	1567	1000-10,000 (76)	1000-10,000 (81)	1000-10,000 (87)

[*1] Data are from 57, 42 and 93 treatment plants in Michigan, England and Wales, and Sweden, respectively. For Michigan, England and Wales, and Sweden, data are derived from those published by Blakeslee (1973), Berrow and Webber (1972), and Berggren and Oden (1972), respectively.

[*2] Percentage of samples within the concentration range indicated are given in parentheses.

10.4.1. As, arsenic

Arsenic compounds are used as paint pigments and in the textile and tanning industries. Also household detergents may contain appreciable amounts (10 to 70 ppm) of As. Because of the high toxicity of most arsenic compounds, the prime source is constituted of all different kinds of pesticides like fungicides, herbicides, insecticides and rodenticides.

The chemistry of arsenic in soils is very similar to that of phosphorus. This may even lead to difficulties in the colorimetric phosphorus determination in which the presence of As may cause interferences (van Schouwenburg and Walinga, 1967; Shukla et al., 1972). Although inorganic arsenicals are no longer widely used in agriculture there is still considerable concern about residual effects of the past high applications (Anastasia and Kender, 1973; Deuel and Swoboda, 1972; Steevens et al., 1972). It is a well-known fact that accumulation may especially occur in the peelings of potatoes. Potatoes grown on foreland soils of the River Rhine, which are flooded once or several times each year with arsenic contaminated river water, may accumulate As in their peelings above the permissible level. This level is established at 2.6 ppm for edible portions of plants. Direct intoxication of man and animals by ingestion of contaminated plants are usually considered as unlikely, since toxicity levels of As for man and animals are also toxic to plants.

Isensee et al. (1973) studied the pathway of several organic arsenicals in a model ecosystem. The accumulation of mercury in the fatty tissue of successively higher components of aquatic food chains must be ascribed to the sufficiently high fat solubility of methylated mercury. Because of the similarity between dimethylmercury and dimethylarsine it was suggested that the latter might also undergo bioaccumulation in the food chain. In these studies the potential for such accumulation was found to be small.

Because of the chemical similarity between P compounds and As compounds it seems evident that the factors influencing the phosphate behavior in soil will also play a dominant role with respect to the behavior of arsenate. Thus Fe, Al and Ca are found to be important for As fixation in soils. The influence of clay content is usually ascribed to the positive correlation between clay content and the amount of reactive Fe and Al. The contribution of Ca in the arsenate fixation is much smaller than that of Fe and Al. This is attributed to differences in solubility products. The solubility product of $FeAsO_4$ is reported as 5.7×10^{-21}, whereas for $Ca_3(AsO_4)_2$ as 6.8×10^{-19}. Thus the As concentration in a solution from pure $Ca_3(AsO_4)_2$ will be a factor 10^6 greater than in a solution from pure $FeAsO_4$ (Woolson et al.,1971). Again similar to phosphates the strong fixation is the main reason that arsenic accumulation in soils is usually found in the top 10 cm. Under specific conditions, however, (i.e. in the absence of bonding or fixing constituents) leaching of arsenic to deeper soil layers may occur (Tammer and De Lint 1969). The same would be true for phosphates but hazards of P are not comparable to those of As. As a result of the competitive behavior of P and As with respect to bonding on soil constituents, As-mobility in soils is increased by phosphate fertilization.

10.4.2. Cd, cadmium

Cadmium is widely used as a coating material. Also in paint pigments and plastic industries extensive use is made of Cd compounds. It is applied in batteries, in photography and as a fungicide. The occurrence of Cd in motor oil and in car tires explains the relative accumulation in roadside soils (Lagerwerff, 1972). The geochemical relationship between Cd and Zn makes that the presences of both elements are usually linked to each other. Thus zinc smelters may cause large emissions of fumes containing both ZnO and CaO. The higher volatility of CdO as compared to ZnO causes a relative enrichment of Cd in these fumes as compared to the original ores. In these ores the Zn/Cd ratio is in the order of magnitude of 900. The presence of Cd in phosphate fertilizers constitutes a very diffuse source of Cd contamination. Cd contents in P-fertilizers vary from 1 - 2 ppm for tertiary calciumphosphate to 50 - 170 ppm for superphosphate.

The most reported illness induced by excessive Cd intake is the development of high blood pressure or hypertension and more generally vascular diseases (Schroeder and Balassa, 1965). Awareness of the health hazards of Cd has made the uptake of Cd by vegetables and agricultural crops a subject of extensive research (Haghiri, 1973, 1974; Jones et al., 1973; Turner, 1973). It is a common finding that cadmium can easily be taken up by most plants. Also in the plant uptake of Cd, an influence of Zn is found. Thus Lagerwerff and Biersdorf (1972) reported that at 100 ppb Cd in solution culture, an increase in Zn from 20 to 400 ppb increased the concentration of Cd in radish leaves by 10%.

The sensitivity of different crops for excess Cd differs strongly. This is shown by figure 10.4, taken from Page et al. (1972), indicating for a number of crops the percentage growth reduction as a function of the Cd concentration in solution.

Cd contents of soil in non-polluted areas are usually below 1 ppm. Yagamata and Shigematsu (1970) described for contaminated rice soils levels up to 50 ppm Cd. For topsoil samples near a zinc smelter values as high as 1,700 ppm have been reported (Buchauer, 1973).

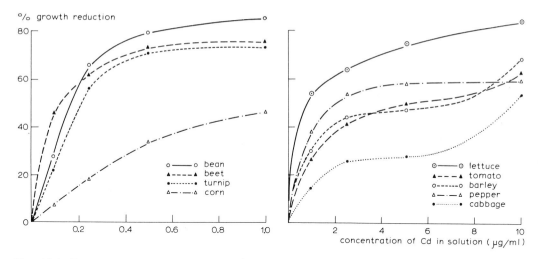

Fig. 10.4. Percentage growth reduction for different plant species as a function of Cd concentration in solution (after Page et al. 1972).

Because of the predominant occurrence of Cd as divalent cation, electrostatic adsorption on exposing adsorption sites probably acts as the main bonding mechanism. Cation exchange studies of Cd on pure clays have been described by Lagerwerff and Brower (1972), and by Bittell and Miller (1974). Using the experimental data of the first mentioned authors, it was shown by Harmsen (1976) that specific adsorption occurs at low Cd concentration. Because of the limited number of adsorption sites with high Cd preference, this specificity remains hidden at higher Cd contents of the clay.

Cadmium may also form complexes with hydroxyl and chloride ions. It was shown by Hahne and Kroontje (1973) that such complexation might contribute to the mobilization of Cd in the environment.

10.4.3. Co, cobalt

Cobalt is used in the production of alloys and in paints, varnishes, enamels and inks. It is a very essential element as well for humans as for other monogastric animals. This need is based on its function as constituent of vitamin B_{12}. Plants that are able to utilize atmospheric nitrogen require cobalt for their root nodule microbial system. Thus more attention has been given in the past on needs of Co and methods to meet the required levels than to toxic concentrations.

Co at high concentrations can be highly toxic to plants (Vanselow, 1966). It may induce Fe deficiency at moderately high concentrations. Most plants require Co solution concentrations not exceeding 0.1 ppm.

Co contents of soil usually do not exceed about 10 ppm, although in specific cases values up to 380 ppm have been reported (Taylor and McKenzie, 1966).

The adsorption of Co^{2+} on clay minerals was studied by Hodgson (1960). It was found that cobalt adsorbed on montmorillonite, in the range of trace adsorption i.e. at a few percent of the total CEC, consisted of two different types, viz. an exchangeable and non-exchangeable part. Even the exchangeable part was adsorbed specifically since 70 - 90% of the total amount adsorbed was exchangeable only with other heavy metals like Cu^{2+} and Zn^{2+} and not with Ca^{2+}, Mg^{2+} or NH_4^+ ions. The non-exchangeable adsorption was attributed to a build-in of Co in the clay mineral lattice. Fixation of cobalt in a comparable way as has been described for potassium and illite may be another possibility (de Haan et al., 1965).

Co is one of the heavy metals that are known to be subjected to chelation in soils (cf. section 10.4.13). There is also a strong association of Co with Mn oxides in soils, probably because Co^{2+} may substitute Mn^{2+} (Lindsay 1973).

Because of the low Co concentrations, even in waste products like sewage sludges, there has been little concern so far on hazardous effects of this heavy metal in soils.

10.4.4. Cr, chromium

Chromium, as metal and in the form of chromates, has its most widespread application as corrosion inhibitor. Many metals, used in household, traffic and industry, are chrome plated to extend their durability. Use in smaller quantities is made in the production of varnishes, inks and dyes.

Cr is an essential element for man and animal. It plays a major role in the so-called glucose tolerance factor, i.e. the return of excess levels of glucose in the blood to normal levels. As such it is probably of high importance to prevent diabetes as its activity is related to that of insulin (Schwartz and Mertz, 1959). This combined with the observation that decreases in Cr level in human tissue are strongly correlated with a reduced dietary intake (Mertz, 1966) has renewed the interest in this element from a health point of view. No damages induced by excessive chromium levels have been reported so far as concerns human beings and animals.

Increased human intake of Cr, if desired, can best be achieved by Cr additions in drinking water and via food industry. Increased Cr content of plants is hard to achieve. Because of immobilization of Cr in most soils, fertilization with Cr-compounds has usually little effect even when using Cr salts of high solubility.

Necessity of Cr for plant growth has not been proved (Pratt, 1966). Toxicity of Cr at high concentration levels has been reported, especially on so--called 'serpentine' soils. These soils are rich in Cr, but also in other heavy metals like Ni and Co. The Cr content of most soils is usually limited to traces only but may reach values up to 46,000 ppm for 'serpentine' soils. Only a very minor part of the total Cr, however, is soluble (0.006 - 0.28% according to Soane and Saunder, 1959), thus strongly restricting plant availability.

The mobility of Cr within the plant is extremely low. It thus remains in that part where the uptake took place (Allaway 1968). Turner and Rust (1971) assume that Cr is taken up by plants as chromate (Cr^{6+}), e.g. after addition of potassium chromate, K_2CrO_4. This then would be the main reason why Cr is not a problem in soils, even not after disposal of Cr containing sludges. At the pH values and redox potentials prevailing in most soils, Cr^{6+} is readily reduced to Cr^{3+}, mainly occurring as the poorly soluble $Cr(OH)_3$. Strijbis and Reiniger (1974) found in experiments with rice that also Cr^{3+} was taken up, although at considerably lower levels than Cr^{6+}. The uptake for both forms was lower under submerged conditions than at field capacity. This is in accordance with the decreased solubility at lower values of the redox potential.

10.4.5. Cu, copper

Prime use of Cu is as wire and brass, and as alloys with a number of different other metals. Most water supply systems consist of copper tubing. Copper in the form of $CuSO_4$ has been used as a fungicide for fruit crops and potatoes. Also in the form of copper sulfate Cu is sometimes used as an additive in swine and poultry feed in order to increase the feed efficiency. Such additions in feed can be made up to a level of 250 ppm Cu. Copper levels up to 750 ppm Cu have been reported for pig manure (Robinson et al., 1971) which means that disposal of this manure on land in excessive rates, or even the application at appropriate nitrogen fertilization levels, may induce an accumulation of Cu in soil.

The effects of high Cu contents of soil on the uptake of certain other micronutrients by plants are well known. High Cu levels may cause iron deficiency which is demonstrated as typical chlorosis features. Also an antagonism between Cu and Zn has been observed. The interaction between Cu and Mo has been described both for human and animal nutrition and for plant nutrition. In a comparable manner as Cu may inhibit Mo uptake, an abundant presence of Mo may in turn induce Cu deficiency, especially when the last is present at relatively low amounts. The proper Cu-Mo balance is thus an important condition for proper plant nutrition.

High levels of Cu can be toxic to microorganisms. Its practical application to control microbial induced diseases is as old as 1882 when the so-called Bordeaux spray, $CuSO_4$, was used for the first time as a fungicide. The presence of Cu in manures, as mentioned above, might thus considerably influence the rate of biodegradation of the organic constituents of these manures.

In view of sewage sludge disposal on land, Chumbley (1971) introduced the 'zinc equivalent factor', to be used as a relative indication of the toxicities of different metals. Very generally, it can be said that copper is about twice as toxic to plants as zinc, whereas nickel is about 8 times as toxic as zinc. The zinc equivalent factor is then defined as:

Zn factor (in ppm) = 1 × ppm Zn + 2 × ppm Cu + 8 × ppm Ni.

Although this factor may be used for a rough relative toxicity indication, it must be kept in mind that the actual relative toxicity and tolerance is governed by these conditions and circumstances which determine the actual availability of Zn versus Cu or Ni; as such the organic matter, phosphate level and pH are the most important (Chaney, 1973).

With respect to crop production it is generally reported that an adverse influence on plant growth results if the Cu concentration in the soil solution exceeds 0.1 ppm. The Cu concentration in drinking water for human consumption is, according to drinking water standards, considered to be safe at levels not exceeding 1.0 ppm Cu. Sheep are known to be sensitive for high Cu levels; concentrations above 20 ppm Cu in feed and forage are reported to be toxic for sheep (Baker, 1974).

Normal Cu contents of soil are around 20 ppm with variations over the range 2 - 100 ppm. At pH values and oxygen pressures which usually prevail in most soils, the predominant ionic form of copper is as divalent cation.

Mobility and displacement of Cu in soils is low. It is strongly bound by organic matter, clay minerals, and even adsorption on pure quartz has been described. As a result of this bonding, downward movement of Cu in silty or clayey soils is almost nil. Even in sandy soils such movement is very small (Jones and Belling, 1967). Peat soils are famous for their copper fixation. After addition of 250 kg Cu per hectare to an acid peat soil only 0.2% was found to be removed from the top 5 cm layer in a time period of 5 years (Lundblad et al., 1949).

The cationic exchange between Cu^{2+} and Ca^{2+} on bentonite has been described by El-Sayed et al. (1970, 1971). Starting with a pure Ca-clay, it was found that the initial Cu-exchange was 'unfavorable', i.e. the clay exhibited as small preference for Ca-ions. With increasing Cu-adsorption, the exchange gradually shifted to 'favorable' exchange, i.e. with preference for Cu. This phenomenon was attributed by the above authors to a condensation of clay platelets with increasing Cu-adsorption, thus excluding a number of adsorption sites for the slightly larger Ca-ions.

Lindsay (1972) suggested for Cu, in a comparable way as for a number of other heavy metal cations, to summarize all Cu reactions in soil by the general equation:

$$Cu^{2+} + soil \rightleftharpoons Cu\text{-}soil + 2H^+$$

According to Norvell and Linday (1969), the log $K°$ of this reaction amounts to -3.2. Such an approach may be useful in the construction of solubility diagrams as a function of pH. As the above equation covers all Cu-reactions that may pertain in soil, no information is gained about the bonding mechanism; this may as well be adsorption on clay minerals or other soil constituents, as precipitation of lowly soluble Cu-salts.

A large number of Cu complexes are known to occur in soils. As will be discussed later, Cu has also been studied intensively with respect to chelation.

10.4.6. Hg, mercury

Mercury is used in measuring and control instruments like thermometers, manometers, etc. It has an application as floating electrode in the electrolysis of chlorine and caustic soda. It is also used as a catalyst in the production of plastics. Agricultural use of mercury, mainly in the form of fungicides e.g. in seed dressings, amounts at present to 5% or less of total industrial mercury consumption. As a fungicide it is also applied for wood preservation and thus products of pulp and paper industry usually contain traces of mercury. Burning of coal and oil is the most important airborn source of Hg emission in the environment. As has been pointed out by Joensuu (1971) the annual natural discharge of mercury in the oceans as a result of chemical and physical weathering of Hg-containing rocks and minerals amounts to an estimated 230 metric tons on worldwide basis.

The hazardous effects of mercury attracted especially attention following the 1950's calamities of Minamata, Japan. Also Swedish researchers called specific attention on mercury behavior in the environment when it became clear that mercury accumulation occurred in food chains. One important Hg source was the use of alkyl mercury in seed dressings. There is a large difference in the toxicity of the different mercury compounds. The aromatic Hg compounds (e.g. phenyl-mercury) and the alkoxyalkyl Hg compounds (e.g. methoxyethyl mercury) are the least toxic. Then follow the inorganic salts of mercury and metallic mercury, $Hg°$, whereas the alkyl compounds (e.g. methyl and ethyl mercury) are the most dangerous. This is due to the high lipid solubility of the last and their high stability. As a combined result of these properties they may easily penetrate and accumulate in human and animal tissue, where they finally may act as a blocking mechanism in the oxygen supply.

Since these differences in toxicity became known, the occurrence of the different compounds and their possible transformations became a major interest. Prediction of mercury behavior is somewhat complicated by the occurrence of several transformation processes. In view of the above it is evident that methylation is the most hazardous of such tranformations (Jacobs and Keeney 1974; for a review on this subject matter see Lexmond et al., 1976).

Types of reactions of mercury in soils greatly depend on the form in which it prevails. One of the most important non-microbial reactions is the oxidation-reduction between mercurous ions and mercuric ions according to the equilibrium:

$$Hg_2^{2+} \rightleftharpoons Hg^{2+} + Hg^{\circ} \quad \text{with } \log K^{\circ} = -1.94$$

This is one of the ways in which both inorganic and organic mercury may transform into metallic mercury.

Cationic mercury forms may easily be adsorbed on soil constituents. Immobilization in the form of lowly soluble mercury phosphate, carbonate and sulfide is another mechanism which prevents translocation in soils. Formation of the highly insoluble HgS from metallic mercury is strongly favored by the presence of sulfate reducing microorganisms.

Anionic mercury species like $HgCl_3^-$ and $HgCl_4^{2-}$ can be adsorbed in a comparable manner as e.g. phosphate ions. This means that for this specific bonding the edges of clay minerals and the positively charged sites of Fe oxides and hydroxides play a predominant role. Also adsorption of molecular mercury compounds like $HgCl_2$ and Hg_2Cl_4 on Mn and Fe hydroxides has been described (Jenne, 1970).

As a result of the strong interactions between mercury compounds and soil constituents, the displacement of mercury in soils in forms other than vapor is usually reported as very low. Poelstra et al. (1973) described the occurrence of Hg in a number of European soils as a function of depth. For Dutch bulb growing soils, where mercury has been applied regularly as a fungicide during the last 50 years, the occurrence with depth was found to be completely determined by soil management. Mercury was found to those depths to which these soils are plowed in order to prevent hard pan formation and to bury disease germs. In soil column studies in which the top layer was labelled with radioactive mercury, ^{203}Hg, they found no translocation in depth after a 6 month leaching period. The rate of water leaching was very high as the artificial precipitation amounted to 24 mm per day. These results were similar for $HgCl_2$, metallic Hg and CH_3HgCl. Figure 10.5 presents results of such leaching experiments for $Hg^°$. The difference in ^{203}Hg at the beginning and after 6 months must be attributed to radioactive decay.

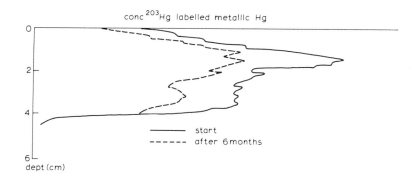

Fig. 10.5. Distribution of labelled metallic Hg as a function of depth at the start of the experiment and after 6 months leaching with 24 mm solution per day (after Poelstra et al., 1973).

Aside from these observations indicating very strong bonding, also the chelation of mercury by soil organic matter compounds has been described (Anderssen and Wiklander, 1965). McLean et al. (1972), when studying the displacement of mercury compounds which were applied on the greens of golf courses, observed both vertical and horizontal translocation. In comparable situations, however, Gilmour and Miller (1973) found that volatilization was the only pathway for Hg disappearance from the soil system.

Hahne and Kroontje (1973) calculated the fractional distribution of total mercury over different complex species with OH and Cl as a function of pH and chloride concentration, respectively. As compared to Cd, Zn and Pb, the relative prevalence of mercury-chloride complexes becomes of importance already at lower chloride levels. This might well be one of the main reasons for the lower mercury accumulation along roadsides as a direct result of increased mercury mobility at the pertaining chloride levels following road salting. As a general conclusion, however, it may be said that mercury transport in soils is low and predominantly limited to transport in the vapor phase.

10.4.7. Mo, molybdenum

Molybdenum is used in steels and alloy production. To a lesser degree also in pigments, electronic tubes and in the production of lubricating materials. Because of its necessity for plant growth it is also applied as a micronutrient fertilizer in agriculture.

Mo deficiency is a well-known cause for decreased plant production. This is ascribed to the Mo function in specific enzymes which have a bearing on the nitrogen assimilation, like nitrate reductase and nitrogenase, which contain Mo. Nitrate reductase is essential in the nitrate assimilation whereas nitrogenase plays a role in the fixation of N_2 to NH_3. As the availability of Mo for plant uptake is strongly influenced by a number of factors, e.g. pH, it is hard to present minimum required levels in soil in terms of general validity.

A number of antagonistic interactions of Mo with other elements are well described in literature. Of these the dual antagonism Cu-Mo has been mentioned already. Uptake of Mo by plants can be impeded by phosphorus and sulfur. Mo itself may have an antagonistic influence on iron uptake.

In animal diets and possibly also in human diets, Mo is required as a constituent of the molybdoproteins xanthine oxidase and aldehyde oxidase. The disease known as molybdenosis, a toxicity effect of too high Mo intakes, can be carried back to a molybdenum-copper imbalance. Ruminants are known to be more sensitive to molybdenosis than monogastric animals.

Normal Mo-contents of plants are around 0.1 ppm on dry matter and, as a result of the above-mentioned function in the nitrogen assimilation, somewhat higher (0.3 - 0.5 ppm) for legumes (Johnson 1966). Mo toxicity symptoms occur when the plant Mo content exceeds 200 - 300 ppm on dry matter. Total Mo-content of soil usually varies in the range 1 - 5 ppm.

The chemistry of molybdenum in soils is fairly complicated. Cationic forms are found only at pH levels below 1.0. This means that in soils only anionic molybdenum compounds occur, molybdates. The valency of the molybdate ions again depends strongly on the pH; below pH 2.5 undissociated H_2MoO_4 predominates, whereas in between pH 2.5 and 4.5 a mixture of H_2MoO_4, $HMoO_4^-$ and MoO_4^{2-} is found. At pH values above 5.0 MoO_4^{2-} will predominate.

At higher concentrations in solution molybdenum may form polymers. As long as the concentration stays below 10^{-4} molar Mo^{6+}, however, polymer formation does not occur (Rohwer and Cruywagen 1964). This means that for all practical purposes polymerization can be left out of consideration since Mo concentrations are reported in the range of $2-8 \times 10^{-8}$ molar under normal soil conditions (Lavy and Barber, 1964).

A number of salts of monomeric molybdic acid are known to occur in soils. Important as such are Ca molybdate, $CaMoO_4$ and ferric molybdate $Fe_2(MoO_4)_3$. Lindsay (1973) indicated the decisive importance of pH on the solubilities of these molybdates. For each unit increase of pH the solubility of calcium molybdate increases by a factor of 100. One of the practical agricultural measures to improve insufficient Mo availability consists of liming; this is in accordance with the above observation on pH-solubility relationship.

The predominant occurrence of molybdenum under normal soil conditions as an anion renders the adsorption on positively charged sites the most pronounced bonding mechanism. Although thus all kinds of positively charged compounds may act as adsorption sites, it is found that especially iron oxides and hydroxides are of prime importance for molybdate adsorption and to lesser degree aluminum oxides, halloysites and kaolinites (Jones, 1957). This may well be one of the reasons that molybdate behavior in soils has been considered as comparable to phosphate behavior. Although this may be true as regards anion adsorption it is certainly not the case as far as solubilities are concerned.

10.4.8. Ni, nickel

The most important use of nickel is in the production of steels and alloys. Ni is also applied in paint pigments, cosmetics and in the production of machinery parts, batteries and electrical contacts. As pointed out by Lagerwerff and Specht (1970) the use of nickled gasoline may be the explanation for the existence of a Ni gradient in soils with respect to their distance to dense traffic lines.

The role of Ni in plant growth is not well understood. To date nickel has not yet been proved to be essential in plants although it can usually be found to be present. Ni was shown to be of importance for growth and development of young chickens (Nielsen, 1970). The fact that Ni is easily taken up by plants when present in soil requires some prudence in Ni applications (e.g. in waste disposals) since Ni is known as highly toxic element to plant growth. For sewage sludge applications this is being taken care of when applying the Zn equivalent factor as has been described before (cf. section 10.4.5). In this factor Ni is considered 8 times as toxic as Zn.

At normal conditions Ni contents of plant do not exceed 1 ppm (dry matter). Total Ni-content of soil may vary from 5 - 500 ppm, with 100 ppm as a rough mean value (Vanselow 1966). 'Serpentine' soils exhibit much higher values.

Being a divalent cation Ni^{2+} must be expected to be adsorbed on the soil complex. It was shown by McLean et al. (1968) that the Ni-Ba exchange on montmorillonite is in accordance with preference expectations. The slight preference for Ba^{2+} over Ni^{2+} is in agreement with the hydrated ion sizes of 5 and 6 Å, respectively. Under normal conditions the total amount of Ni adsorbed is very small only because of the low solution concentrations. This is not so for 'serpentine' soils in which the Ni-concentration may be up to 300 - 700 ppm in the soil solution.

A first decrease in the Ni toxicity of high Ni soils can be obtained by the addition of phosphates. The formation of lowly soluble nickel phosphates like $Ni_3(PO_4)_2 \cdot 8H_2O$ and $Ni_3(PO_4)_2 \cdot 2NiHPO_4$ induces a considerable decrease of the Ni concentration in the soil solution (Pratt et al., 1964). This mechanism does, however, not explain the normal Ni levels in soil which are around 0.005 to 0.050 ppm Ni in solution. Trinickelphosphates would allow a Ni concentration of 1 ppm and up, at pH values of 7 and below. Probably silicate ions govern the Ni concentration in soil solution, immobilization of Ni thus being caused by the formation of nickel silicate minerals. If so, the abundance of silicate ions in soils provides an almost infinite storage capacity when time is available for the formation of these nickel solids. Like Cd^{2+} and Cu^{2+}, also Ni^{2+}, however, is susceptible to chelation which may considerably affect its displacement in soils.

10.4.9. Pb, lead

Most lead is used in the automobile industry, in the production of batteries and as the anti-knock gasoline additives tetraethyl lead and tetramethyl lead. To a lesser degree, the application of lead containing pesticides in agriculture causes a Pb burden on the environment. Pb application in pigments and plumbing is relatively small.

A distribution of total industrial lead consumption over different categories in the USA is presented in Table 10.3.

TABLE 10.3.

Percentual distribution of industrial lead consumption in the USA.

Branch	Percent
batteries	37
gasoline additives	23
metal products	19
ammunition	7
solder	6
pigments	4
brass and bronze	4

Although the total amount used is highest in batteries this type of lead application has little bearing only on Pb dispersion in the environment. It is especially the gasoline additives that cause major discharges.

Gasoline combustion primarily causes an air pollution problem. The lead particles, however, reach the soil surface, especially in precipitation, and thus a soil pollution problem is induced. When it became apparent that Pb may cause health hazards, research about Pb effects on the environment has initially been centered around dense traffic areas. Figure 10.6 presents average values on lead contents in topsoil samples in Illinois, USA. The area of relatively high Pb content coincides with the main traffic lines between Chicago and St. Louis.

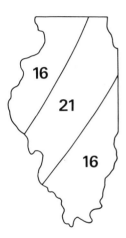

Fig. 10.6. Lead content in ppm of surface soils in Illinois, USA (Alexander, 1971)

Directly along a roadside in the Los Angeles metropolitan area Pb contents in soil as high as 2,400 ppm have been reported (Lagerwerff and Specht 1970).

In a similar way as was done for Cd it could be shown that the ingestion of Pb by people living in the vicinity of a Lead-Zinc smelter is at least 50% higher than normal. Too high Pb intake by man (and animals) may lead to toxic effects following accumulation in liver, kidneys and bones. Largest intake under normal conditions is by food, especially in the form of meat and vegetables. Daily intake levels from food are on the average around 300 μg, from air 10 - 100 μg. It is assumed that the solid food lead intake should not exceed 600 μg per day (Kehoe 1966). This makes the Pb content in plants and the Pb uptake by plants of direct importance for human health conditions. The mechanism of this plant uptake is still not clear and disagreement exists as to the proportion that is being taken up from the soil system and from the air. Translocation in the plant is usually small.

The Pb-content of most plant species is normally in the range 0.5 - 3 ppm. For certain species, however, the Pb toxicity level is very high. This may create a rather dangerous situation because such plants may show no toxic symptoms and apparently look healthy at Pb levels which are hazardous for human consumption. Alloway (1969) reported Pb contents in radish as high as 498 ppm in roots and 136 ppm in tops. Difference in lead tolerances was also described by John and van Laerhoven (1972) for lettuce and oats. They found lettuce to be much more susceptible for Pb accumulation and thus pointed to the necessity of plant selection for the cultivation of Pb contaminated soils. Baumhardt and Welch (1972) found for corn a significant increase of lead in the forage but not in the grain. This would indicate that use of corn grown on lead contaminated soils for silage could be more dangerous than use after grain harvesting.

Not much is known about Pb chemistry in soils. Gasoline combustion being the main Pb source most of the lead will be deposited as the soluble halides, lead chlorobromide, PbCl.Br. Singer and Hanson (1969) described how the danger of excess lead is probably decreased after deposition on the soil, due to the formation of relatively insoluble compounds like $PbCO_3$, $Pb_3(PO_4)_2$ and to a lesser degree $PbSO_4$. Because of the formation of these solids and of the adsorption of Pb when present as divalent cation, lead displacement in soils is mostly small. Slight downward movement by leaching was mentioned by Tso (1970). As has been discussed by Lindsay (1973), soils of high pH may release fixed lead when becoming acidic, especially if $PbCO_3$ is involved in the lead immobilization.

Bittel and Miller (1974) measured the exchange of Pb^{2+} against Ca^{2+} on montmorillonite, illite and kaolinite. In these experiments Pb^{2+} was preferentially adsorbed, resulting in selectivity coefficients of about 2 - 3.

In a comparable manner as was done for several heavy metals mentioned before, Hahne and Kroontje (1973) calculated the distribution of total Pb over different complexes with OH and Cl.

10.4.10. Se, selenium

Major use of selenium is in electronics and electrical industry, e.g. in the production of voltage regulators and photoelectric cells. It is also applied in paints and inks, in cosmetics and in paint removers.

Se is known to be required in the animal diet. Deficiency symptoms, indicated as white muscle disease (WMD) or muscular distrophy are found in livestock, especially lambs and calves, and in poultry. Although the Se function is not completely understood a relationship with vitamin E has been observed. MWD is found if the Se content in the feed is less than 0.03 to 0.1 ppm (Muth and Allaway, 1963). Also Se toxicity has been observed, namely if the Se concentration in the diet exceeds 5 ppm.

In view of the Se requirements considerable attention has been given to the geographical distribution of soils low in selenium, and on measures to adjust the Se content of plants used in animal feeding. Problems met in this respect are caused by the toxicity at too high Se availability, i.e. selenium adjustment must be regulated within fairly narrow limits.

The occurrence of possible selenium forms in soils has extensively been described by Geering et al. (1968). They discussed the criteria for the prevalence of Se in different oxidation states, namely:

+6, selenates, e.g. K_2SeO_4
+4, selenites, e.g. K_2SeO_3
0, elemental, $Se^°$
−2, Se^{2-} e.g. in H_2Se and $CuSe$.

The probability of the occurrence of these different forms is primarily governed by the redox potential. Concentrations found in the soil solution depend on solubility products and pH of the system. Solubility products of ferric selenites were found as low as 10^{-33} for $Fe_2(SeO_3)_3$ and $10^{-62.7}$ for $Fe_2(OH)_4SeO_3$. Amendment of Se in selenium deficient soils may best be obtained by using selenites (Cary and Gissel-Nielsen, 1974). Due to the low solubilities the efficiency is relatively low, but selenate applications would induce toxic levels in crops as a result of their high solubility.

Se is usually found at very low concentrations only in sewage sludges. This makes it unlikely that this element would induce hazards as a result of build-up in the soil, except following frequent disposals of sludges which would contain Se at ppm level.

10.4.11. V, vanadium

Vanadium is used in steels and nonferrous alloys. It is also applied in catalyst compounds, in photographic developers and in the processing of paints and varnishes.

Although there is no conclusive evidence that V is essential for plants, many beneficial effects of small amounts of V on plant growth have been reported. The V influence is specifically manifested via an increased nitrogen supply which might be an indication that V is required for the development of microorganisms which are involved in plant nitrogen nutrition. Also toxic effects of V on plants have been described. Toxicity may, depending on plant species and vanadium form, start at concentrations as low as 0.5 - 1 ppm. As described by Hopkins and Mohr (1971) vanadium is an important factor in controlling the blood cholesterol level of animals. The same might be true for man. Toxicity for man and animal may occur at 5 ppm levels and higher.

Behavior of V in soils has not yet been studied. It may be reasoned that its expected behavior will depend on the form in which it prevails. Four different oxidation states are known which can be listed as:

+5, vanadates, VO_3^-; e.g. NH_4VO_3
+4, vanadyls, VO^{2+}, e.g. $VOCl_3, VOSO_4$
+3, V^{3+}; e.g. VCl_3, $VOCl$
+2, V^{2+}; e.g. VCl_2, VSO_4

Which of these forms is present and stable in soils depends again on the redox potential. Since under normal conditions V is predominantly present as a cation it seems warranted to expect a low degree of V displacement in soils.

10.4.12. Zn, zinc

The most important use of zinc is as a metal coating and in alloys. Galvanized pipes are used in domestic water supply systems whereas galvanized metals used to be an important material for gutters and rainpipes. The usually observed increased Zn level in waste waters in highly industrialized areas is not necessarily caused by direct Zn discharges in the sewer system. The increased acidity of the precipitation in such areas may induce an increased corrosion of Zn containing building materials. No doubt, however, that specific industries like ore smelters constitute a very pronounced direct Zn pollution source. Zinc is also applied in a great number of various articles like inks, copying paper, cosmetics, paints, rubber and linoleum.

Zn constitutes an essential element for animals and plants. In animal metabolism it is of importance as a constituent of a number of different enzymes and hence deficiencies appear as results of insufficient enzyme activities. Also in plant growth Zn is essential for enzymes like dehydrogenases and peptidases.

Zn toxicity has also been reported, though at relatively high concentrations only. Since waste products, especially sewage sludges, may be very high in Zn (up to 50,000 ppm; Chaney, 1973), the possible accumulation of zinc in soils after disposal of such waste materials and its consequences deserve special attention. Zinc toxicity is usually caused by interactions in plant uptake of other essential elements like phosphorus and iron. According to

Chapman (1966) Zn deficiency symptoms usually start for most plant species at Zn levels lower than 20 - 25 ppm. Toxic levels are at about 400 ppm and up. Zinc toxicity in animals starts if the Zn content of the diet exceeds roughly 1,000 ppm. Most plants have been severely injured at such high Zn levels. This higher toxicity sensitivity of plants as compared to animals serves as an automatic protection against Zn accumulation in the food chain. Total Zn levels in soil are at normal conditions in the 10 - 300 ppm range, with 30 - 50 ppm as a rough average value.

Chaney (1973) pointed to the fact that the toxicity level of Zn, in a similar way as for Cu and Ni strongly depends on several factors which control the Zn availability in soils. These factors may differ considerably for different soils. One of the major factors controlling Zn availability is the pH. For the same Zn addition level the yield reduction and the zinc contents of chard leaves were found to be much higher at pH 5.3 than at pH 6.4. pH was also considered by Chumbley (1971) as the main controlling factor when he suggested that not more of toxic elements should be applied to soils in a 30--year period than 250 ppm Zn equivalent if the pH of the soil would be equal to or higher than 6.5. Leeper (1971) suggested a maximum toxic metal addition up to 5 percent of the cation exchange capacity of the soil. Chaney (1973), reasoning that this percentage would still be damaging when the pH would drop below 6.5, combined both recommendations in stating that the toxic metal addition should not exceed Zn equivalent levels equal to 5 percent of the CEC at pH $\leqslant$ 6.5. This would indeed allow minor additions of a large number of sewage sludges only but safeguard soils for toxic metal pollution.

Zn predominantly occurs in soil as a divalent cation and as such it may be adsorbed on the adsorption complexes of clays and organic matter. Competition for adsorption by other cations like Ca and Mg, which are present in relative abundance, prevents substantial zinc adsorption, in a comparable manner as for other heavy metals. As seems to be quite similar for all heavy metals, also zinc may preferentially be adsorbed at low concentration level.

As shown by Lindsay (1973) the predominant zinc species in soils below pH 7.7 is Zn^{2+} and at higher pH values the neutral zinc hydroxide, $Zn(OH)_2$. Large pH influence on zinc solubility is indicated by the fact that the zinc activity in solution at equilibrium conditions decreases 100-fold for each unit increase in pH.

Prediction of Zn behavior in soil is complicated by the occurrence of a great number of complexes which have a cationic character like e.g. $Zn(OH)^+$, or an anionic character like, e.g. the zincate anions $Zn(OH)_3^-$, $HZnO_2^-$ and ZnO_2^{2-}. Moreover, chelation may influence Zn-soil interactions.

10.4.13. Chelation and metal mobility

Interest in the formation of chelates of metal cations in soils originates from the finding that crop supply with micronutrient metals can be improved by adding the deficient metal in chelated form. This practice has already been applied for several decades and was started for iron. Apparently the chelated iron remained in solution at higher concentration levels than were to be expected from solubility considerations of solid iron phases. Thus chelation increases the mobility and consequently the plant availability of metal cations.

In view of the present concern about heavy metals in the environment the mobilization as mentioned is of interest with respect to heavy metal transport processes in soil and their soil accumulation or distribution in ground and surface water. If chelation concerns a substantial proportion of the total amount of a certain heavy metal present in soil, this would mean that displacement is not governed by adsorptive interactions and precipitation-dissolution considerations only.

The influence on mobility is basically caused by a change in ionic behavior. After chelation the former metallic cation is completely enveloped by the chelating agent. This may transform the cation into an anionic complex with the corresponding enhanced opportunities for displacement and leaching in soils. For a specific chelating agent a series of chelates may be formed with a certain metal cation, depending on the number of H^+ ions and OH^- ions that are involved. As an example, EDTA (ethylenediaminetetraacetic acid) can be symbolized as L^{4-} when dissociation of H^+ at each of the four acetic acid groups takes place. Chelation of Fe^{3+} by EDTA may thus result in the following possibilities:

FeL^-, $FeHl$, FeH_2L^+, $FeOHL^{2-}$, $Fe(OH)_2L^{3-}$, $Fe(OH)_3L^{4-}$

The work of Lindsay and coworkers (1967, 1969) has greatly contributed to the basic understanding of the chemistry of chelates in soils. Norvell (1972) provided a procedure for the prediction and estimation of chelate stabilities. The two principal points in the thermodynamic approach to describe chelation concern in the first place the availability of the metallic cation in order to compete with other cations for a bonding with the chelating agent and secondly the stability of the chelates involved. For the more abundant cations in soil like H^+, Ca^{2+}, Mg^{2+}, Al^{3+} and Fe^{3+}, the equilibria with chelating agents may be expressed by means of stability diagrams. In such diagrams the mole fraction of the chelating agent that is associated with these cations is related to the pH in the system Departing from a number of realistic assumptions for soil systems and from known solubility products of solid phases, values for the most probable activities of the cations as a function of pH can be calculated (cf. chapter 6). Construction of chelate stability diagrams also requires values for chelate formation constants as provided by Sillen and Martell (1971). In this way Norvell (1972) calculated the stability diagrams for eleven different chelating agents. Such diagrams allow a comparison between different cations for one specific chelating agent, and between the different chelates of one specific cation. They are as such indispensable in the prediction of most probable chelation features.

The above author pointed to the complications met when applying the same approach for heavy metals, caused by uncertainties and lack of knowledge about the metallic cation concentration. By introducing in this case the ratio between chelated and free metallic cation, the influence of chelation as a function of pH can be expressed as the most probable distribution of total metal over chelated and non-chelated form, which information is of prime importance in transport considerations. Norvell (1972) showed in this manner for Zn^{2+}, Cu^{2+}, Mn^{2+}, Ca^{2+} and Ni^{2+} that a large difference exists between the stabilities of the different chelating agents.

Gradual progress is made in the understanding of the mechanism of chelation and the factors that influence chelation. Considerations as the above concern, however, chelating agents which are very well defined and synthetically producible. Nature may bring about a vast series of chelating agents which have not yet been identified. This is of practical importance in relation to disposal of organic matter in combination with toxic metal containing wastes, like sewage sludges and hog manure. Following the incorporation of the organic matter in the soil matrix and the inherent formation and conversion of all kinds of organics, the probable occurrence of chelation requires increased precautions toward such disposals.

10.5. ORGANIC PESTICIDES IN SOIL

In the rapid development during the last decade(s) of the research on the fate and behavior of environmental contaminants, no subject matter has attracted so much attention as organic pesticides. This must be ascribed to the broad awareness about the hazards involved. Also the continually increasing number of pesticides has undoubtedly contributed to this situation. At present a number of active compounds far exceeding 60,000 is known. Large scale application, however, is limited to less than 50.

Specific compounds will be referred to only for explanation and demonstration purposes, without implying a measure of the extent of the use of such compound in practice. Sources considered here are confined to purposeful applications. No attention is given to industrial discharges, accidents and carelessness although these may constitute incidental and non-neglectable point sources for environmental pollution with pesticides. Also no distinction will be made as regards the type of biological activity exhibited by these compounds, thus treating fungicides, herbicides, insecticides, nematicides, miticides, rodenticides and slimicides as one group. Non-pesticide organic chemicals like defoliants, soil conditioners, growth regulators and fertilizer extenders are left out of consideration in this text as their level of toxicity tends to be different from the pesticides.

As in the previous sections main emphasis will be placed on the interactions of organic pesticides with soil, but preceding that discussion some general remarks will be made on the environmental hazards involved. The latter have been discussed to some extent by Nicholson (1968). The most spectacular effects are fishkills following excessive pesticide concentrations in surface water, but observed effects of this nature have fortunately been incidental and were then usually traceable to particular local discharges of a preventable nature. Sublethal concentrations in open water are probably much more common than calamities of the above type and many of these may have been unnoticed. This refers to a second problem namely the exposure of aquatic life, wildlife and higher animals, man inclusive, to subacute doses of pesticides and the not well-understood long-term effects of such exposure. A third and evident problem is given by the possible contamination of drinking water supply.

The main pathways for the dispersion of pesticides from soils on which they were applied into the environment are again via runoff from the soil surface and by leaching out of the soil profile. Here an important difference may be noted when comparing pesticides with other pollutants discussed in the previous sections. On the one hand pesticides are added purposely to soil at a killing level for certain organisms. Thus direct runoff of excess surface water into adjacent open water constitutes indeed a hazard, albeit local, incidental and temporary.

In contrast, N and P, if present in runoff are never an immediate threat to life, and thus in that case the significance of runoff must be measured in terms of the mean effect during long time periods. Also in case of heavy metals, runoff is only one aspect of the dispersion pattern, as these are - as a rule - not added to soil at a killing level, and thus again the effect must be averaged over a longer period.

With regard to the displacement by leaching, however, one finds often a very effective retention of pesticides by soil, because of the combination of strong adsorption and biodegradation acting on relatively small amounts. In contrast again, part of the N moves relatively fast (cf. section 10.3.5), while neither P nor the heavy metals are subject to actual biodegradation. Granted that exceptions to the above must be carefully followed it may nevertheless be stated that runoff and its control are of prime importance with respect to the dispersion of pesticides into the environment.

It is interesting to note that a 'universal soil-loss equation' as presented by Wischmeier and Smith, (1967) - although a very useful tool in guiding farm conservation planning for hilly areas - is of restricted value for the prediction and control of pesticide contamination by runoff. This is mainly caused by the fact that the above-mentioned equation relates the erosion controlling factors like rainfall, soil erodibility, slope length and gradient, cropping management and conservation practice to the total amount of soil solids that is lost on a yearly basis. As has been described for symmetric triazines by Hall et al. (1972) and Hall (1974), the relative contribution of water dissolved pesticide in runoff is much higher than that of solid-adsorbed pesticide. This is not because the herbicides involved were not adsorbed on the solid phase (concentration in runoff solids was much higher than in runoff water) but because the amount of runoff water exceeded many times the amount of eroded solids. Besides this imbalance between the amounts of water and of solids transported during erosion events, the second limitation of the erosion equation is that one is, in pesticide contamination, not so much interested in yearly contributions or average values on a yearly basis. Especially the occurrence of specific erosion events and the prediction of such events is of importance. In other words, the probability of sudden concentrated contaminations, is of major concern.

Bailey et al. (1974), in discussing a model for the prediction of pesticide runoff, pointed to the important practical implications of a better understanding of runoff events. It may well be possible that such predictive modelling contributes to the formation of guidelines in order to prevent runoff as much as possible. Such guidelines should not necessarily be confined to practical farming but may be extended to the manufacturing of the organic compounds in order to introduce properties which minimize their mobility. The last mentioned factor is of more importance with respect to in-profile movement than to displacement due to erosion. Strong adsorption on soil compounds may, however, also limit environmental pollution by runoff.

When discussing the movement of pesticides in soil the same approach as applied for inorganic compounds can be followed. This means that the transport and accumulation may be described with the same, or comparable, mathematical relationships. There is, however, an important difference in the final elaboration of such approach. When describing e.g. phosphate transport, the adsorption may be accounted for by the introduction of adsorption isotherms, whereas dissolution-precipitation reactions can be built in by introducing a 'production' factor. Considering pesticide movement the picture becomes much more complicated due to the biodegradation of pesticides in soil. Indeed, biodegradation is a very important feature from a pollution point of view as decomposition prohibits accumulation. This biodegradation may be described as a 'negative production' factor, representing the change in concentration with time as a function of a number of controlling factors. Unfortunately these degradation processes are often complicated, that is they may involve many different steps which are not known in detail, let alone that the relevant rate parameters are known. Differences with precipitation reactions are brought forward by actual disappearance of the compound from the system

In the following paragraphs the two main factors controlling the fate of pesticides once present in soil, viz. the interaction with the solid phase and the decomposition in the soil environment, will be discussed. It is noted that these factors have already been attracting considerable attention for many years, as the ensuing mobility and perseverance of the pesticide in the soil profile determines largely the method of application and thus the efficiency of the pesticide as a controlling agens for certain plagues.

Mobility of a pesticide may determine whether the compound will reach the intended zone of uptake or activity when not applied at that specific zone. This is specifically so in soil fumigation, in which case the transport is predominantly governed by the soil vapor phase (Leistra, 1972). Moreover, the rate of decomposition of the chemical affects its efficiency in the control of a disease or organism. In this respect the term dosage has been introduced. This is the concentration toward which the organism is being exposed integrated over the exposure time. It is self-evident that high degradability (corresponding to low persistency) and high mobility require an increase of dosage.

Environmental concern has recently added another dimension to this long-existing professional interest in the mobility and decomposition of pesticides in soil. The following sections point to the main factors that control these phenomena. For more detailed information the reader has to be referred to the literature in this field.

10.5.1. Bonding by soil constituents

The soil constituents that will be considered here are clay minerals and organic matter. It is realized that also oxides and hydroxides play a role in pesticide bonding which may be of special importance if such compounds predominantly constitute the soil adsorbing complex. This usually is the case in low organic matter, aluminum and iron containing sandy soils and especially in lateritic soils. For all practical purposes these specific soil constituents are left out of consideration. Their adsorbing characteristics are only known in a qualitative way. They predominantly act as anion adsorbers whereas the adsorption process is strongly influenced by the pH of the system (Breeuwsma, 1973). The pH is actually the all-important factor controlling the pesticide interaction with clays and particularly with organic matter. This must be attributed to the fact that the pH usually influences the adsorbing properties of both adsorber and adsorbate.

Results of adsorption measurements are commonly expressed by means of adsorption isotherms. Regularly used mathematical descriptions for adsorption isotherms are the Freundlich equation and the Langmuir equation (cf. also section 10.3.8). The Freundlich adsorption equation reads in general terms:

$$x/m = K \cdot c_o^{1/n} \tag{10.1}$$

in which:

x/m = quantity of compound or ions adsorbed per unit weight of adsorber
c_o = equilibrium concentration of the adsorbing compound once adsorption has been established.
K and n are constants, which may be derived from the experimental adsorption data.

The Freundlich adsorption equation is of empirical nature and has the (unrealistic) consequence that the amount adsorbed should increase infinitely with increasing concentration.

The Langmuir adsorption isotherm has originally been derived for the adsorption of gases on adsorptive surfaces, and may thus have its limitations in the extrapolation to adsorption of other compounds. A number of assumptions, which are valid when considering gaseous adsorption, are not necessarily equally valid for other compounds. The Langmuir adsorption equation reads as follows:

$$x/m = \frac{Kb \cdot c_o}{1 + K \cdot c_o} \tag{10.2}$$

in which:

x/m = amount adsorbed per unit weight of absorbent
c_o = equilibrium concentration of the adsorbing compound once adsorption has been established
K = constant relating to the bonding energy
b = adsorption maximum or the amount adsorbed when the adsorbent is completely saturated

The Langmuir adsorption isotherm has the advantage that both the K and b values may be determined from relatively simple laboratory experiments. Once these values are known, the amount that will be adsorbed at any particular input may be estimated. As such the adsorption isotherm takes a key position in the prediction of the fate of compounds in soil.

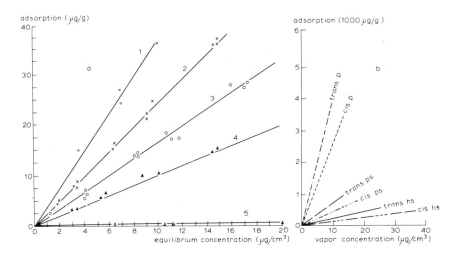

Fig. 10.7. Adsorption isotherms for two organic chemicals.

a: lindane adsorption on different soil fractions (after Kay and Elrick,1967).
  1. clay, ($<2\mu$)
  2. silt ($2\text{-}50\mu$) including organic matter
  3. silt after removal of organic matter
  4. sand ($>50\mu$) including organic matter
  5. sand after removal of organic matter
b: adsorption of cis- and trans- 1,3-dichloropropene on peat (p), peaty sand (ps) and humic sand (hs). (after Leistra, 1972).

As the propagation of the nematicide 1,3 dichloropropene predominantly occurs via the vapor phase, not the concentration in the soil solution but in the soil gaseous phase has been plotted in figure 10.7.b.

In the above cases a linear relationship was found over the entire concentration range. The proportionality factor between amount adsorbed and concentration in solution is then simply given by the constant slope of the adsorption isotherm Such a linearity of the isotherm is not uncommon when the concentrations do not exceed the values that actually occur in the soil solution after pesticide application (several ppm range). In adsorption measurements the concentrations are usually increased to much higher values resulting in sloping isotherms as have been reported for e.g. atrazine adsorption (Armstrong et al.,1967)

As has been discussed by Hamaker and Thompson (1972) the Freundlich equation (10.1) and the Langmuir equation (10.2) both result in one simplified adsorption equation for specific conditions. These conditions are that

$n = 1$ in the Freundlich equation and that c_o is very small, allowing the substitution of 1 for $1 + Kc_o$ in equation (10.2). The simplified adsorption equation then reads:

$$x/m = K_d c_o \qquad (10.3)$$

in which K_d is the adsorption proportionality factor, corresponding to the slope of the linear adsorption isotherm. K_d, indicated as the distribution coefficient, presents a very important practical characteristic for the adsorptive behavior. Its value actually reflects the distribution of the organic chemical over the different soil phases according to:

$$K_d = \frac{\text{amount adsorbed per unit weight of soil}}{\text{amount in solution (or gas phase) per unit volume of liquid (or gaseous) phase}} \qquad (10.4)$$

The so-called distribution ratio, (cf. R_D, chapter 4) is then found as K_d/W with W = liquid content (or air content for gaseous adsorption) in ml per gram of soil.

Values of K_d may vary widely for different chemicals in accordance with their adsorption characteristics, e.g. from <1 for picloram to $>1 \times 10^5$ for DDT. Since the organic matter plays a dominant role in the adsorption of organic chemicals in soil, K_d referring to the soil as a whole may for practical reasons be substituted by K_{om} (organic matter) or K_{oc} (organic carbon). The relationship between these factors is given by:

$$K_{om} = \frac{K_d}{\text{\% organic matter}} \times 100 \qquad (10.5)$$

$$K_{oc} = \frac{K_d}{\text{\% organic carbon}} \times 100 \qquad (10.6)$$

The factors which affect the magnitude of the distribution coefficients can best be discussed when considering the different adsorption mechanisms that may prevail, in combination with the different types of adsorbers that may occur in soil.

10.5.1.1. Pesticide adsorption mechanisms

The different bonding mechanisms that may occur between pesticides and soil constituents have been reviewed by Bailey and White (1970) and by Stevenson (1972). The most important interactions are:

a. Physical bonding due to London- van der Waals forces
b. Electrostatic bonding by ion exchange and protonation
c. Hydrogen bonding
d. Coordination bonding or ligand exchange

These bonding mechanisms do not necessarily occur separately. Adsorption in the soil will usually be governed by a combination of two or more of the above mechanisms.

a. Physical bonding

The London- van der Waals forces which induce this type of bonding result from different interactions between molecules and ions in a system. Actually the London- van der Waals forces are the combined result of dipole-dipole interactions, ion dipole interactions, dipole-induced dipole interactions, and induced dipole-induced dipole interactions. It is a well-known fact that London van der Waals forces decrease extremely rapidly with increasing distance. They are inversely proportional to the seventh power of the distance, in contrast with electrostatic forces between charged particles which are inversely proportional to the second power of the distance between the effecting components.

Due to this close-distance action the size and shape of the molecule or ion may considerably influence the strength of this type of bonding. Small and regularly shaped molecules allow closer contact with the adsorber than bulky and irregular compounds. On the other hand extended molecules may have a larger area of contact per molecule. The prediction of the relative magnitude of London- van der Waals forces in a complicated system as soil is difficult. Hance (1965) attributed the decrease in adsorption of urea herbicides in the order

neburon $>$ linuron $>$ diuron $>$ monuron $>$ fenuron

to the greater contribution of London- van der Waals forces with increasing chain length. The energy of London- van der Waals adsorption is commonly found to increase with increasing molecular size, especially if such increase is accompanied by an increase in double or triple bonds. The overall result of increased sizes on the bonding depends on which of the factors dominates, as the increase in bonding energy and the distance effect are both to be considered. Physical adsorption is usually considered to be of minor importance in the overall pesticide bonding in soil. Interlayer bonding in clay minerals can often be neglected because of the rather large molecular size of many pesticides. Thus the physical adsorption is limited to external surfaces (Bailey and White, 1970).

b. Electrostatic bonding

This bonding mechanism may occur when the pesticide is present as a cation or becomes a cation after protonation. The pesticides diquat and paraquat are applied in cationic form and may as such directly be adsorbed on the cation exchange complex of the clay minerals or the soil organic matter. Figure 10.8 presents an example how the divalent diquat may be adsorbed to two carboxyl groups of organic matter which are negatively charged following proton dissociation.

The dissociation constants of the functional groups in the organic matter molecules thus determine the extent of dissociation at a certain pH level and consequently the possibilities for this type of adsorption on organic matter.

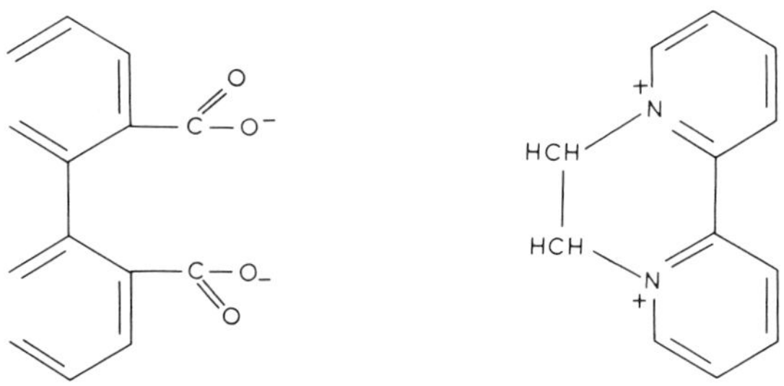

Fig.10.8. Electrostatic adsorption of diquat on part of an organic matter molecule (after Stevenson,1972)

A large number of pesticides may be transformed to cations after association of one or more protons as is demonstrated in figure 10.9 for symmetric triazines.

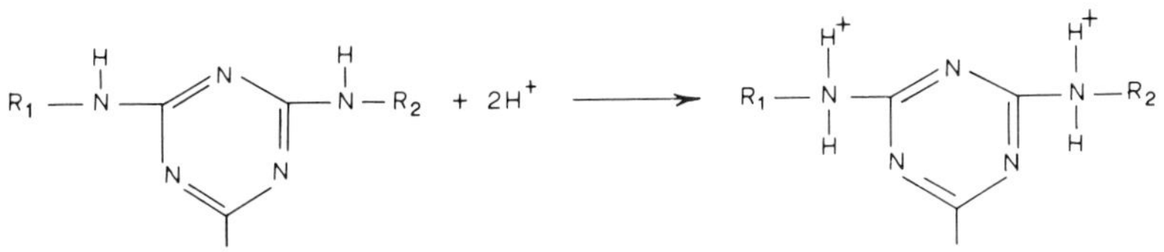

Fig. 10.9. Conversion of an s-triazine molecule into a divalent cation after association of two protons.

The pK_D (cf. chapter 2) values of the pesticides determine the degree of proton association and consequently of their cationic behavior. Thus it may be seen here that the pH of the soil system in relation to the values of the dissociation constants of the organic matter components on the one hand and the values of the proton association constants for the pesticides at the other hand, completely govern the pesticide-organic matter interaction. Stevenson (1972) presented a general picture which allows the prediction of the degree of adsorption of different s-triazines on two types of functional groups in fulvic acids.

The correct evaluation of the pH influence on this type of bonding is considerably complicated by the fact that the proton concentration in the close vicinity of a soil colloid usually is essentially higher than in the equilibrium soil solution. This then means that for a certain pH value the degree of proton association in a soil system is different from that to be expected for the same pH value in aqueous solution (Bruggenwert, 1972). As a result pesticides are sometimes found to be adsorbed at pH values where they should be expected not to adsorb according to pH - pK_D considerations.

In the prediction of the adsorptive behavior it must be kept in mind that for compounds which exert proton dissociation and association about 10 percent occurs in associated form at a pH value of 1 unit higher than the pK_D whereas 90 percent prevails in associated form at a pH value of 1 unit below the pK_D. Complete association and dissociation will occur if the pH is two units lower and higher, respectively, than the pK_D value.

c. Hydrogen bonding

This mechanism can be looked upon as an incomplete protonation, i.e. there is only a partial charge transfer between the electron donor and the electron acceptor. This charge transfer is complete in the case of actual protonation (Hadzi et al.,1968).

Hydrogen bonding is very common if the pesticide contains an N-H group as is the case in e.g. phenylcarbamates and substituted ureas. The hydrogen bond may occur with an oxygen of the clay surface or of organic matter compounds. In the latter case this may as well be oxygen of organic acid groups as phenolic oxygen. Examples of hydrogen bonds are presented in figure 10.10.

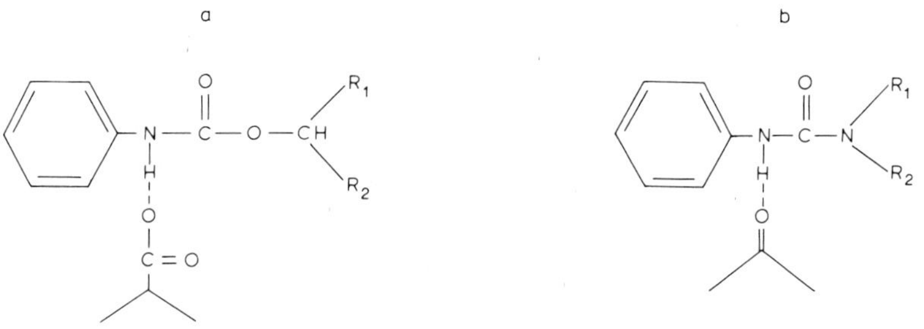

Fig. 10.10. Examples of hydrogen bonding
a. between phenylcarbamate and organic carboxyl group
b. between substituted urea and phenolic OH-group

d. Coordination bonding

Bonding by means of coordination compounds may be explained as the result of various complex formations. Such complexation is similar to the one described in the heavy metal section. Actually, the complex is formed via the donation of electron pairs by the ligand and the acceptance of these electrons by the metal (atom or ion). This makes that a central metal atom or ion is surrounded by a cluster of ligands. The known complexing agents as well as organic chemicals or clay minerals may act as ligands. Since one or more of the complexing agents may be substituted by pesticides the term 'ligand exchange' for this type of bonding is obvious. As an example coordination bonding of symmetric triazines is presented in figure 10.11.

The metal ion thus forms a bridge between the pesticide and the soil constituent (clay or organic matter). The following metal ions are known to act in the above coordination

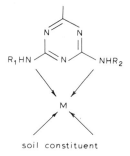

Fig. 10.11. Coordination bonding of s-triazines.

bonding of pesticides: Co^{2+}, Cu^{2+}, Fe^{2+}, Mn^{2+} and Ni^{2+}. An evident prerequisite for the occurrence of ligand exchange is that the metal ions have a greater affinity for the pesticide than for the substituted ligand.

A bonding mechanism similar to the above coordination bonding, although less complicated, is given by the salt bridge formation. In that case a non-monovalent cation acts as a bridge between soil constituents and carboxyl groups of pesticides.

The above mechanisms all pertain to positive adsorption or bonding of the pesticide compound. Depending on pesticide characteristics, also indifferency or exclusion may occur e.g. in the neutral pH range for 2,4-D; 2,4,5-T and picloram.

10.5.1.2. Pesticide adsorbents in soil

Main adsorptive surfaces in soils of temperate regions are provided by clay minerals and organic matter. In accordance with the magnitude of the cation exchange capacity the most common clay minerals kaolinite, illite and montmorillonite usually show an increase in pesticide adsorption capacity in this order.

Frissel (1961) calculated for a number of herbicides the distribution between liquid phase and solid phase as a function of pH on the basis of the pK_a values (negative logarithm of the association constants) of the pesticides.

In these calculations it was assumed that the herbicides are distributed homogeneously over a furrow slice with a thickness of 20 cm and that no adsorption on other compounds than clay minerals were to take place. A bulk density of 1.5 was chosen and a moisture content of 20%. Use was made of adsorption isotherms as measured for the different herbicides.

The results are given in table 10.4, together with the application rates, indicating again the great influence of pH on the adsorption.

TABLE 10.4

Calculated concentrations in soil solution and percentages adsorbed for a number of herbicides as a function of pH. Predominant clay minerals: I, illite; K kaolinite; and M, montmorillonite (after Frissel, 1961).

Compound	Application kg/ha	Clay	Conc. in solution (ppm) pH			Percentage adsorbed pH		
			5.5	6.5	7.3	5.5	6.4	7.3
DNC	4	I	0.07	0.19	6.70	99.0	97.0	0
		K	2.50	6.70	6.70	63.0	0	0
		M	0.06	0.18	6.70	99.1	97.0	0
dinoseb	1	I	0.02	0.05	1.70	99.0	97.0	0
		K	0.63	1.70	1.70	63.0	0	0
		M	0.02	0.04	1.70	99.1	97.0	0
2,4-D		I	0.05	0.09	1.70	97.0	95.0	0
2,4,5-T	1	M	1.70	1.70	1.70	0	0	0
MCPA								
monuron	1	I	0.07	0.07	0.08	96.0	96.0	95.0
diuron		M	0.03	0.03	0.03	98.0	98.0	98.0
trietazine		I	0.01	0.02	0.04	99.6	99.4	99.0
simazine	1.5	K	0.07	0.14	0.14	97.0	95.0	95.0
chlorazine		M	0.00	0.01	0.01	99.8	99.7	00.6

The above pH values refer to solution pH. As has been mentioned before the concentration of H-ions in the vicinity of the adsorbent may greatly exceed the concentration in the equilibrium solution.

The data of table 10.4 are in agreement with the practical observation that soils of high adsorptive capacity require higher dosages of most herbicides in order to obtain the intended biological effect, than soils of low adsorptive capacity.

In experiments on saturation of the adsorption complex with pesticides, values of amounts adsorbed in accordance with exchange capacity have been reported, as well as indicating both oversaturation or undersaturation. Cation exchange undersaturation is usually ascribed to a steric hindrance. Especially in cases of high surface charge density the average space per charge can easily be too small to allow complete neutralization of all electrical charges by the relatively large organic chemical ions.

Oversaturation of CEC has been explained in several ways. For methylene blue adsorption, this oversaturation was attributed to physical bonding by London- van der Waals forces in addition to the electrostatic adsorption by Coulomb forces. In other cases the formation of 'hemi-salts' was supposed to occur; in such hemi-salts a single proton is being shared by two basic molecules as may be visualized according to the reaction:

$$\text{clay} - RNH_3^+ + RNH_3Cl \rightarrow \text{clay} - RNH_3^+(RNH_3Cl) \rightarrow$$

$$\rightarrow \text{clay} - RNH_3^+(RNH_2) + HCl$$

The decrease of pH according to this equation has actually been observed.

The influence of organic matter on the adsorption may for instance be seen from the isotherms of figure 10.7. The term 'organic matter' is nothing but a noun for the vast array of compounds that constitute the soil organic matrix. As was thoroughly described by Stevenson (1972), pesticide interaction with organic matter can best be discussed when following the distinction in humic acids and fulvic acids, and considering the dissociation, polymerization and functional group characteristics of these acids. Increased insight in the structure and composition of organic matter in soil is a demand for a quantitative interpretation of pesticide adsorption.

From the foregoing it will be clear that for most pesticides no general value for the distribution ratio can be presented. Such values depend entirely on the prevailing conditions of which organic matter content and pH are the most important.

As a very generalized rule it may be said that most pesticides are tightly bound to soil constituents at practically prevailing pH ranges, thus preventing pesticide movement, except for pure sandy soils.

10.5.2. Decomposition of pesticides in soil

The main processes causing degradation of pesticides in the soil system can be categorized as photodecomposition, purely chemical degradation and microbial degradation. These features have in this order an increasing contribution on the overall breakdown of pesticides in soil. Only the last mentioned mechanism may be indicated as biodegradation; in that case the decomposition is caused by breakdown processes of biological nature. The interaction between microorganisms and pesticides is dual: microbes may govern pesticide concentrations, but pesticides may also greatly influence the microbial composition and activity of soil.

The above three decomposition mechanisms may occur separately. Degradation under field conditions, however, usually is the combined result of two or all three mechanisms acting simultaneously. It may be hard to distinguish between the relative contribution of each. Comparison of degradation rates under sterilized and non-sterilized conditions has become a standard procedure to measure the relative contribution of microbial decomposition. It must be realized, however, that sterilization may also considerably influence chemical decomposition processes like hydrolysis.

10.5.2.1. Photodecomposition

Degradation of pesticides in soil by direct influence of light has been the least investigated. The relative contribution must usually be small since this type of degradation is completely limited to the soil surface. This means that only those compounds may be attacked which are not soil-incorporated and those which have moved to the soil surface following capillary water rise during drying (Helling et al., 1971).

Photolysis has been studied particularly in aquatic environments under conditions considerably differing from radiation at the soil surface, i.e. ultraviolet light of relatively high intensity is mainly used as energy source. Nevertheless the rate of photodecomposition is usually found to be low. Probably the most important contribution of photolysis is provided by the increased microbial degradability of compounds after photolytic pretreatment. Such an influence has been shown for organics of low microbial degradability like lignin compounds (Park et al., 1972; Rockhill et al., 1972).

10.5.2.2. Chemical degradation

A large number of pesticides are subject to degradation processes in which purely chemical decomposition constitutes an important part. The degradation of malathion may be mentioned as an example involving organophosphorus insecticides (Walker and Stojanovic, 1973, 1974). Following the division as given by Helling et al. (1971) the chemical degradation processes can be categorized as reactions catalyzed and not catalyzed by soil constituents.

Non-catalyzed chemical degradation reactions consist of hydrolysis, oxidation, isomerization, ionization and salt formation. Of these, hydrolysis and oxidation are the most important. The previously mentioned organophosphates are subjected to alkaline hydrolysis. Acidic hydrolysis is known for the compound 2,4-DEP. The phenylcarbamate proximpham can hydrolyse at acid, neutral and alkaline conditions.

Soil constituents may considerably catalyze chemical decomposition reactions. This has usually been attributed to the increased H-concentration near clay minerals. In addition to this, some specific influence must also be exerted by the soil compounds since reactions are mostly found to be more effective in the presence of a soil colloid than with an equivalent amount of acid alone. Also specific soil constituents like iron oxides and amorphous alumina are found to catalyze degradation. On the other hand the presence of organic matter may retard chemical decomposition.

The mechanisms of soil catalysis are not clearly understood. Catalytic effects greatly depend on the nature of the pesticide. A soil factor which acts catalytic in the degradation of the one compound can be a retardation factor in the chemical degradation of another pesticide.

10.5.2.3. Biodegradation

This pathway of pesticide decomposition is controlled by microorganisms, which makes it obvious that the same factors influencing normal microbial activity in soil also govern pesticide biodegradation. It is well known that these main factors are soil temperature, appropriate moisture content and presence of organic matter.

In the microbial metabolism of pesticides the organic compounds are used as an energy source for other metabolic processes. As pointed out by McNew (1972) microorganisms gradually developed capabilities of decomposing organic residues from plants and animals. Most of the pesticides are new for the soil microorganisms. This causes lack of adaptation on the part of the microflora and consequently incapability for biodegradation. Bioadaptation usually proceeds slowly as a result of low pesticide concentrations in soil. Moreover, applications are irregular and infrequent. It seems probable that an accelerated adaptation of microorganisms to newly introduced molecules and compounds provides one of the ways to prevent undesired pesticide accumulation.

Requirements for uptake of pesticide molecules by microbial cells have extensively been discussed by Meikle (1972). One of such prerequisites is solubility of the compound. Different mechanisms, like diffusion and specific carrier transport, may be responsible for the transport through the cytoplasma membrane.

The compound which has probably received most attention with respect to degradation is DDT. Like most highly chlorinated compounds, DDT is very resistant to degradation. Decomposition of DDT results in the compound DDE which is also extremely persistent. Both compounds are undesirable from an environmental point of view because of storage in fat and because of food-chain magnification. This example clearly demonstrates that not only the microbial attack of the originally applied pesticide is of concern but also the metabolites generated during biodegradation. A similar continuation of toxic activity after biodegradation is given by the oxidation of aldrin, resulting in dieldrin. Dieldrin has insecticidal properties comparable to those of aldrin.

Turnover of DDT into DDD was found to be rapid under anaerobic conditions, i.e. only 1 to 2 percent of applied DDT was left after three months anaerobic incubation (Guenzi and Beard, 1968). Under aerobic conditions a very slow conversion to DDE takes place. At comparable incubation conditions except for aerobic environment most of the DDT (75 percent) appeared still to exist after half a year whereas at maximum 4 percent DDE could be detected. With aerobic conversion only trace amounts of DDD were found. Kearny et al. (1969) have suggested to use the relatively rapid anaerobic DDT decomposition by applying flooding as a means for the reclamation of DDT contaminated soils.

It should be kept in mind that degradation rates as mentioned above were measured in laboratory experiments. This makes the extrapolation of such results to practical conditions speculative, as field conditions are usually much less favorable than those maintainable in a laboratory.

A large number of microorganism species have been shown to possess the capability of DDT decomposition. Although not with the same efficiency, and as a matter of fact all with low efficiency as compared to many other pesticides, there are at least 28 microorganisms which can rapidly convert DDT to DDD. For the conversion aldrin-dieldrin at least 80 pure cultures showed some capacity.

The most important chemical reactions in the pesticide biodegradation are:
a. Oxidation; especially the β-oxidation is an important feature in the chain shortening of alifatic compounds; also oxidation of the terminal C atom (with or without a double bond) occurs frequently.
b. Ring cleavage; this can be ortho cleavage or meta cleavage; it is the prime pathway in which the aromatic nucleus is attacked after catechol formation by bonding of two hydroxyde groups.
c. Reduction; this may result in e.g. a double bond in case of alifatic compounds, conversion of alcohols from aldehydes and hydroxide formation from carboxyl groups.
d. Hydrolysis; organophosphorus compounds can be subjected to enzymatic hydrolysis; breakage of the phosphorus bond is strongly influenced by the pH.

Actual processes in the biodegradation consist of a combination of two or more of the above principal reactions, preceded or followed by specific reactions for specific chemical structures. Dehalogenation is an example in the case of halogen compounds. Moreover the chemical attack may start at different places on the molecule.

A thorough compilation of chemical pathways involved in the microbial attack of pesticides has been presented by Meikle (1972). The fact that biodegradation mostly involves quite complicated organic chemistry places its discussion beyond the scope of this text.

Application of a certain pesticide can also affect nontarget microorganisms, animals or plants. The degree to which these influences occur depends completely on the specificity of the compound. These side effects have long been recognized and it may be said that soil pollution aspects of pesticide use provide a collective indication of part of these side effects. Increased specificity and decreased persistence as major targets in the development of new pesticides, combined with biological control of diseases and development of disease resistant crop varieties, provide the most effective ways to protect the soil against pesticide pollution.

A somewhat specific problem in addition to pesticide accumulation in agricultural areas is caused by the disposal of surplus and waste pesticide materials. Here again soil is usually considered as the appropriate sink for such wastes. As discussed by Stojanovic et al. (1972) for a large number of pesticides the soil may indeed be used as a natural biological incinerator. Many others, however, have to be disposed of in combination with different treatments like incineration or chemical treatment.

10.6. MISCELLANEOUS SOIL POLLUTION SOURCES

Sources of soil pollution as discussed in the preceding sections are predominantly diffuse, except for the typical disposal systems as for example raw sewage water and sewage sludge disposals. Some other point sources are brought about by oil spills and oil sludge disposal, gas leakages and sanitary landfills. These sources deserve some consideration here.

10.6.1. Oil spills and oil sludge disposal

Due to the huge quantities of petroleum oils that are handled and used, the potential of soil and surface water contamination with oil is great. Contamination may result from accidents in transport as well as from deliberate disposals of oily wastes. Especially drinking water supply areas have to be protected against oil contamination since extremely low concentrations of oil components may adversely affect the smell and taste of water. In principle there is no difference between the composition of oil sludges which result from refinery operations and those from spillages of crude oils and refinery products. Oil sludges may be handled by incineration. Sludges resulting from spillages may involve large amounts of oil polluted soil which may make incineration very expensive.

Decomposition of oil residues following spillages on soil appeared very well possible when appropriate measures are taken. Thus it became a consideration to treat all types of oily sludges, also the industrial waste sludges originating from refineries, by disposal on land (Dotson et al., 1972). This approach is known as land spreading. Aerobic hydrocarbon decomposition is much faster than anaerobic degradation. Thus one of the prime measures in land spreading of oily sludges concerns the oxygen supply by cultivation of the soil. After mixing the sludge with the top soil layer of the disposal area, rotary mixing and discing at regular time intervals may provide the aerobic microorganisms with the required oxygen.

Microorganisms responsible for oil decomposition are well identified. They predominantly belong to the genera *Arthrobacter*, *Corynebacterium*, *Flavobacterium*, *Nocardia* and *Pseudomonas* (Carlson et al, 1961). According to microbial counts the number of organisms present appeared independent of the temperature. The microbial activity, however, is strongly influenced by the soil temperature resulting in considerably higher decomposition rates at increasing temperature. This effect is demonstrated in figure 10.12, taken from Kincannon (1972). Initial decomposition may be accelerated by seeding the spill or sludge disposal with the appropriate microorganism species. This can be achieved by using soil from an old oilspill location.

There are a number of factors, besides the soil temperature and aeration conditions, that influence the decomposition rate. One of these is the fertility status. The C:N:P ratio for optimal microbial activity may be adjusted by nitrogen and phosphorus fertilization of the spillage or disposal area. The indications 1, 2 and 3 in figure 10.12 refer to fertilizer applications of 1100, 550 and 0 kg N per ha, respectively, and of 220, 110 and 0 kg P_2O_5 per ha, respectively. The increase of the oil decomposition rate with higher fertilization level is clearly demonstrated.

Also the chemical composition of the oil affects its degradation. It is well known that aromatic compounds usually can be much better attacked and more rapidly decomposed than other hydrocarbons. Of the long chain hydrocarbons the paraffin groups appear the most readily decomposable.

An important practical measure in case of an oil spillage is to allow the oil to spread itself as much as possible. The oil should not be confined. Infiltration of oil into deeper layers of the soil may greatly hamper proper treatment of the contaminated soil. This is the more so because the soil pore space will not be completely saturated with oil. It will stay at the so-called residual oil saturation value which is considerably lower (Blokker, 1971).

10.6.2. Gas leakages

A less complicated hydrocarbon energy source is brought into the soil near leaks of natural gas, which consists for about 80 - 95% of methane. A change-over from town or mine gas to natural gas is often accompanied by severe injury to trees which grow in the vicinity of the subsurface distribu-

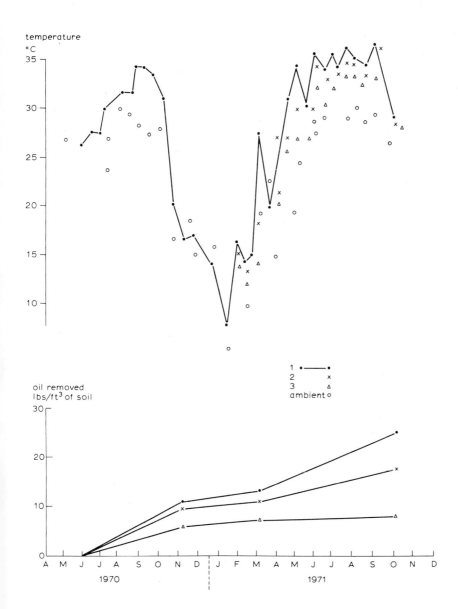

Fig. 10.12. Influence of temperature on decomposition rate of oil in soil. 1, 2 and 3: heavily, medium and non-fertilized, respectively (after Kincannon, 1972).

tion system. After the change-over the number of leaks strongly increases due to the combined effect of increased gas pressure and desiccation of lead-oakum joints. Lead-oakum joints usually prevail in the oldest parts of the distribution system, the town centres, and hence damages to vegetation are largest there (Hoeks, 1971). For Dutch conditions about 60 - 90% of total mortality among street trees in town centres was found to result from gas leakages.

Hoeks (1972) has presented a complete description of this phenomenon. A number of reasons for plant damage near gas leaks may exist. The injury can be caused by one or more of the following facts: low oxygen, high carbondioxide, possibly the presence of ethylene, toxic concentrations of reduced compounds (e.g. Mn^{2+}). The above author showed that the distribution of the main soil gas components CH_4, O_2 and CO_2 near gas leaks can be described very well for a number of situations by using transport equations for gaseous transport. An example of such distribution is given in figure 10.13 for winter and summer circumstances, thus showing the temperature influence. This temperature influence exerts itself on the methane consumption which decreases at falling temperatures due to a decline in the microbial activity. As a consequence, the concentration distributions of oxygen and carbondioxide are also changing with temperature. Several different

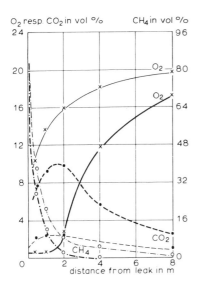

Fig. 10.13. Distribution of CH_4, O_2 and CO_2 as measured around an artificial gas leak for summer (thick lines) and winter (thin lines) conditions. (after Hoeks, 1972).

zones can be distinghuished in the soil around a gas leak. These are the anaerobic zone, which is closest to the leak and in which the highest concentration of methane in the soil gas phase is found. Next to the anaerobic zone the methane oxidation zone occurs. There both oxygen and methane are present thus allowing the oxidation of methane to CO_2 and H_2O. Finally a so-called oxygen transit zone may be distinguished which serves as the transport medium of oxygen from the unaffected soil part to the methane oxidation zone.

Regeneration of injured vegetation can be achieved - if the injury is not too severe- by improvement of soil aeration. Also after repair of the leak, poor aeration conditions would continue for a number of months if no measures are taken. Accelerated improvement of soil aeration can be obtained by installment of ventilation channels and by the use of air compressors.

10.6.3. Sanitary landfills

With increasing population density, handling and treatment of municipal solid wastes may become a very serious problem. Several approaches can be applied in such treatment. They are incineration, composting, and disposal on land. The last is the most attractive from an economic point of view. Consequently, the majority of municipal solid wastes are stored on or in the soil (about 60% for the Netherlands and 90% for the USA).

Many of these disposal systems are still nothing but open dumps. As such they cause a deterioration of landscape and form continuous sources for fires, odors, and other nuisances. Because of increasing concern about environmental conditions, open dumps are gradually replaced by so-called sanitary landfills. In a sanitary landfill nuisances and hazards are prevented as much as possible by coverage of the refuse with soil. This coverage must be performed at least daily. A side aspect of sanitary landfills is the endeavor to let the fill have an appropriate function in the landscape at the termination of the refuse disposal.

In the interior of a sanitary landfill a large number of processes occur which cause decomposition of the organic part of the refuse material. Circumstances like degree of compaction, water content, and sizes of fill determine the possibilities for oxygen intrusion and thus whether these conversions will be aerobic or anaerobic in nature.

In a comparable manner as for soils, sanitary landfills usually exhibit spots and zones of anaerobic circumstances alternating with aerobic conditions. Depending on which predominate the leachate of the fill will be characterized by anaerobic or aerobic decomposition products. This is of importance from a pollution point of view. Products of aerobic decomposition are stable like CO_2, H_2O, NO_3 and SO_4 and a relatively stable form of organic matter. The

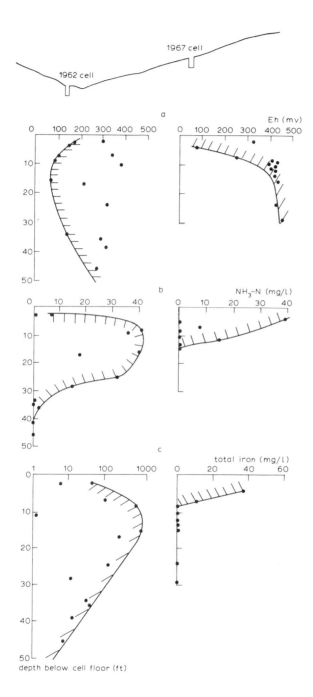

Fig. 10.14. Distribution of three characteristics of groundwater as a function of depth at two observation points under a sanitary landfill; a: redox potential; b: NH_3-N concentration; c: total iron concentration. (after Apgar and Langmuir, 1971).

leachate resulting from anaerobic decay usually has a high oxygen demand since reduced or only partially oxidized compounds are formed during such decomposition, like for example methane and organic acids. Because of the predominance of anaerobic conditions in most sanitary landfills or most parts of the fills, pollution of groundwater with organic compounds in the proximity of refuse piles is probable. This is the more so because the sites used are usually natural land depressions or soil excavations. The distance between bottom of the fill and water table is therefore commonly small. Sometimes this bottom is even situated below groundwater level thus enhancing direct contact between groundwater and refuse or leachate from the refuse.

The anaerobic conditions as mentioned may induce increased mobility of metals like manganese and iron. Also pH changes may considerably affect the metal concentrations in the leachate. If, however, also sulfate is being reduced (which requires amongst others the presence of the microorganisms *Desulfovibrio desulfuricans*) the formation of low soluble metal sulfides will occur.

Apgar and Langmuir (1971) measured the course of several groundwater characteristics as a function of depth below a sanitary landfill. Results for redox potential, NH_3-N and total iron concentrations are presented in figure 10.14. Two locations in the fill were chosen, which have an age difference of 5 years. Also the geographical situation of the two observation points is different. The 1967 cell (see figure 10.14) lies more upslope and at the freshly filled edge of the refuse fill. This makes that oxidation conditions are considerably better here, whereas also dilution of the leachate with uncontaminated groundwater may occur. These differences are clearly reflected in figure 10.14, representing analysis data of 1970. All three variables confirm for the 1962 cell that the most anaerobic condition in the leachate occurs at about 10-15 feet below the bottom. Apparently the 1962 cell has been depleted of biodegradable organic compounds, thus allowing again more aerobic conditions in the leachate.

The area in which the groundwater may be affected by the presence of a sanitary landfill completely depends on geophysical conditions. Velocity of groundwater flow, direction of such flow, occurrence of highly permeable layers like creviced bedrocks, etc. are all considerations. Exler (1972) reports detectable influences at distances as large as 1 kilometer. These have been measured at extremely high groundwater flow velocities. Hoeks (1973) describes in a literature review on this subject matter that the influence is usually confined to a few hundred meters. Thus maintenance of a minimum distance of at least 1-2 kilometers between a sanitary landfill and the nearest groundwater well can usually be considered as safe.

10.7. POSITIONING THE PRESENT TREATISE WITH RESPECT TO ADJACENT AREAS OF INTEREST

In line with the scope of a text on soil chemistry, the preceding discussion concentrated on the chemical and physico-chemical interactions between soil constituents and specific compounds which could give rise to malfunctioning of soil. As was pointed out in the introductory section of this chapter, a thorough knowledge of these interactions constitutes a necessary prerequisite for predictions of such malfunctioning and thus occupies a central position within the totality of scientific know-how on which a program of pollution control could be based. In fact the soil chemist's effort in the field of soil pollution is directed toward specifying the concentration (c.q. activity) of particular compounds and their reaction products as a function of time and of position in the soil profile, ultimately in the form of a mathematical model. Accordingly the factual information contained in this chapter must be brought to use within the broader context of the chemical (and physical) processes discussed in previous chapters.

Particularly in those cases where organic compounds are involved, the construction of such models requires also a specification of the microbially controlled transformations in terms of relevant rate coefficients. In order to obtain such information there is a great need for close cooperation between soil microbiologists and soil chemists.

A practical complication in acquiring such a concerted effort of these two disciplines lies in a difference of purpose of the two fields as grown historically. Thus the soil chemist is in the present case basically concerned with the rate of disappearance of a particular compound in about the least 'pure' system that one could imagine from a microbiological standpoint, i.e. a system in which almost any microbial species is potentially present. In contrast the microbiologist has generally been oriented toward studying the life cycle of a specific microorganism in pure culture in reaction to different substrates and varying environmental conditions.

A second point which should be mentioned once more is the interpretation of the terms 'functioning' and 'malfunctioning' of the soil introduced in section 10.1. The soil ecosystem is an intricate system comprising abiotic and biotic 'compartments' which are continually interacting with each other. In the present approach main emphasis was put on processes in the abiotic portion of soil and the functioning or malfunctioning was tied in the first place to soils ability to 'support' the growth of certain crops. It should be realized here that regular agricultural practice already implies a certain degree of manipulation of the soil ecosystem. Thus certain processes in the biotic part of soil may become disturbed in comparison to the truly natural soil ecosystem. Accordingly, the term 'malfunctioning' has been used here in a limited sense and was supposed to be based on the judgement of a crop ecologist and not so much on that of a 'natural' ecologist. Admittedly, disturbances in the biotic compartments of soil, although for the time being not noticeable in terms of crop yields, could in the long run still prove undesirable. Here the recent awakening of widespread interest in soil as a natural ecological unit may prove to be of great value in spotting such disturbances at an early stage.

In contrast to the above, malfunctioning of soil as a filter is usually judged against requirements with respect to the composition of open water as put forward by ecologists and in some cases of groundwater as source for supply of drinking water (or industrial water use). As was pointed out locally such requirements tend to be more severe than those for irrigation water.

Somewhat opposed to the study of soil pollution phenomena with the purpose of protecting an important environmental component is the consideration of soil as a recipient for waste materials. Such waste disposal may comprise flooding with poor quality water as e.g. raw sewage water or effluents of water treatment plants. Although in principle unwarranted from a soil protection point of view, it must be realized that in regions of high human activity many wastes are produced and the need for disposal, or preferably purification, must be weighed against the risks of polluting the soil. In such cases it is a challenge to environmentalists and soil scientists together to manipulate the soil system in such a manner that an optimal combination of its mentioned functions may be arrived at without overestimating the long-term capacity of soil for regeneration.

LITERATURE

Aldrich, S.R., Oschwald, W.R. and Fehrenbacher, J.B., 1970. Implications of crop production technology for environmental quality. *AAAS Meeting, Chicago, Ill.*

Alexander, J.D., 1971. Lead in Illinois agriculture. *Univ. Ill. Urbana:* 46-48.

Allaway, W.H., 1968. Agronomic controls over the environmental cycling of trace elements. *Adv. Agron.* 20: 235-274.

Allison, F.E., 1965. Evaluation of incoming and outgoing processes that affect soil nitrogen. *Soil Nitrogen*, Madison, Wisc., Am. Soc. Agron.: 573-606.

Alloway, B.J., 1969. *The soil and vegetation of areas affected by mining for non-ferrous metalliferous ores with special reference to Cd, Cu, Pb and Zn.* Ph.D. Thesis, Univ.Wales.

Anastasia, F.B. and Kender, W.J., 1973. The influence of soil arsenic on the growth of lowbush blueberry. *J. Envir. Qual.* 2: 335-337.

Anderson, G. and Arlidge, E.Z., 1962. The adsorption of inositol phosphates and glycerophosphates by soil clays, clay minerals and hydrated sesquioxides in acid media. *J.Soil Sci.* 13: 216-224.

Anderssen, A. and Wiklander, L., 1965. Something about mercury in nature. *Grundforbattring* 18: 171-177.

Apgar, M.A. and Langmuir, D., 1971. Ground-water pollution potential of a landfill above the water table. *Ground Water*, 9: 76-96.

Armstrong, D.E., Chesters, G. and Harris, R.F., 1967. Atrazine hydrolysis in soil. *S.S.S.A. Proc.* 31: 61-66.

Bailey, G.W., 1968. *Role of soils and sediment in water pollution control.* U.S. Dept. Int., F.W.P.C.A. Southeast Water Lab. ; 90 pp.

Bailey, G.W. and White, J.L., 1970. Factors influencing the adsorption, desorption and movement of pesticides in soil. *Residue Reviews* 32: 29-92.

Bailey, G.W., Swank, R.R. and Nicholson, H.P., 1974. Predicting pesticide runoff from agricultural land. A conceptual model; *J. Envir. Qual.* 3: 95-102.

Baker, D.E., 1974. Copper: soil, water, plant relationships. Ecological problems of high level nutrient feeding. *Fed. Proc.* 33: 1188-1193.

Baumhardt, G.R. and Welch, L.F., 1972. Lead uptake and corn growth with soil-applied lead. *J. Envir. Qual.* 1: 92-94.

Beek, J. and de Haan, F.A.M., 1974. Phosphate removal by soil in relation to waste disposal. *Proc. Int. Conf. Land Use Waste Manag.*, Ottawa, Oct. 1973: 77-87.

Beek, J., de Haan, F.A.M. and van Riemsdijk, W.H., 1976. Phosphates in soils flooded with sewage water. 1 Fractionation of accumulated phosphates. *J. Envir. Qual.* (in press).

Beggren, B. and Oden, S., 1972. *Analys resultat rorande fungmetaller och klorerade kolväten i rötslam fran Svenska remingsverk '69-'71.* Lantbrukshögskolan 75007, Uppsala, Sweden.

Bernstein, F., 1960. Distribution of water and electrolyte between homoionic clays and saturating NaCl solution. *Clays and Clay Minerals*, 8: 122-128.

Berrow, M.L. and Webber, J., 1972. Trace elements in sewage sludges. *J. Sci. F. Agr.* 23: 93-100.

Bittell, J.E. and Miller, R.J., 1974. Lead, cadmium and calcium selectivity coefficients on a montmorillonite, illite and kaolinite. *J. Envir. Qual.*, 3: 250-253.

Black, C.A., 1968. Phosphorus; Chapter 8 in *Soil-Plant relationships.* Wiley and Sons Inc., New York, 558-653.

Blakeslee, P.A., 1973. *Monitoring considerations for municipal wastewater effluent and sludge application to the land.* U.S.E.P.A.; U.S.D.A.

Blokker, P.C., 1971. *Migration of oil in soil.* Rep. 9/71, Concawe, The Hague, 16 pp.

Bollag, J.M., 1973. *Nitrate and nitrite volatilization by microorganisms in laboratory experiments.* EPA-660/2-73-002, 65 pp.

Bouwer, H., Rice, R.C. and Escarcega, E.D., 1974a. High-rate land treatment, I: Infiltration and hydraulic aspects of the Flushing Meadows Project. *Journ. WPCF*, 46: 834-843.

Bouwer, H., Lance, J.C. and Riggs, M.S., 1974b. High-rate land treatment, II: Water quality and economic aspects of the Flushing Meadows Project. *Journ. WPCF*, 46: 844-859.

Breeuwsma, A., 1973. *Adsorption of ions on hematite (α-Fe_2O_3)*. Ph.D. Thesis, Wageningen

Bremner, J.M., 1965. Organic nitrogen in soils. in *Agronomy 10, Soil Nitrogen*. Am. Soc. of Agronomy, Madison, Wisc. pp 93-149.

Briggs, G.G. and Dawson, J.E., 1970. Hydrolysis of 2-4-Dichlorobenzonitrile in soils. *J. Agr. Food Chem.* 18: 97-99.

Broadbent, F.E., 1960. Factors influencing the reaction between ammonia and soil organic matter. *Trans. 7th Int. Congr. Soil Sci.*, Madison, Wisc.

Broadbent, F.E. and Stevenson, F.J., 1966. Organic matter interactions. *Agricultural anhydrous ammonia, technology and use;* Madison, Wisc. Am. Soc. Agr. pp 169-187.

Bruggenwert, M.G.M., 1972: *Adsorptie van Al-ionen aan het kleimineraal montmorilloniet* Ph.D. Thesis, Wageningen.

Buchauer, M.J., 1971. *Effects of zinc and cadmium pollution on vegetation and soils.* Ph.D. Thesis, Rutgers Univ., New Brunswick, N.J.

Buchauer, M.J. 1973. Contamination of soil and vegetation near a zinc smelter by zinc, cadmium, copper and lead. *Envir. Sci. Technol.* 7: 131.

Burge, W.D. and Broadbent, F.E., 1961. Fixation of ammonia by organic soils. *SSSA Proc.* 25: 199-204.

Carlson, V., Bennett, E.O. and Rowe, J.A., 1961. Microbial flora and their relationship to water quality. *Soc. Petr. Eng. J.* 1: 71-80.

Chaney, R.L., 1973. Crop and food chain effects of toxic elements in sludges and effluents. *Recycling municipal sludges and effluents on land;* Champaign, Ill. pp. 129-141.

Chen, Y.S.R., Butler, J.N. and Stumm, W., 1973. Kinetic study of phosphate reaction with aluminum oxide and kaolinite. *Envir. Sci. Techn.* 7: 327-332.

Chumbley, C.G., 1971. *Permissible levels of toxic metals in sewage used on agricultural land*. A.D.A.S. Advis. paper 10, 12pp.

Commoner, B., 1968. Treats to the integrity of the nitrogen cycle: Nitrogen compounds in soil, water, atmosphere and precipitation. *Am. Assoc. Adv. Sci. Meeting, Dallas, Texas.*

Cook, S.F. and Heizer, R.F., 1965. *Studies on the chemical analysis of archeologic sites.* Univ. Cal. Publ., Anthrop. 2.

Cosgrove, D.J., 1963. The chemical nature of soil organic phosphorus; I. Inositol phosphates. *Austr. J. Soil Res.* 1: 203-214.

Cosgrove, D.J., 1964. An examination of some possible sources of soil inositol phosphates. *Plant Soil*, 21: 137-141.

Cosgrove, D.J., 1967. Metabolism of organic phosphates in soil. Chapt. 9 in *Soil Biochemistry*, pp. 216-228.

Dahnke, W.C., Malcolm, J.L. and Menendez, M.E., 1964. Phosphorus fractions in selected soil profiles of El Salvador as related to their development. *Soil Sci.* 98: 33-37.

Deuel, L.E. and Swoboda, A.R., 1972. Arsenic toxicity to cotton and soybeans. *J. Envir. Qual.* 1: 317-320.

Diest, A. van and Black, C.A., 1959. Soil organic phosphorus and plant growth. II Organic phosphorus mineralized during incubation. *Soil Sci.* 87: 145-154.

Dotson, G.K., Dean, R.B., Kenner, B.A. and Cooke, W.B., 1972. Land spreading, a conserving and non-polluting method of disposing of oily wastes. *Proc. 5th Int. Wat. Poll. Res. Conf.* 1, II: 36/1-36/15.

Ellis, B.G. and Erickson, A.E., 1969. *Movement and transformation of various phosphorus compounds in soils.* Soil Sci. Dept., Mich. St. Univ., 35 pp.

El-Sayed, M.H., Burau, R.G. and Babcock, K.L., 1970. Thermodynamics of Copper II - Calcium exchange on bentonite clay. *SSSA Proc.* 34: 397.

El-Sayed, M.H., Burau, R.G. and Babcock, K.L., 1971. Reaction of Copper tetrammine with bentonite clay. *SSSA. Proc.*, 35: 571.

Enfield, C.G. and Shew, D.C., 1975. Comparison of two predictive non-equilibrium one dimensional models for phosphorus sorption and movement through homogeneous soils. *J. Envir. Qual.* 4: 198-202.

Exler, H.J., 1972. Defining the spread of groundwater contamination below a waste tip. *Proc. Conf. Groundwater Poll.*, Univ. Reading.

Flaig, W., 1966. The chemistry of humic substances. *The use of isotopes in soil organic matter studies.* Pergamon Press, New York, pp 103-127.

Fried, M. and Broeshart, H., 1967. *The soil-plant system in relation to inorganic nutrition* Ac. Press, New York.

Frissel, M.J., 1961. *The adsorption of some organic compounds, especially herbicides on clay minerals.* Agr. Res. Rep. 76, 3, 54 pp..

Furukawa, H. and Kawaguchi, K., 1969. Contributions of organic phosphorus to the increase of easily soluble phosphorus in water-logged soil, especially in relation to phytin phosphorus (inositol hexaphosphate). *Soil Sci. Pl. Nutr.* 15: 243.

Garman, W.H., 1973. Agriculture's place in the environment; Considerations for decision making. *J. Envir. Qual.* 2: 327-333.

Geering, H.R., Cary, E.E., Jones, L.H.P. and Allaway, W.H., 1968. Solubility and redox criteria for the possible forms of selenium in soils. *SSSA Proc.* 32: 35-40.

Gilbert, R.G., Robinson, J.B. and Miller, J.B., 1974. The microbiology and nitrogen transformations of a soil recharge basin used for waste water renovation. *Proc. Conf. Land Use Waste Management*, Ottawa, Oct. 1973: pp 87-97.

Gilmour, J.T. and Miller, M.S., 1973. Fate of a mercuric-mercurous chloride fungicide added to turfgrass. *J. Envir. Qual.*, 2: 145.

Goring, C.A.I. and Bartholomew W.V., 1952. Adsorption of nucleotides, nucleic acids and nucleoproteins by clay. *Soil Sci.* 74: 149-164.

Guenzi, W.D. and Beard, W.E., 1968. Anaerobic conversion of DDT to DDD and aerobic stability of DDT in soil. *SSSA Proc.* 32: 522-524.

Haan, F.A.M. de, 1965. *The interaction of certain inorganic anions with clays and soils.* Agr. Res. Rep. 655, Wageningen, 167 pp.

Haan, F.A.M. de, Bolt, G.H. and Pieters, B.G.M., 1965. Diffusion of Potassium-40 into an illite during prolonged shaking. *SSSA Proc.*, 29: 528-530.

Haan, F.A.M. de, 1972. *Results from loading the soil with great quantities of waste.* Note 657, Inst. Land Water Management Research, Wageningen, 56 pp.

Haan, F.A.M. de, Hoogeveen, G.J. and Riem Vis, F., 1973. Aspects of agricultural use of potato starch waste water. *Neth. J. Agr. Sci.* 21: 85-94.

Haan, F.A.M. de, 1973. Maintenance of sound soil conditions. *Planning and Development in the Netherlands.* VII - II; 98-110.

Haan, F.A.M. de, 1975. The effects of long term accumulation of heavy metals and selected organic compounds in municipal wastewater on soil. *Proc. Conf. on Renovation and Recycling of wastewater through aquatic and terrestrial systems*, Bellagio, Italy.

Haan, F.A.M. de, 1975. On the acceptability of sewage sludge application to soil. *Proc. Conf. on Renovation and Recycling of wastewater through aquatic and terrestrial systems.* Bellagio, Italy.

Haan, F.A.M. de, 1975. Interaction mechanisms in soil in relation to soil pollution and groundwater quality. *Versl. Mededel. Cie Hydr. Onderzoek T.N.O.*

Hadzi, D., Klofutar, C. and Oblak, S., 1968. Hydrogen bonding in some adducts of oxygen bases with acids. Part IV. Basicity in hydrogen bonding and in ionization. *J. Chem. Soc. A.* : 905-909.

Haghiri, F., 1973. Cadmium uptake by plants. *J. Envir. Qual.* 2: 93-96.

Haghiri, F., 1974. Plant uptake of cadmium as influenced by cation exchange capacity, organic matter, zinc and soil temperature. *J. Envir. Qual* 3: 180-183.

Hahne, H.C. and Kroontje, W., 1973. Significance of pH and chloride concentration on behavior of heavy metal pollutants: Mercury, Cadmium, Zinc and Lead. *J. Envir. Qual.* 2: 444-450.

Hall, J.K., Pawlus, M. and Higgins, E.R., 1972. Losses of atrazine in runoff water and soil sediment. *J. Envir. Qual* 1: 172-176.

Hall, J.K., 1974. Erosional losses of s-triazines herbicides. *J. Envir. Qual.* 3: 174-180.

Hamaker, J.W. and Thompson, J.M., 1972. Adsorption. Chapt. 2 in *Organic chemicals in the soil environment* (ed. Goring and Hamaker), Dekker, New York.

Harmsen, K., 1976. Ph.D. Thesis. Agricultural State University. Wageningen, Netherlands.

Harmsen, G.W. and Kolenbrander, G.J., 1965. Soil inorganic nitrogen in *Soil Nitrogen*, Agronomy 10, Madison, Wisc., pp 43-92.

Helling, C.S., Kearny, P.C. and Alexander, M., 1971. Behavior of pesticides in soil. *Adv. Agr.* 23: 147-240.

Henkens, Ch. H., 1972. Fertilizer and the quality of surface water. *Stikstof* 15: 28-40.

Hodgson, J.F., 1960. Cobalt reactions with montmorillonite. *SSSA Proc.* 24: 165-168.

Hoeft, R.G., Keeney, D.R. and Walsh, L.M., 1972. Nitrogen and sulfur in precipitation and sulfur dioxide in the atmosphere in Wisconsin. *J. Envir. Qual.* 1: 203-208.

Hoeks, J., 1971. *De verbetering van de bodemluchtsamenstelling by straatbomen.* SIAB rep. 5, The Hague, 19 pp.

Hoeks, J., 1972. *Effect of leaking natural gas on soil and vegetation in urban areas.* Agr. Res. Reports, 778, Wageningen, 120 pp.

Hoeks, J., 1973. *Verontreiniging van bodem en grondwater bij vuilstortplaatsen.* (Pollution of soil and groundwater near refuse dumps; a review). Note 737 Inst. Land Water Management Research, Wageningen, 35 pp.

Hopkins, L.L. and Mohr, H.E., 1971. Effect of vanadium deficiency on plasma cholesterol of chicks. *Fed. Proc.* 39: 462.

Hsu, P.H. and Rennie, D.A., 1962. Reactions of phosphate in aluminum systems; I. Adsorption of phosphate by X-ray amorphous aluminum hydroxide. *Can.J. Soil Sci.* 43: 197-209.

Isensee, A.R., Kearney, P.C., Woolson, E.A., Jones, G.E. and Williams, V.P., 1973. Distribution of alkyl arsenicals in a model ecosystem. *Envir. Sci. Techn.* 7: 841-845.

Jackman, R.H. and Black, C.A., 1951. Solubility of iron, aluminum, calcium and magnesium inositol phosphates at different pH values. *Soil Sci.* 72: 179-186.

Jacobs, L.W. and Keeney, D.R., 1974. Methylmercury formation in mercury-treated river sediments during in-situ equilibration. *J. Envir. Qual.*, 3: 121-126.

Jenne, E.A., 1970. Atmospheric and fluvial transport of mercury; Mercury in the environment. *Geol. Surv. Paper* 713: 40-45.

Joensuu, O.I., 1971. Fossil fuels as a source of mercury pollution. *Science* 172: 1027-1028.

John, M.K. and van Laerhoven, C., 1972. Lead uptake by lettuce and oats as affected by lime, nitrogen and sources of lead. *J. Envir. Qual.* 1: 169-171.

Johnson, C.M., 1966. Molybdenum. Chapter 20 in *Diagnostic Criteria for plants and soils* p 286.

Jones, G.B. and Belling, G.B., 1967. Movement of copper, molybdenum and selenium in soils as indicated by radioactive isotopes. *Austr. J. Agr. Res.* 18: 733-740.

Jones, L.H., 1957. The solubility of molybdenum in simplified and aqueous soil suspensions. *J. Soil Sci.* 8: 313-327.

Jones, R.L., Hinesly, T.D. and Ziegler, E.L., 1973. Cadmium content of soybeans grown in sewage sludge amended soil. *J. Envir. Qual.* 2: 351-353.

Kardos, L.T., Sopper, W.E., Myers, E.A., Parizek, R.R. and Nesbitt, J.B., 1974. *Renovation of secondary effluent for reuse as a water resource.* Envir. Prot. Tech. Series, 660/2-74-016.

Kay, B.D. and Elrick, D.E., 1967. Adsorption and movement of lindane in soils. *Soil Sci.* 104: 314-322.

Kearney, P.C., Woolson, E.A. and Plimmer, J.R., 1969. Decontamination of pesticides in soil. *Residue Review*, 29: 137-149.

Kehoe, R.A., 1966. Under what circumstances is ingestion of lead dangerous ? *Symp. Envir. Lead Cont. Public Health Service*, No. 1440, 51-58.

Kincannon, C.B., 1972. *Oily waste disposal by soil cultivation process.* EPA-R2-72-110, 115 pp.

Klausner, S.D., Zwerman, P.J. and Scott, F.W., 1971. Land disposal of manure in relation to water quality. *Agr. Wastes Principles and Guidelines for practical solution;* Cornell Univ., 36-46.

Kolenbrander, G.J., 1972. The eutrophication of surface water by agriculture and the urban population. *Stikstof* 15: 56-67.

Kononova, M.M., 1966. *Soil organic matter.* Pergamon Press, 2nd ed.,New York.

Kunishi, H.M., Taylor, A.W., Heald, W.R., Gburek, W.J. and Weaver, R.N., 1972. Phosphate movement from an agricultural watershed during two rainfall periods. *J. Agr. Food Chem.* 20: 900-905.

Lagerwerff, J.V. and Specht, A.W., 1970. Contamination of roadside soil and vegetation with cadmium, nickel, lead and zinc. *Envir. Sci. Techn.* 4: 583-586.

Lagerwerff, J.V., 1972. Lead, mercury and cadmium as environmental contaminants. Chapt. 23 in *Micronutrients in agriculture*, pp 593-636.

Lagerwerff, J.V. and Biersdorf, G.T., 1972. Interaction of zinc with uptake and translocation of cadmium in radish. *Proc. 5th Ann. Conf. Trace Subst. and Envir. Health*, 515-522.

Lagerwerff, J.V. and Brower, D.L., 1972. Exchange adsorption of trace quantities of cadmium in soils with chlorides of aluminum, calcium, and sodium. *SSSA Proc.* 36: 734-737.

Lagerwerff, J.V., Brower, D.L. and Biersdorf, G.T., 1973. Accumulation of cadmium, copper, lead, and zinc in soil and vegetation in the proximity of a smelter. *Trace Substances in Environmental Health*, VI; Univ. Miss., Columbia, 71-78.

Lagerwerff, J.V. and Brower, D.L., 1975. Effect of a smelter on the agricultural conditions in the surrounding environment.Agr. Envir. Qual. Inst. ; Beltsville, Md.

Lance, J.C., 1972. Nitrogen removal by soil mechanisms. *Journ. WPCF*, 44: 1352-1361.

Lance, J.C. and Whisler, F.D., 1972. Nitrogen balance in soil columns intermittently flooded with secondary sewage effluent. *J. Envir. Qual.* 1: 180-186.

Lance, J.C. and Whisler, F.D., 1974. Nitrogen removal during land filtration of sewage water. *Proc. Int. Conf. on Land Use for Waste Management*, Ottawa, Oct. 1973, pp 174-183.

Lavy,T.L. and Barber, S.A., 1964. Movement of molybdenum in the soil and its effect on availability to the plant. *SSSA Proc.* 28: 93-97.

Leeper, G.W., 1972. *Reactions of heavy metals with soils with special regard to their application in sewage wastes.* U.S. Dept. Army, DACW 73-73-C-0026, 70 pp.

Leistra, M., 1972. *Diffusion and adsorption of the nematicide 1,3-dichloropropene in soil* Agr. Res. Rep. 769, 105 pp.

Lexmond, Th.M., de Haan, F.A.M. and Frissel, M.J., 1976. On the methylation of inorganic mercury and the decomposition of organo-mercury compounds, a review. *Neth. J. Agr. Science* , 24,1.

Lindsay, W.L., Hodgson, J.F. and Norvell, W.A., 1967. The physico-chemical equilibrium of metal chelates in soils and their influence on the availability of micronutrient cations. *Int. Soil Sci. Soc. Trans. Comm.* II, IV: 305-316.

Lindsay, W.L. and Norvell, W.A.,1969. Equilibrium relationships of Zn^{2+}, Fe^{3+}, Ca^{2+} and H^+ with EDTA and DTPA in soils. *SSSA Proc.* 33: 62-68.

Lindsay, W.L., 1972. Inorganic phase equilibria of micronutrients in soils. Chapter 3 in *Micronutrients in Agriculture*, pp 41-58.

Lindsay, W.L., 1973. Inorganic reactions of sewage wastes in soils. *Recycling municipal sludges and effluents on land;* Champaign, Ill., pp 91-96.

Lundblad, K., Svanberg, O. and Ekman, P., 1949. The availability and fixation of Cu in Swedish soils. *Plant Soil*, 1: 277-302.

Mattson, S. and Koutler-Andersson, E., 1942. The acid-base condition in vegetation, litter and humus. *Lantbrukshogskoles Annåler*, 10: 284-286.
Mc.Lean, G.W., Pratt, P.F. and Page, A.L., 1966. Nickel-barium exchange selectivity coefficients for montmorillonite. *SSSA Proc.* 30: 804-805.
Mc.Lean, A.J., Stone, B. and Cordukes, W.E.,1972. Amounts of mercury in soil of some golfcourse sites. *Can. J. Soil Sci.*, 53: 130.
McNew, G.L., 1972. Interrelationships between agricultural chemicals and environmental quality in perspective. *J. Envir. Qual* 1: 18-22.
Meikle, R.W., 1972. Qualitative relationships of decomposition. Chapt. 3 in *Organic Chemicals in the Soil Environment;* Dekker, New York, pp 147-252.
Mertz, W., 1966. Chromium in our food. *Food Nutr. News.* 38: 1-4.
Muth, O.H. and Allaway, W.H., 1963. The relationship of white muscle disease to the distribution of naturally occurring selenium. *J. Am. Vet. Med. Ass.* 142: 1379-1384.
Nicholson, H.P., 1968. Pesticides, a current water quality problem. *Trans. Kans. Ac. Sci.* 70: 39-44.
Nielsen, F.H., 1970. Symptoms of nickel deficiency in the chick. *Fed. Proc.* 29: 696.
Norvell, W.A. and Lindsay, W.L., 1969. Reactions of EDTA complexes of Fe, Zn, Mn and Cu with soils. *SSSA Proc.* 33: 86.
Norvell, W.A., 1972. Equilibria of metal chelates in soil solution. Chapt. 6 in *Micronutrients in agriculture*, pp 115-138.
Olsen, S.R. and Watanable, F.S., 1957. A method to determine a phosphorus adsorption maximum of soils as measured by the Langmuir isotherm. *SSSA Proc.* 21: 144-149.
Oniani, O.G., Chater, M. and Mattingly, G.E.G., 1973. Some effects of fertilizers and farmyard manure on organic phosphorus in soil. *J. Soil Sci.* 24: 2-9.
Omotoso, T.I. and Wild, A., 1970. Content of inositol phosphates in some English and Nigerian soils. *J. Soil Sci.* 21: 216-223.
Owens, M. and Wood, G., 1968. Some aspects of the eutrophication of water. *Water Res.* 2: 151-159.
Page, A.L., Bingham, F.T. and Nelson, C., 1972. Cadmium absorption and growth of various plant species as influenced by solution cadmium concentration. *J. Envir. Qual.* 1: 288-291.
Page, A.L., 1974. *Fate and effects of trace elements in sewage sludge when applied to agricultural lands.* EPA-670/2-74-005, 98 pp.
Park, J.E., Rockhill, R.C. and Klein, D.A., 1972. Photooxidative approach to annual ryegrass; Straw modification with corresponding increases in microbial response. *J. Envir. Qual.* 1: 298-300.
Patrick, W.M., 1964. Extractable iron and phosphorus in a submerged soil at controlled redox potentials. *8th Int. Congr. Soil Sci.* IV: 605-609.
Patrick, W.H. and Tusneem, M.E., 1972. Nitrogen loss from flooded soil. *Ecology* 53:735--737.
Patrick, W.M. and Khalid, R.A., 1973. *Phosphate release and sorption by soils and sediments: Effect of aerobic and anaerobic conditions.* Louis. St. Univ., 9 pp.
Peperzak, A., Caldwell, A.C., Hunziker, R.R. and Black, C.A., 1959. Phosphorus fractions in manures. *Soil Sci.* 81: 293-302.
Poelstra, P., Frissel, M.J., v.d. Klugt, N. and Tap, W., 1973. Behaviour of mercury compounds in soils: accumulation and evaporation. *Symp. Comp. Asp. Food Envir. Cont.*, Helsinki.
Pratt, P.F., Bair, F.L. and McLean, G.W., 1964. Reactions of phosphate with soluble and exchangeable nickel. *SSSA Proc.*, 28: 363-365.
Pratt, P.F., 1966. Chromium. Chapter 9 in *Diagnostic criteria for plants and soils.* p 136-141.
Riemsdijk, W. van, Weststrate, F. and Bolt, G.H., 1975. The reaction rate of phosphate with aluminum hydroxide with evidence for the formation of a new phase. *Nature* 257: 473-474.

Riemsdijk, W. van, Weststrate, F. and Beek, J., 1976. Phosphates in soils flooded with sewage water. 2. Kinetic studies on the reaction of phosphate with aluminum compounds. *J. Envir. Qual.* (in press).

Robinson, K., Draper, S.R. and Gilman, A.L., 1971. Biodegradation of pig waste: Breakdown of soluble nitrogen compounds and the effect of copper. *Envir. Poll.* 2: 49-56.

Rockhill, R.C., Park, J.E. and Klein, D.A., 1972. Photooxidative degradation of ligninsulfanate to substrates enhancing microbial growth. *J. Envir. Qual.* 1: 315-317.

Rohwer, E.F.C.H. and Cruywagen, J.J., 1964. The first protonation constant of monomeric molybdic acid. *J. South Afr. Chem. Inst.* 17: 145-148.

Sawyer, C.M., 1947. Fertilization of lakes by agricultural and urban drainage. *J. Engl. Wat. Works Assoc.* 61: 109-127.

Schroeder, H. and Balassa, J.J., 1965. Influence of chromium cadmium and lead on rat aortic lipids and circulating cholesterol. *Am. J. Physiol.* 209: 433-437.

Schwartz, K. and Mertz, W., 1959. Chromium III and the glucose tolerance factor. *Arch. Biochem. Biophys.* 85: 292.

Schouwenburg, J. Ch. van and Walinga, I., 1967. The rapid determination of phosphorus in presence of arsenic, silicon and germanium. *Anal. Chim. Acta.* 37: 271-274.

Shukla, S.S., Syers, J.K. and Armstrong, D.J., 1972. Arsenic interference in the determination of inorganic phosphate in lake sediments. *J. Envir. Qual.* 1: 292-295.

Sillen, L.G. and Martell, A.E., 1971. *Stability constants of metal-ion complexes.* The Chem. Soc., London, 754 pp plus supplement.

Singer, M.J. and Hanson, L., 1969. Lead accumulations in soils near highways in the Twin Cities metropolitan area. *SSSA Proc.* 33: 152-153.

Smith, G.E., 1971. *Hearings before Senate Committee on Public Works.* Subcomm. on Air and Water pollution. Kansas City, No. 92-H, 11, p 2527-2540 and 2941-3048.

Soane, B.D. and Saunder, D.H., 1959. Nickel and Chromium toxicity of serpentine soils in Southern Rhodesia. *Soil Sci.* 88: 322-330.

Stanford, G., Frere, M.H. and Schwanzinger, D.H., 1973. Temperature coefficient of soil nitrogen mineralization. *Soil Sci.* 115: 321-323.

Stanford, G. and Smith, S.J., 1972. Nitrogen mineralization potentials of soils. *SSSA Proc.* 36: 465-472.

St.Amant, P.P. and McCarty, P.L., 1969. Treatment of high nitrate waters. *J. Am. Water Works Assoc.* 61: 659-662.

Steevens, D.R., Walsh, L.M. and Keeney, D.R., 1972. Arsenic phytotoxicity on a Plainfield sand as affected by ferric sulfate and aluminum sulfate. *J. Envir. Qual.* 1: 317-320.

Stevenson, F.J. and Wagner, G.H., 1970. Chemistry of nitrogen in soils. Chapter 8 in *Agricultural Practices and Water Quality,* Iowa State Univ. Press, Ames, Iowa, pp 125--141.

Stevenson, F.J., 1972. Organic matter reactions involving herbicides in soil. *J. Envir. Qual.* 1: 333-343.

Stevenson, F.J., 1974. *Nitrogen transformations accompanying the application of livestock wastes to soil.* Statement prepared for Ill. Poll. Contr. Board Hearings, Amboy, Ill.

Stojanovic, B.J., Kennedy, M.V. and Shuman, F.L., 1972. Edaphic aspects of the disposal of unused pesticides, pesticide wastes and pesticide containers. *J. Envir. Qual.* 1: 54-62.

Strijbis, K. and Reiniger, P., 1974. *Growth of rice plants on a chromium contaminated soil.* Report Biology Group, C.E.C., Ispra, Italy, 39 pp.

Tammer, P.M. and de Lint, M.M., 1969. Leaching of arsenic from soils. *Neth. J. Agr. Sci.* 17: 128-132.

Taylor, R.M. and McKenzie, R.M., 1966. The association of trace elements with manganese minerals in Australian soils. *Austr. J. Soil Res.* 4: 29-39.

Taylor, A.W., Edwards, W.M. and Simpson, E.C., 1971. Nutrients in streams draining woodland and farmland near Coshocton, Ohio. *Wat. Resource Res.* 7: 81-89.

Tso, T.C., 1970. Limited removal of Po and Pb from soil and fertilizer by leaching. *Agr. J.* 66: 663-664.

Turner, M.A. and Rust, R.H., 1971. Effect of chromium on growth and mineral nutrition of soybeans. *SSSA Proc.* 35: 755-758.
Tusneem, M.E. and Patrick, W.H., 1971. *Nitrogen transformations in water-logged soils.* Bull. 657, Dept. of Agron., Louisiana State Univ., 75 pp.
Vanselow, A.P., 1966. Cobalt; Chapter 10 in *Diagnostic criteria for plants and soils.* Univ. Cal., pp 142-156.
Vanselow, A.P., 1966. Nickel, Chapter 21 in *Diagnostic criteria for plants and soils. p 302-309.*
Vollenweider, R, 1968. *Les bases scientifiques de l'eutrophisation des lacs et des eaux courantes sous l'aspect particulier du phosphore et de l'azote comme facteurs d'eutrophisation.* report O.E.C.D.
Walker, G.W., Bouma, J., Keeney, D.R. and Magdoff, F.R., 1973. Nitrogen transformations during subsurface disposal of septic tank effluents in sand; I: Soil transformations. *J. Envir. Qual.* 2: 475-480.
Walker, G.W., Bouma, J., Keeney, D.R. and Olcott, P.G., 1973. Nitrogen transformations during subsurface disposal of septic tank effluents in sands; II. Ground water quality. *J. Envir. Qual.* 2: 521-525.
Walker, W.W. and Stojanovic, B.J., 1973. Microbial versus chemical degradation of malathion in soil. *J. Envir. Qual.* 2: 229-232.
Walker, W.W. and Stojanovic, B.J., 1974. Malathion degradation by an Arthrobacter species. *J. Envir. Qual.* 3: 4-10.
Welch, L.F., 1972. More nutrients are added to soil than are hauled away in crops. *Illinois Research*, 14: 3-4.
Wetselaar, R., 1962. Nitrate distribution in tropical soils; III Downward movement and nitrate accumulation in the subsoil. *Plant Soil* XVI, 1: 19-31.
Whittaker, R.H., 1970. *Communities and ecosystems.* MacMillan, London.
Williams, J.D.H., Syers, J.K., Walker, T.W. and Rex, R.W., 1970. A comparison of methods for the determination of soil organic phosphorus. *Soil Sci.* 110: 13-18.
Winton, E.F., Tardiff, R.G. and McCabe, L.J., 1971. Nitrate in drinking water. *J. Am. Water Works Assoc.* 63: 95.
Woldendorp, J.W., 1972. Nutrients limiting algal growth. *Stikstof* 15: 16-27.
Woolson, E.A., Axley, J.H. and Kearney, P.C., 1971. The chemistry and phytotoxicity of arsenic in soils; I. Contaminated field soils. *SSSA Proc.* 35: 938-943.
Yamagata, N. and Shigematsu, I., 1970. *Cadmium pollution in perspective.* Bull. Inst. Publ. Health, 19, Tokyo.
Zwerman, P.J. and de Haan, F.A.M., 1973. Significance of the soil in environmental quality improvement. *The Science of Total Environment*, 2: 121-155.

SUBJECT INDEX

Accumulation of sesquioxides, 156
acid-base equilibria, 30—32
acid neutralizing compounds, 86
acids, acid strength, 31
—, thermodynamic activity constant, 30
—, thermodynamic dissociation constant, 31, 34
acid soils, 62, 63
acid sulfate soils, 62, 166
actinomycetes, 202
activity (chemical), 14—16
—, mean activity of neutral electrolytes, 17
—, of a pure compound, 24
—, of dissolved ions, 28, 29
activity coefficients, 16, 17
—, at high ionic strength, 20
—, Debye-Hückel equation, 16, 17
—, in aqueous solutions, 16, 28, 29
—, mean activity coefficient, 17, 20, 28
—, mean salt method, 20
—, of single ions, 19, 28, 29
—, of uncharged species, 29
adsorption capacity, 55
adsorption complex, *see* also cation, anion adsorption
—, composition of, 56—63, 82
adsorption of anions, *see* anion adsorption
adsorption of cations, *see* cation adsorption
adsorption of H-ions, 76—81
—, by organic matter, 78—80
—, by oxides, 80, 81
—, non selective, 76, 77
adsorption properties of inorganic components, 2—8
aggregates, 53
albic horizon, 156
albite, 3
—, hydrolysis constant, 148
—, solubility diagram, 147
Albolls, 165
Al-clay, properties of, 80—85
Alfisols, 144
aliphatic acids, 155
alkaline sodic soils, 185
alkaline soils, 167—169, 175
alkalinity (concentration), 105, 162, 175, 183, 184

—, and reduction processes, 162
alkalinization, 172, 183—186
—, hazard, 181, 185
—, relation with sodication, 186
—, under irrigation, 183—186
alkalization, *see* sodication
aluminum, adsorption by clays, 77, 78
—, chelates, 155, 156
—, $\bar{G}_f^0$ of Al^{3+}, 25
—, hydroxocomplexes, 115, 123
—, mobilities in soils, 146
—, phosphates, 118, 119, 123
aluminum oxides/hydroxides, 1, 114, 141, 144, 154
—, formation from kaolinite, 116—118
—, molybdate adsorption by, 231
—, organic pesticides adsorption by, 242
—, organic phosphate adsorption by, 214
—, PO_4 adsorption by, 95
—, solubility, 114—116, 122
—, some common forms, 3
amides, 202
amines, 202
amino acids, 9, 201, 202, 207
amino sugars, 202
ammonia and ammonium,
—, fixation of, 74
—, nitrogen, 203—205
—, solubility in water, 123
ammonification, 203, 204
ammonium nitrite decomposition, 207
analcime, 167, 168
anion adsorption, 91—95
—, deficit of anions, 64, 65
—, negative (exclusion), 60, 91, 92
—, net adsorption, 91, 93
—, positive, 60, 61, 91
—, selectivity, 93
anion exchange capacity, 80, 92, 93
anion exclusion, 60, 64, 65, 91—95
—, effective distance of, 92
—, estimation of, 91, 92
annite, 148
anorthite, hydrolysis constant, 148
—, solubility diagram, 147
antigorite, 8
apatite, 3, 4, 118
Aqualfs, 165

Aquults, 165
aragonite, 96
argillic horizon, 141, 163
Aridisols, 144, 172
aromatic acids, 155
arsenic in soil, 219, 221, 222
atrazine adsorption, 244
augites, 3, 141

Basalt, 145, 149
base, *see* acid—base equilibria
base saturation, 81, 82
basin irrigation, 177, 178
bayerite, 23—25
—, $\bar{G}_f^0$, 25
—, solubility product, 24, 122
beidellite, 153, 166
bioaccumulation of pollutants, 221
biodegradation of organic pesticides, 226, 241, 253—255
biological agents, 155
biological homogenization, 141
biological oxidation, 155
biomass, 8, 202
biosphere, 141, 142
biotite, 3, 152, 148
boehmite, 3, 122, 214
Boltzmann equation, 47—49, 67, 91
Boltzmann factor, 91
border irrigation, 177, 178
boron, 173
brucite, 4
buffering capacity of soil, 193, 194
bulk solution in soil, 43, 48

Cadmium in soil, 122, 219, 222, 223
calcite, 2, 3, 96, 167, 168
—, solubility diagram, 147
—, solubility product, 100, 122
calcium, Ca-carbonate complexes, 104, 123
—, Ca-phosphate complexes, 123
—, Ca-phosphates, 118, 119, 122
—, $CaSO_4^0$-complex, 25, 29, 123
—, $\bar{G}_f^0$ of Ca^{2+}, 25
calomel electrode, 75, 87—89
carbonate equilibria, 96—105
—, calculations, 97, 98
—, conditions, 96, 97
—, solubility diagrams, 103
—, total dissolved concentrations, 92

carbonates, 3, 141, 145
carbon dioxide, as weathering agent, 142
—, concentration in the gas phase, 11, 12
—, CO_2—H_2O systems, 96—99, 123
—, transport of, 126
carbonic acid, 96, 97
—, activity in solution, 97
—, protolysis reactions, 97, 123
carnallite, 3
cat clays, 166
catenas, 144
cation adsorption, 54—90
—, by clay minerals, 54—75
—, by organic matter ligands, 75, 76
—, determination methods of, 56
—, excess adsorbed, 60, 61, 64, 65
—, of trace amounts, 71, 72, 135
—, selectivity of, 66, 67, 72, 73
cation exchange (adsorption), 54, 127—132
—, complex, 60, 172
—, equations, 65—69
—, —, Gapon, 68
—, —, Kerr, 66
—, equilibrium, 65
cation exchange capacity, 55
—, calculation of, 58
—, effect of pH, 77—81
—, effect of salt level, 76—81
—, methods of determination of, 56
—, of clay minerals, 56
—, of organic matter, 56, 75
cation exchange chromatography, 127—132
cation exchange isotherm, favorable, 134
—, linear, 133
—, non favorable, 134
—, normalized, 132, 133
cation exchange reaction, 54, 55, 65—72, 153
—, heterovalent, 67, 68
—, homovalent, 66, 67
—, rate of, 55, 65
cation fixation, 73
cesium fixation, 74, 75
chelate hypothesis, 155, 156
chelate stability diagrams, 238, 239
chelation, 75, 76, 155, 238, 239
chemical activities, *see* activities (chemical)
chemical equilibria, 13—42
—, conditions for, 13, 14, 22
—, equilibrium constant, 14

—, general rules, 13, 14
chemical potential, 14, 15, 23
—, at standard state, 15, 23
—, concentration dependent part, 15
—, standard, 21
chernozems, 144, 145
chlorazine, 250
chlorites, 7, 8, 153, 154
chromatography, exchange, 127—131
—, precipitation, 139
chromium in soil, 224, 225
chrysotile, 148, 153
citric acid, 155, 156
clay minerals, adsorption of cations, *see* cation adsorption
—, adsorption of organic pesticides, 242, 249, 250
—, adsorption of phosphates, 94, 95, 214
—, expanding lattice, 7
—, formation, 141
—, lattice structure, 4—8, 44
—, pH-dependent charge, 145
—, physico-chemical behavior, 5
—, solubility, 2, 3, 148
—, some common, 3, 8
—, specific surface area, 4—8, 55
—, substitution charge, 44
—, surface density of charge, 43—45, 49
clino-enstatite, 148, 153
C/N-ratio of soil, 9, 209
cobalt in soil, 76, 219, 223, 224
coions, 48, 49
complex formation, 36—38
congruent dissolution (of minerals), 145—147, 149
consumptive use of water by plants, 175, 176
convection flux, 127
coordination bonding of organic pesticides, 248
copper in soil, 76, 225—227
correction of soil pH, 86, 87
counterions, 44—49
crandallite, 118

DDT, 245, 253, 254
Debye-Hückel equation, 16, 17
—, Davis extension, 17, 18
decay of organic matter, 155
decomposition of oil residues, 255, 256
decomposition of organic pesticides, 251—255

deficit of anions adsorbed, *see* anion adsorption
degradability of organic pesticides, 195, 242
denitrification (microbial), 203, 208, 209
—, conditions, 208
—, rate of, 208
denitrifiers, 208, 209
depth of penetration, of a liquid feed, 130
—, of a solute front, 128, 132
diaspore, 3
1,3-dichloropropene, 244
diffuse (electric) Double Layer, 47—53
—, charge, 48
—, coion concentration, 48, 49
—, counterion concentration, 47—49
—, electric capacity, 50
—, electric potential, 48
—, extent, 47—50, 92
—, formation, 45—48
—, influence on soil properties, 52, 53
—, model considerations, 63—65
—, truncated, 50, 51
diffusion (and dispersion) of solutes, 126, 127
—, flux, 127
—, influence on solute front, 134, 135
diffusion of gases, 142
dinosep, 250
di-octahedral clays, 4
diopside, 148
diquat, 246, 247
dispersion of clay particles, 53
dispersion of solutes, *see* diffusion
displacement of soluble components, 126—140
—, in case of complete exchange, 128—130
—, in case of incomplete exchange, 130, 131
—, influence of exchange isotherm, 131, 132
distribution coefficient, 245
distribution ratio of ions, 61, 62, 129
diuron, 246, 250
dolomite, 3, 96, 122, 167, 185
drainage water composition, 53, 175, 176

Electrical conductivity, 173
electric double layer, *see* diffuse D.L.
electrode potential, 34—36
electromotive force (EMF), 34—36
electron transfer, 32—34

—, donor, acceptor, 32
—, free electrons, 33
—, relative electron activity, 32, 34, 158
eluvial A_2-horizon, 156
equilibrium, acid—base, 30—32
—, constant, 14, 21
—, overall, 153, 154
—, oxidation—reduction, 32—34
—, partial, 153
—, solution composition, 26—29
—, thermodynamic constants, 14, 21, 23—26, 40, 41, 122—124, 148, 160
equivalent fraction adsorbed, 59
eu-polytrophic water, 199
eutrophication, 95, 196—199, 210, 211
—, classification of, 199
excess of cations (adsorbed by soil), 64, 65
Exchangeable Sodium Percentage, 175, 180
—, lowering of, 186—189
exchange, see cation, anion exchange (adsorption)
expanding lattice clays, 7
expulsion of anions, see anion exclusion

Faraday constant, 34
favorable exchange, 134
fayalite, 146
feed solution in chromatography, 132
—, step increase of concentration of, 132—134
—, volume of, 127
feldspars, 141, 153
—, weathering of, 145, 146, 149
felsic rocks, 167, 169
fenuron, 246
ferric oxides/hydroxides, 1, 3, 4, 105, 106, 141, 144, 154
—, PO_4-adsorption by, 95, 214
—, solubility, 3, 4, 36—38, 105—113, 122
—, some common, 3
—, specific surface area, 3, 4
ferrolysis, 165, 166
ferrous compounds, 105, 106
filtering capacity of soil, 193
fixation of cations, 73
fixation of phosphate, 94, 95
flocculation of clays, 94
florencite, 118
fluorapatite, 122, 216
fluorite, 123

flux, 126
—, autonomous flux, 127
—, convective flux, 126, 127
—, total flux of solutes, 126, 127
—, volume flux of soil solution, 126, 127
forsterite, 145, 147, 148
—, solubility diagram, 147
—, weathering of, 145
free enthalpy, 21, 22
—, of formation, 21, 25, 40, 41
—, partial molar, 22, 23
—, standard free enthalpy of a reaction, 14, 21, 23, 26, 34
—, standard partial molar, 23
Freundlich adsorption equation, 243
fulvic acids, 10, 202
fumigants, 126
fungicides, 221, 222, 226, 227, 240

Garnet, 3
gas leakages in soil, 197, 256—259
gibbsite, 3, 23—27, 80, 81, 152—154
—, formation, 146
—, $\bar{G}_f^0$ of, 25
—, hydrolysis constant, 148
—, solubility diagram, 24, 115, 116, 122, 147
—, stability diagram, 151, 152
—, surface charge, 44
Gibb's phase rule, 102
glass electrode, 35, 87—89
gley, 163, 164
gley soils, 144
goethite, 3, 105, 149, 152, 153
—, solubility product, 122
gorceixite, 118
gypsum, 2, 3, 27—29, 141, 167—169, 178
—, $\bar{G}_f^0$ of, 25
—, lowering ESP by, 186, 187
—, solubility diagram, 147
—, solubility product, 123, 127

Halides, 2, 3
halite, 3, 167
halloysite, 8
hausmannite, 160
H-clays, 77, 82, 83
—, aged, 78
—, titration curve, 82, 83
heavy metals in soil, 76, 197, 218—239
—, chelates, 238, 239

—, mobility, 219, 238, 239
—, sources, 219—237
hematite, 3, 80—85, 105, 111—113, 149
—, reduction of, 107—110
—, solubility product, 122
—, stability diagrams, 107, 108, 110, 112, 114
Henry's law, 96
herbicides, see organic pesticides
homoionic clay, 56
hornblende, 3, 141, 152
humic acids, 10, 202
humification, 9, 155, 202, 203
humus, 9
—, adsorption of organic pesticides by, 242, 246, 247, 251
—, cation exchange capacity, 56, 75, 79, 81
—, H-ion adsorption by, 9, 10, 78—80
—, metalic organic complexes, 10, 75, 76, 155, 226, 238, 239
—, structure, 9, 10
—, surface charge density, 45, 75, 79
—, titration curve, 84, 85
hydrodynamic dispersion, 177
hydrogen, adsorption of, 76—81
—, by clays, 76—78, 81
—, by organic matter, 78—81
—, by oxides, 80, 81
hydrogen bonding, 248
hydrogen ions, $\overline{G}_f^0$, 25
hydrological cycle, 10, 11
hydromorphic characteristics, 164
hydromorphic soils, 163—166
hydrosphere, 142
hydrous micas, 6
hydroxyapatite, 120, 122
hydroxy benzoic acid, 155
hydroxylamine, 203

Illite, 3, 8, 152—154
—, adsorption of cations by, 67
—, adsorption of organic pesticides by, 249, 250
—, adsorption of phosphate by, 94
—, cation exchange capacity, 56
—, fixation of cations, 73
—, hydrolysis constant, 148
—, 'open' illite, 74
—, specific surface area, 7, 55
—, substitution charge, 7
—, surface density of charge, 55

illuvial B-horizon, 156
immobilization (microbial), 201—203
Inceptisols, 144
incongruent dissolution (of minerals), 145, 146
indurated horizons, 145
inosilicates, 3
inositol phosphates, 213, 214
—, adsorption by soil components, 214
insecticides, see organic pesticides
interstitial water, 143
ionic mobilities, 173
ionic strength, 17
iron accumulation, 164
iron, see also ferric and ferrous
—, chelates, 76, 155, 156, 238
—, elemental, 108
—, hydroxy complexes, 37—39, 105, 106, 123
—, mobility, 146, 155, 156
—, oxidation—reduction, 159—161
—, phosphates, 118, 119, 123
iron oxide accumulation, 169, 170
irreversible reactions, in soil formations, 154
irrigation water, 175—178
—, anionic composition, 182
—, calculation of total amount, 178
—, classification, 181
—, excess amount, 178
—, quality, 181

Jarosite, 166

Kaolinite, 3, 8, 152—154, 166
—, adsorption of organic pesticides by, 249, 250
—, formation of, 144, 145
—, hydrolysis constant, 148
—, lattice structure, 4, 5
—, solubility diagram, 117, 147
—, specific surface area, 5, 55
—, stability diagrams, 149—152
—, surface density of charge, 55
K-beidellite, 148, 153
—, hydrolysis constant, 148
—, stability diagram, 149, 150, 152
K-feldspar, 152
—, stability diagram, 149—152
—, weathering, 145, 146
K-fixation, 73—75

K-mica, stability diagram, 149, 150

Landscape model, 143—145
latosols, 152
leaching, 153, 186, 200
—, efficiency factor, 177, 178
—, rate, 154
—, requirement, 178, 179
lepidocrocite, 105
ligands in organic matter, 10, 75, 78—80
lignin compounds, 9
lime, lowering ESP-value by, 186
—, neutralizing power, 87
—, potential, 81, 82
—, requirement, 87
liming, 86
—, factor, 87
—, material, 87
limonite, *see* ferric hydroxide

Mafic rocks, 167, 169
maghemite, 105
magnesite, 96, 122, 167
magnesium, carbonates complexes, 123
—, ions in solution, 144
magnetite, 105, 111—113, 152, 153, 160
—, reduction, 108—110
—, stability diagram, 107, 108, 110, 112, 114
malfunctioning of soil, 192, 196, 210
manganese, oxidation—reduction, 159—161
—, oxides, 141, 162
manganese accumulation, 164
manganite, 160
manure, 205, 213, 214
mass action principle, 14
mean depth of penetration of a cation, 128, 129, 131
mean diameter of solvated ions, 17, 18
mean salt method, 20
mercury in soil, 126, 227—229
metal-organic complexes, 10, 75, 76, 155, 156, 238, 239
methane, 256, 258
Mg-beidellite, hydrolysis constant, 148
—, solubility diagram, 147
—, stability diagram, 151, 152
Mg-chlorite, hydrolysis constant, 148
—, stability diagram, 151, 152

micas, 7, 141
microbial immobilization, 202
microcline, hydrolysis constant, 148
—, stability diagram, 149, 150
mineralization, of organic matter, 155, 201—204
—, of wastes, 197
minerals, congruently dissolution, 145, 146
—, hydrolysis, 148
—, incongruent dissolution, 145, 146
—, primary, 141, 147, 153
—, residual, 3, 146
—, secundary, 141, 153, 154
—, solubility, 2, 3, 122, 123, 146, 148
—, some common, 3
—, specific surface area, 4—8, 55
—, weathering, 145—158
—, weathering rate, 156
minor elements, 3, 218, 230
Mollisols, 144, 145
molybdenum in soil, 230, 231
montmorillonite, 3, 8, 144, 153
—, adsorption of organic pesticides by, 249, 250
—, adsorption properties, 7, 68
—, PO_4-adsorption by, 94
—, specific surface area, 7, 55
—, surface density of charge, 55
—, swelling properties, 7
monuron, 246, 250
muscovite, 3, 8, 149, 153
—, hydrolysis constant, 148
—, solubility diagram, 147
—, specific surface area, 7
—, stability diagram, 150

Nahcolite, 167
neburon, 246
negative adsorption of anions, 60, 91, 92
nematicides, *see* organic pesticides
Nernst's law, 32, 34—36
nesosilicates, 3
nickel in soil, 76, 226, 231, 232
nitramide, 203
nitrate-nitrogen, distribution in soil, 206
—, health hazards, 198
—, reduction of, 159, 161, 203
—, standard in drinking water, 198
nitrates (mineral), 2, 3
nitre, 3
nitrification, 203, 205—208

nitrite-nitrogen, 203, 207, 208
nitrogen, assimilation, 230
—, immobilization, 201—203
—, inorganic forms in soil. 203. 204
—, mineralization, 201—203
—, mobility in soil, 197
—, organic forms in soil, 201—203
—, overfertilization, 198
—, oxides, 203, 207, 208
—, pathway through soil, 204—210
—, pollution effects, 198, 199, 218
—, pollution hazards of, 197—199
—, sources, 199—201
non favorable exchange, 134
normative mineral composition, 152

Oil spills, 197, 255, 256
—, residues of, 255, 256
oligo-mesotrophic water, 199
olivine, 3
organic acids, 155, 156
organic compounds in soil (*see also* humus), 8—10
—, adsorption of organic pesticides, 242, 246, 247, 251
—, as weathering agent, 142
—, C:P:N:S-ratio, 202
—, decay of, 8, 9, 155
—, mineralization, 155
—, oxidation, 141, 159—161
—, resynthesis of, 9, 155
organic nitrogen, *see* nitrogen
organic pesticides in soil, 12, 239—256
—, adsorption by clay minerals, 249, 250
—, adsorption by ferric oxides, 242
—, adsorption by organic matter, 251
—, adsorption mechanisms, 245—249
—, biodegradation, 240—242, 251, 253—255
—, bonding by soil constituents, 242—249
—, chemical degradation, 251—253
—, degradability, 242
—, dosage, 242
—, environmental hazards of, 240
—, hemi salts, 251
—, persistency, 242
—, photodecomposition, 251, 252
—, transport and accumulation of, 240, 241
organic phosphates, 118, 213, 214
organic soils, 9
orthoclase, 3

'overall' distrubution ratio, 62, 131
oxalic acid, 155
oxidation in soils, 144
oxidation potential, 34—36
oxidation—reduction equilibria, 32—34, 106, 124
—, in soil, 158—164
oxides, cation exchange capacity, 80, 81
—, H-ion adsorption by, 80, 81, 83
—, hydrolysis constants, 148
—, some common, 3
—, surface charge, 44, 45, 50, 80, 81
—, surface potential, 44, 50
Oxisols, 144, 152, 154, 163
oxygen, concentration in gasphase, 11, 12
—, transport, 126

Paraquat, 246, 247
parent material (of soil constituents), 3, 141, 153, 154
partial molar free energy, *see* free enthalpy
peat soils, 9, 144
pe (in soil), 33, 158—161
—, determination, 35, 36
penetration of solute front, 132—135
—, influence of diffusion and dispersion, 134, 135
—, influence of exchange isotherm, 132—134
peptization (of clay suspensions), 94, 180, 181
pesticides, *see* organic pesticides
phenylcarbamates, 248
pH (in soil), 32
—, as a function of P_{CO_2} and Alk, 183, 184
—, correction, 86, 87
—, determination, 35, 36, 88, 89
—, salt concentration effect on, 77—85
phosphate, adsorption, 94, 215—218
—, distribution in soil, 216—218
—, eutrophication due to, 196, 199, 210, 211
—, fixation, 93, 95, 213
—, mobility in soil, 198
—, organic forms, 212—214
—, pollution effect of, 210—213, 218
—, pollution hazards of, 197, 198
—, some common minerals, 3, 118
—, sources in soil, 211—213
phosphoric acid, 118, 123
photodecomposition of organic pesticides, 251, 252
p-hydroxy benzoic acid, 156

phyllosilicates, *see also* clay minerals
—, some common, 3, 8
picloram, 245, 249
plagioclase, 152
Planosols, 165
podzols, 144, 155, 156
pollutants, *see also* pollution of soil
—, adsorption, 71, 72, 75, 76, 135, 192—271
—, penetration into soil, 137—139
pollution of soil, 192—271
—, biological indicators of, 194
—, by gas leakages, 256—259
—, by sanitary landfills, 259—261
—, definition, 192, 193
—, point sources of, 255
—, recognition and prediction, 194—197
—, with heavy metals, 218—239
—, with nitrogen, 197—210
—, with oil spills, sludge disposal, 255—257
—, with organic pesticides, 239—242
—, with phosphates, 197, 198, 210—218
—, with raw sewage water disposals, 219, 255
—, with sewage sludge disposals, 219, 255
—, with solid wastes, 219, 259—261
polysaccharides, 9
polytrophic water, 199
pore volume, 11
positive adsorption of anions, *see* anion adsorption
potassium fixation, 73—75
potential determining ion, 144
precipitation-chromatography, 139
pressure filtration, 92
primary minerals, *see* minerals
proton hydration, 32
proton transfer, 30
proximity effect of surface charges, 77, 78
pseudogley, 163, 164
pseudogley soils, 144, 165
pyrite, 3
—, oxidation, 146, 157
pyrochroite, 160
pyrolusite, 160
pyroxenes, 153

Quartz, solubility diagram, 123, 147, 148

Rate of salinization, 176

rates of chemical reactions, 1, 13
ratio law, 68
reclamation of Na-soils, *see* sodic soils
redox potential, 32—34, 158—161, 260, 261
redox reactions, 32—34, 106, 158—161
—, thermodynamic equilibrium constants, 32—34, 123, 160
red tropical soils, 105
reduced ratio (of concentrations or activities), 68, 172
reducing conditions, 109
reduction and oxidation in soil, *see* soil reduction and oxidation
reduction of soils, 142—144
residual alkalinity, 168
residual minerals, *see* minerals
Residual Sodium Carbonate, 185
reverse osmosis, 92
reverse weathering, 166—169
rubidium fixation, 74
runoff, 200
—, of pesticides, 240—242

Saline soils, 2, 11, 61, 91
—, agricultural problems, 172, 173
—, chemical characterization, 173—175
—, classification, 174
—, formation, 144, 175—177
—, reclamation, 186—189
salinity, 166
salinization, 168, 169, 171, 172
—, following irrigation, 175—179
—, hazard, 181
—, rate, 176
—, zonal, azonal, 171
salt balance, 175, 176
salt concentration effect on pH, 77—81, 83
salt effect on crop growth, 174
salt free zone, 92
salt management, 171
—, leaching, 171
salt profiles, 177
salt-sieving effect, 53, 92
sanitary landfills, 197, 259—261
saturated paste, 183, 184
saturation extract, 173—175
—, chemical conductivity, 173—175
—, osmotic pressure, 174
—, salt concentration, 173—175
saturation percentage, 174, 176

seasonal flooding, 144
secundary minerals, see minerals
secundary phases in soil formation, 141
selectivity coefficients, high selectivity, 72, 73
—, of anions, 93
—, of cations, 66, 67
selenium in soils, 234, 235
semi-arid soils, 144, 171
sepiolite, 148, 167
septic tanks effluents, 205
serpentine, 153
—, soils, 225, 232
sesquioxides/hydroxides, see ferric and aluminum oxides/hydroxides
sewage farm, 216—218
siderite, 105, 111—113
—, solubility, 122
—, stability diagram, 112, 114
silicates, 3, 141
—, hydrolysis constants, 148
—, some common, 3
—, stability diagram, 150, 151
—, weathering, 145, 146
silicic acid, 123
silicon dioxide, 3, 4, 122, 141, 144
—, hydrolysis constant, 148
—, solubility, 3, 4, 122
—, stability diagram, 150—153
—, surface charge, 44
smectite, 144, 147, 154, 167
sodication, 135, 137—139
—, hazard, 181, 182
—, rate, 172
—, upon irrigation, 179—183
sodic soils, 2, 49, 53, 62, 63, 168, 171—173
—, agricultural problems, 172, 173
—, chemical characterization, 173—175
—, formation (see also sodication), 144
—, reclamation, 135—137, 186—189
Sodium Adsorption Ratio, 180
sodium saturation, 172
soil as an environmental component, 192—194
soil catenas, 144
soil formation, 1, 141, 142, 152, 154
—, factors, 141—145
soil horizon, 141
soil pollution, see pollution of soil
soil reduction, 142, 143, 158—162
soil solution, composition, 10, 11, 155, 156

—, concentration under field conditions, 11, 179
—, dielectric constant, 43
—, interaction with solid phase, 43—53, 141—170
solid phase components, 2—10
solid residues, 146
solod, 183
Solodic Planosols, 165, 166
solodised solonetz, 183
solonetz, 183—185
solubility, 40, 41, 143, 146
—, and stability relationships, 146—153
—, diagrams, 36—39
—, equilibria, 122, 123, 148
—, of inorganic compounds (see minerals), 2—4
—, product, 24, 27, 36
solute flux, see flux
solute front, 132—135
—, influence of diffusion and dispersion, 134, 135
—, infleunce of exchange isotherm, 132—134
—, magnitude of spreading effects, 135
—, modulation, 132—135
—, rate of displacement, 132, 133
—, shape, 132, 134
solute transport, see displacement of soluble components; flux
solution flux, see flux
specific surface area, 2, 55
—, determination, 55, 92
spinell, 153
spodic horizon, 156
Spodosols, 144, 155, 156
stability diagram, 107, 149—151
standard free enthalpy, see free enthalpy
standard states, 14—16
strengite, 118, 122
strontium, 138
structure decay, 187
sulfate ions, $\bar{G}_f^0$, 25
sulfates (minerals), 2, 3
sulfate—sulfide equilibrium, 159—161
sulfides (minerals), 3
surface (of solid phase), charge density, 43—45, 55, 65, 76—81
—, chemistry, 43
—, dissociation of surface ions, 43, 44
—, phase, 43
—, structure, 40, 41, 43
suspension effect, 87—89

swelling pressure, 50, 51
—, influence on soil properties, 52, 53
sylvine, 3

Talc, 3, 148
tectosilicates, 3
thermodynamic equilibrium (constant),
  see equilibrium
titration curves, of Al-clay, 84, 85
—, of H-clay, 82—84, 85
—, of organic matter, 84, 85
—, of oxides, 84, 85
—, of soil samples, 85
—, of soil samples characteristics, 84
toposequence, 143
total dissolved carbonate concentration,
  see alkalinity
tourmaline, 3
toxic effects of heavy metals, 218, 219
—, of nitrate in drinking water, 198
trace components, see also heavy metals
—, adsorption, 71, 72, 135
—, penetration into soil, 137—139
transition metals, 75
transition metals, see heavy metals
transport, see also flux
—, in the gas phase, 126, 127
—, in the liquid phase, 126, 127
—, of trace components, 137—139
—, with the liquid phase, see displacement
  of soluble components
triazines, 240, 247, 249
tridymite, 3
trietazine, 250
tri-octahedral clays, 4, 8
trona, 167
truncated diffuse double layer, 50, 51
—, concentration distribution in, 51
—, swelling pressure of, 50, 51

Ultisols, 144, 154
ultra oligotrophic water, 199
universal soil-loss equation, 241
urea herbicides, 246

Vanadium in soil, 235, 236
Van 't Hoff equation, 21
variscite, 118, 120, 122
vermiculite, 3, 8, 74
—, expanding lattice, 74
—, K—Mg-adsorption, 74
—, specific surface area, 7, 8
Vertisols, 144, 155, 166, 167
vivianite, 3, 118, 122

Waste disposal (systems), 193, 196, 201
—, recycling of nitrogen in, 208
water, analysis, 142, 143
—, balance equation, 175, 178
—, dissociation constant, 25, 123
—, interstitial, 143
—, management, 171
—, oxidation, reduction, 105, 106
—, standard free enthalpy of formation,
  25
waterlogging in soils, effect of, 208,
  160—166
wavellite, 118
weatherability, 149
weathering,
—, agents, 142
—, of primary minerals, 141
—, phenomena, 142
—, process, reactions, 1, 4, 142, 143
—, products, 154
—, rates, 156
—, residues, 149
—, reverse, 166—169
—, sequences, 152
—, under seasonally reduced conditions,
  164—166

Yield-value (of ESP), 179

Zeolites, 3, 167, 168
zero-point of charge, 45
zinc equivalent factor, 226
zinc in soil, 219, 236, 237
—, carbonate, 122
—, phosphates, 119, 122